PHYSICS RESEARCH AND TECHNOLOGY

AN ESSENTIAL GUIDE TO ELECTRODYNAMICS

PHYSICS RESEARCH AND TECHNOLOGY

Additional books and e-books in this series can be found
on Nova's website under the Series tab.

PHYSICS RESEARCH AND TECHNOLOGY

AN ESSENTIAL GUIDE TO ELECTRODYNAMICS

NORMA BREWER
EDITOR

Copyright © 2019 by Nova Science Publishers, Inc.

All rights reserved. No part of this book may be reproduced, stored in a retrieval system or transmitted in any form or by any means: electronic, electrostatic, magnetic, tape, mechanical photocopying, recording or otherwise without the written permission of the Publisher.

We have partnered with Copyright Clearance Center to make it easy for you to obtain permissions to reuse content from this publication. Simply navigate to this publication's page on Nova's website and locate the "Get Permission" button below the title description. This button is linked directly to the title's permission page on copyright.com. Alternatively, you can visit copyright.com and search by title, ISBN, or ISSN.

For further questions about using the service on copyright.com, please contact:
Copyright Clearance Center
Phone: +1-(978) 750-8400   Fax: +1-(978) 750-4470   E-mail: info@copyright.com.

NOTICE TO THE READER

The Publisher has taken reasonable care in the preparation of this book, but makes no expressed or implied warranty of any kind and assumes no responsibility for any errors or omissions. No liability is assumed for incidental or consequential damages in connection with or arising out of information contained in this book. The Publisher shall not be liable for any special, consequential, or exemplary damages resulting, in whole or in part, from the readers' use of, or reliance upon, this material. Any parts of this book based on government reports are so indicated and copyright is claimed for those parts to the extent applicable to compilations of such works.

Independent verification should be sought for any data, advice or recommendations contained in this book. In addition, no responsibility is assumed by the Publisher for any injury and/or damage to persons or property arising from any methods, products, instructions, ideas or otherwise contained in this publication.

This publication is designed to provide accurate and authoritative information with regard to the subject matter covered herein. It is sold with the clear understanding that the Publisher is not engaged in rendering legal or any other professional services. If legal or any other expert assistance is required, the services of a competent person should be sought. FROM A DECLARATION OF PARTICIPANTS JOINTLY ADOPTED BY A COMMITTEE OF THE AMERICAN BAR ASSOCIATION AND A COMMITTEE OF PUBLISHERS.

Additional color graphics may be available in the e-book version of this book.

Library of Congress Cataloging-in-Publication Data

ISBN: 978-1-53615-705-5

Published by Nova Science Publishers, Inc. † New York

CONTENTS

Preface		vii
Chapter 1	The Organic Electron and the Periodic Table of the Elements *Peter J. Fimmel*	1
Chapter 2	Maxwell's Electromagnetic Equations, Elementary Introduction *George J. Spix and V. M. Red'kov*	27
Chapter 3	Maxwell Electromagnetic Equations in the Uniform Medium *E. M. Ovsiyuk, V. Balan, O. V. Veko, Ya. A. Voynova and V. M. Red'kov*	101
Chapter 4	Hidden Aspects Of The Electromagnetic Field *Ivanhoe B. Pestov*	131
Chapter 5	Radiation of Electromagnetic Waves Induced by Electron Beam Passage over Artificial Material Periodic Interfaces *Yuriy Sirenko, Petro Melezhik, Anatoliy Poyedinchuk, Seil Sautbekov, Alexandr Shmat'ko, Kostyantyn Sirenko, Alexey Vertiy and Nataliya Yashina*	169
Chapter 6	The Cornell Potential in Lee-Wick Inspired Electrodynamics *Anais Smailagic and Euro Spallucci*	207
Chapter 7	Electrodynamics in Uniformly Accelerated/Rotating Frames *A. Sfarti*	221
Index		297
Related Nova Publications		301

PREFACE

The opening chapter of *An Essential Guide to Electrodynamics* describes a new theory of the electron, from which derives a fully deductive explanation of the chemical inertness of the group 18 elements of the periodic system.

The authors propose that there is a need to present the detailed mathematical steps that are required to prove the equations of Maxwell textbooks and course instruction to help students gain a firm grasp of the equations and their applications.

Additionally, this compilation examines the wave equation for the electromagnetic 4-potential, which has a form that explicitly involves the 4-velocity vector of a moving frame. Hence, Minkowski electrodynamics implies the absolute nature of mechanical motion in medium.

Next, the authors represent the electromagnetic field from different and unknown points of view, and the duality of natural time is considered.

Chapter five is focused on an accurate and profound investigation, interpretation and explanation of resonant and anomalous phenomena in radiated electromagnetic fields that arises due to the passage of charged particle beams over arbitrary-shaped periodic interfaces of natural or artificial material.

Later, it is shown that a suitable modification of the Lee-Wick idea can also lead to linear potential at large distances. For this purpose, the authors study an Abelian model that "simulates" the quantum chromodynamics confining phase while maintaining the Coulomb behaviour at short distances.

In the final chapter, the authors present a generalization of the transformation of the electromagnetic field from the frame co-moving with an accelerated particle into an inertial frame of reference and from an inertial frame into the frame co-moving with a moving particle.

Chapter 1 describes a new theory of the electron, from which is derived a fully deductive explanation of the chemical inertness of the group 18 (VIIIA) elements of the periodic system. The authors show that the oscillation of the Dirac electron, when coupled with the Aristotelian doctrine of the actualisation of immaterial potential, is sufficient for a description of charged-particle kinematics which is quantum mechanical and dependent upon the rules of special relativity. The principle of the one-electron system guides the evolution naturally, first, into the electron pair and then the many-electron plasma, and finally when in combination with the proton, the electron pair evolves into a global electromagnetic interaction for the helium atom, which evolves further into each in turn of the single-atom interactions of the remaining members of the group 18 elements. Each of the natural elements has a one-atom electromagnetic interaction which engages all the electrons and protons of each atom. Each step in the evolution proceeds in accordance with a single principle: the actualisation of immaterial potential. The electromagnetic interaction is one of discrete, direct interparticle action; there are no valence electrons or inert electrons close to the nucleus, or geometric relations among charged particles, because all the electrons and protons are continually active, on a time scale of zeptoseconds. The kinematics is restricted naturally to the antisymmetrical two-particle ensemble, thus fulfilling the requirement of Pauli's two-valuedness, and the concept of field is not required. The six members of the group 18 elements are uniquely inert among the natural elements, because their electron numbers (viz. 2, 10, 18, 36, 54 and 86) together with their nuclear protons, each form a self-isolating, and therefore inert, electromagnetic interaction between an actual and a potential treble of charged particles, mediated by virtual photons and electron pairs; the latter behave as composite bosons. Single-atom group 18 electron collectivities of the theory mirror the orbital electron configurations for the same elements. The theory offers new insights into molecular bond formation.

As explained in Chapter 2, there is a need to present the detail mathematical steps that are required to prove the equations of Maxwell Text books and course instruction do give students a firm grasp of the equations and their applications. What is often missing is the step by step exposition leading from the basic experiments to the established equations. The following book fills this gap admirably. These are presented for students and erstwhile students who are interested in how the great physicists derived and proved their equations.

All of the math and physics presented here are more than covered in the four year college engineering course. However, students with a passing interest in high school math can easily understand all that is contained in these three papers. For some readers, there will be the revisiting of old friends. Other readers will get the "Ah, Ha!" experience. The examples in using the equations are presented in great detail and are easy to follow.

Two known, alternative to each other, forms of presenting the Maxwell electromagnetic equations in a moving uniform medium are discussed in Chapter 3. The commonly used Minkowski approach is based on two tensors, and the constitutive equations relating these tensors change their form under Lorentz transformations, so that Minkowski equations depend upon the 4-velocity of the moving inertial reference frame. In this approach, the wave equation for the electromagnetic 4-potential has a form which explicitly involves this 4-velocity vector of a moving frame. Hence, the Minkowski electrodynamics in a sense implies the absolute nature of mechanical motion in medium. An alternative formalism may be constructed, when the Maxwell equations are written with the use of only one tensor. This form of Maxwell equations exhibits symmetry under modified Lorentz transformations, in which instead of the vacuum speed of light c one uses the speed of light $c' < c$ in the medium. Due to this symmetry, the formulation of Maxwell theory in this medium becomes invariant under the mechanical motion of the reference frame, while the transition velocity must follow modified Lorentz formulas. The transition to 4-potential leads to a simple wave equation which does not contain any additional 4-velocity parameter; also this equation describes waves which propagate in space with light velocity kc, which is invariant under the modified Lorentz transformations. In connection with these two alternative schemes, an essential issue must be stressed: it seems more reasonable to perform the Poincaré-Einstein clock synchronization in the uniform medium with the use of real light signals, which leads us to modified Lorentz symmetry.

In Chapter 4 the goal is to represent the electromagnetic field from different and unknown points of view. The duality of natural Time is considered. It is established that there are two different Times in nature and the dual Time is tightly connected with natural rotation. On this ground, equations of dual electrodynamics are established and a new theoretical approach to the world of leptons and dual particles (quarks) is formulated, which gives a simple and evident explanation of lepton-quark symmetry, quark confinement and baryon number conservation.

It is shown that the electromagnetic field emerges as a ground state of the simplest form of energy called the generalized electromagnetic field, and the formulated theory of this field predicts the existence of the shadow world that consists of particles (heavy photons) which do not interact at all with any detectors and the influence of which becomes apparent only in the form of gravity and a new form of an energy flow.

The authors show that the ground state of the electromagnetic field itself has a hidden structure which takes a form of a complex scalar field associated with a one-dimensionally extended object called a charged string. The theory which predicts the existence of these objects is formulated.

On the basis of the principles of general relativity and the natural concept of Time the motion of a charged massive point particle in the external gravitational and electromagnetic fields is considered. The general covariant and reparametrization invariant Newton equations are derived from the equations of geodesic motion. The general covariant and reparametrization invariant expressions for the physical momentum, velocity and energy of a massive and charged point particle are presented. It is marked that the change of the orientation of a path and the operation of the charge conjugation are connected. The existence of the gravitational force that is defined by the momentum of the gravitational field and is not trivial only for the non-static gravitational fields is predicted. An adequate solution to the well known problem of zero Hamiltonian is manifested as well.

Chapter 5 is focused at accurate and profound investigation, interpretation and explanation of resonant and anomalous phenomena in radiated electromagnetic field that arises due to the passage of charged particles beams over arbitrary shaped periodic interface of natural or artificial material including smartmaterials and metamaterials. Reliability of the results is assured by the fact that the study is based on rigorous accurate solutions to electromagnetic boundary and initial boundary value problems and corresponding robust numerical algorithms.

Two types of structures are considered in theory: (i) infinite arbitrary profiled periodic interfaces of conventional or artificial materials with a priori given dispersion law, their consideration is based on frequency domain (FD) methods of analytical regularization; and (ii) infinite structures constructed of periodic arrays of various materials, their consideration is based on solutions to the corresponding electrodynamic problems, which are developed with a help of the method of exact absorbing conditions (EAC) enabling the consideration of the problems both in time domain (TD) and FD.

In the seventies, Lee and Wick proposed an interesting modification of classical electrodynamics that renders it finite at the quantum level. At the classical level, this modified theory leads to a regular linear potential at short distances while also reproducing the Coulomb potential at large distances. In Chapter 6, it is shown that a suitable modification of the Lee-Wick idea can also lead to a linear potential at large distances. For this purpose, the author study an Abelian model that "simulates" the QCD confining phase while maintaining the Coulomb behavior at short distances.

Chapter 6 is organized in three parts. In the first part, it presents a pedagogical derivation of the static potential in the Lee-Wick model between two heavy test charges using the Hamiltonian formulation. In the second part, the authors describe a modification of the Lee-Wick idea leading to the standard Cornell potential. In the third part, they consider the effect of replacing a point-like charge with a smeared Gaussian-type source, that renders the electrostatic potential finite as $r \to 0$.

In Chapter 7 the authors present a generalization of the transforms of the electromagnetic field from the frame co-moving with an accelerated particle into an inertial frame of reference and from an inertial frame into the frame co-moving with a moving particle. The solution is of great interest for real time applications, because Earth-bound laboratories are inertial only in approximation. Real life applications include accelerating and rotating frames more often than the idealized case of inertial frames. Our daily experiments happen in the laboratories attached to the rotating, continuously accelerating Earth. Many books and papers have been dedicated to transformations between particular cases of rectilinear acceleration and/or rotation and to the applications of such formulas. In Chapter 7 the authors are presenting the equations of electrodynamics in an accelerated frame as viewed from the accelerated frame, as viewed from an inertial frame and as viewed from a uniformly rotating frame. There is also great interest in producing a general solution that deals with arbitrary orientation of acceleration in the case of rectilinear motion, so they produced the equations for the general case as well. The main idea is to generate a standard blueprint for a general solution that gives equivalent of the Lorentz transforms for the case of the transforms between an inertial frame and a uniformly accelerated/uniformly rotating frame. The authors conclude by deriving several applications: the general form of the relativistic Doppler Effect and of the relativistic aberration formulas for the case of accelerated/uniformly rotating motion. They also give an explanation of the Mossbauer effect as viewed from a rotating frame. Chapter 7 concludes with an application to the general explanation of the Bremsstrahlung Effect (electromagnetic radiation due to particle acceleration).

In: An Essential Guide to Electrodynamics
Editor: Norma Brewer

ISBN: 978-1-53615-705-5
© 2019 Nova Science Publishers, Inc.

Chapter 1

THE ORGANIC ELECTRON AND THE PERIODIC TABLE OF THE ELEMENTS

Peter J. Fimmel
Gooseberry Hill, Western Australia

Abstract

This chapter describes a new theory of the electron, from which is derived a fully deductive explanation of the chemical inertness of the group 18 (VIIIA) elements of the periodic system. We show that the oscillation of the Dirac electron, when coupled with the Aristotelian doctrine of the actualisation of immaterial potential, is sufficient for a description of charged-particle kinematics which is quantum mechanical and dependent upon the rules of special relativity. The principle of the one-electron system guides the evolution naturally, first, into the electron pair and then the many-electron plasma, and finally when in combination with the proton, the electron pair evolves into a global electromagnetic interaction for the helium atom, which evolves further into each in turn of the single-atom interactions of the remaining members of the group 18 elements. Each of the natural elements has a one-atom electromagnetic interaction which engages all the electrons and protons of each atom. Each step in the evolution proceeds in accordance with a single principle: the actualisation of immaterial potential. The electromagnetic interaction is one of discrete, direct interparticle action; there are no valence electrons or inert electrons close to the nucleus, or geometric relations among charged particles, because all the electrons and protons are continually active, on a time scale of zeptoseconds. The kinematics is restricted naturally to the antisymmetrical two-particle ensemble, thus fulfilling the requirement of Pauli's two-valuedness, and the concept of field is not required. The six members of the group 18 elements are uniquely inert among the natural elements, because their electron numbers (viz. 2, 10, 18, 36, 54 and 86) together with their nuclear protons, each form a self-isolating, and therefore inert, electromagnetic interaction between an actual and a potential treble of charged particles, mediated by virtual photons and electron pairs; the latter behave as composite bosons. Single-atom group 18 electron collectivities of the theory mirror the orbital electron configurations for the same elements. The theory offers new insights into molecular bond formation.

Keywords: Dirac electron, potential and actual, periodic table of the elements

Introduction

In the first 30 years of the twentieth-century quantum mechanics were created as part of the project to understand the atom, particularly the smallest and simplest of them all, the hydrogen atom. And we still have no explanation from first principles of the well-known periodicity of the natural elements; the hallmark of which is the noble gases of the periodic table. Spectroscopy and quantum chemistry together have revealed much about the chemical energies of atomic charged-particle states, but Mendeleev's periodic law still lacks a properly deductive explanation [1]. The large complexities that attend quantum calculations have led to the necessity of employing approximation methods.

At a deeper level, there is no unanimity on the completeness or otherwise of quantum mechanics [2], or the appropriateness of its mathematics [3], or whether it is a theory of elementary particle physics or epistemology [4]. The quantum formalism deals essentially with probabilities, and it does so without illuminating the real nature of the system, its behaviour, or interactions; they remain a closed book [5]. What are the components of the atom doing that makes them chemically reactive, quantum mechanical and conform with special relativity? Where is the ontology? Einstein was not satisfied either.

> 'It is to be expected that behind quantum mechanics there lies a lawfulness and a description that refer to the individual system. That it is not attainable within the bounds or concepts taken from classical mechanics is clear' [6].

In the quantum mechanical scheme of things, the electron exhibits "behaviour" such as entanglement and superposition of states and non-local change of position, etc. all of which are at or beyond the fringe of human understanding. Current chemical theory works around the quantum cloud of mystery and impossibly complex calculations. In the chemistry domain the electron has an essentially passive existence. It obeys Newton's laws, responding predictably to forces and field effects that impinge on it. Since the late nineteenth century celestial mechanics has been a significant guide for models, theories and discussion of atomic structure, particularly the electron component of the atom. The concept of electron shells is still in use.

Like several of the creators of quantum physics, Dirac was acutely aware of the difficulties that stood in the way of understanding the microscopic world, which was slowly emerging at that time.

> There are, at present, fundamental problems in theoretical physics … the solution of which … will presumably require a more drastic revision of our fundamental concepts than any that have gone before. Quite likely, these changes will be so great that it will be beyond the power of human intelligence to get the necessary new ideas by direct attempts to formulate the experimental data in mathematical terms [7].

It was Heisenberg who, borrowing from the ancients, suggested the actual–potential concept as an aid to the acceptance of the non-classical, counterintuitive behaviour of quantum systems [9]. The origin of the actual–potential distinction lay with Aristotle, whose doctrine of material objects includes the postulate: "everything is both potential and actual, but not at the same time" [10]. It is probable that there is no standard approach to the question: what is it about elementary particles that we are missing? Dirac's answer, described above is: 'a ... drastic revision of our fundamental concepts'. Nature is not inert, and neither is it waiting for the next Newtonian impetus to induce a little variety into existence, even at the scale of the electron. Nature is active and above all it is organic.

Here we describe the theory of the actualisation of potential, as it applies to the electron. Even at the scale of the electron there is an organic tinge, a hint of reproduction and other organic features. One consequence of which is that the electron takes on a rudimentary organic character. And unlike classical mechanics, the impetus for much of its action is internal to itself. The general ideas that form the background to the present theory are those of both A. N. Whitehead [8] and Aristotle of Stagira, who agreed generally that the universe emerges organically from the actions of its own history. This emergence both embraces and re-realises aspects of the past.

The distinction between potential and actual lay with the pre-Socratics, who were the source for Aristotle. His doctrine of change and particularly material objects includes the postulate "everything is both potential and actual, but not at the same time" [10]. Such a state of matter seems to be a natural partner for the oscillation of the Dirac electron; despite its pre-scientific origins, actualisation and the modern Dirac oscillation together suggest a level of harmony.

In the present chapter, we begin the description of the new theory with an outline of the concepts of 'potentia' and its contrast, 'actual' as they are used here. Then follows the oscillation of the electron. A quote from Dirac, its creator, reveals something of the surprise that the appearance of the oscillation from the equation caused. Then follows the coupling of the concept of potential with the subject of the oscillation. Next is the photon-mediated interaction among electrons and finally the interaction of protons and electrons in the atomic sector. We explain the rôles of the particle vacuum, the photon and Pauli exclusion. The oscillation applies not only to the electron, but to all the low energy subatomic particles, including the photon, and to groups of particles.

The application of the process to small numbers of atomic charged particles leads naturally to the global interaction of single atoms. A number of features of the Periodic Table of the Elements are explained by the theory. Foremost among them is the unique chemical inertness of the group 18 elements and with it the length of the periods. The uniformly high first ionisation energy of the Group 18 elements.

Potential and Actual

The concept of potential has a long history. The pre-Socratic Greeks used the concepts of 'potential' and 'actual', but it was Aristotle who later brought it to prominence and applied it widely in debates on the subjects of change and the natural world generally. Aristotle regarded potential and actual as equally important aspects of reality, they were ontologically equal analogues of one another. Underlining that neither were a more significant form of reality than the other. And importantly, neither made sense without the other. Potentia are entirely immaterial[1]. It is the necessary precursor to the physical entities that abound. They

---

[1] The classical use of the term 'potential' should not be confused with its use in contemporary physics. Potential energy, for example, is not the kind of potential used here.

become actual as the result of some process, which is here referred to simply as 'actualisation'.

From the perspective of present-day physics, the concept of potentia has the limitation that it does not readily lend itself to the quantitative analysis of data. However, the ancients had not sufficiently advanced the application of mathematics to natural science, a milestone which had to wait for over a thousand years, until the Oxford calculators and their contemporaries came to prominence. Thus prior to the 13th century potentia was a normal mode of analysis.

By the Late Middle Ages the utility of the concept of potentia had been in decline for some time. It was finally supplanted by the 13th and 14th centuries with the advent of the Oxford calculators of Merton College, who were in search of more precise means of representing motion. Quantitative methods, whose real virtue was precision, were about to displace the merely qualitative methods of the ancients.

Ordinary language discourse found potential a comfortable analytical tool; one of its virtues being its qualitative nature, which well-suited philosophical discussion, but it did not lend itself to quantitative analysis of data, especially numerical data. Its eclipse in the 14th century was completed by the rise to prominence of the Oxford calculators of Merton College. Thomas Bradwardine, William Heytesbury, Richard Swineshead and John Dunbleton, were searching for the mathematical means that would enable quantification and thereby introduce precision to the study of motion, and especially acceleration, through space. They built on the earlier work of Walter Burley and Gerard of Brussels and effectively began the revolution which saw the mathematical expression of physical quantities displace the imprecise, qualitative description of the manifestations of change. What would become classical physics spawned a mathematical edifice that prevails over 600 years later.

Galileo, Newton and Einstein, serially, with other mathematicians, made calculation an art form for physical modelling which was to achieve, by the time of Niels Bohr and quantum mechanics, the transition to post classical physics of wave/particles, superposition of states, entanglement and the rest of the quantum mysteries.

Werner Heisenberg introduced Aristotle's concept of potentia into quantum mechanics, which had brought about a crisis in epistemology in the first decades of the twentieth century. This time, the introduction of potential was sudden by comparison with its early development by the ancients. Quantum mechanics, according to Bohr, needed no deeper knowledge than that the formalism worked - the need to understand the details of the behaviour of quantum systems was little more than folly. Bohr's theory of complementarity and Heisenberg's uncertainty principle were all that were required, according to Bohr. Many of Heisenberg's contemporaries were simply not impressed with the use of the ancient concept of potential and its actualisation in Nature, they thought it a regressive step. Heisenberg was not put off; like Aristotle, he believed in an immaterial ontological phase of normal matter, and applied it to quantum mechanics.

Following the physics successes of the first 30 years of the twentieth century, it was very difficult to interest the scientific community in the qualitative analysis of data. However, some quantum physicists have followed Heisenberg's lead and applied the concept of potentia to the quantum problem. A detailed analysis is developed in the work of Michael Epperson [11]. Others include: Robert Griffiths, Murray Gell-Mann, James Hartle and Roland Omnès.

Aristotle, Heisenberg and Epperson and the present author all agree that potential and actual are equally important elements of reality; they are ontologically sound. A potential electron is as ontologically real as an actual electron.

The Oscillation of the Dirac Electron

Dirac's comments on the interpretational difficulties of the physical consequences that follow from his equation for the electron include the following:

> "It is found that an electron which seems to us to be moving slowly, must actually have a very high frequency oscillatory motion of small amplitude superposed on the regular motion which appears to us." [12]

It is at this stage of the narrative that the Dirac electron and the Aristotelian doctrine of the 'actualisation of potential' are coupled. We postulate that the oscillatory motion of the electron is a continual physical process of actualisation of the immaterial potential of an individual, ideal electron. As depicted in Figure 1, the oscillation begins in a state of pure formless, immaterial, potential at t_0. Actualisation follows gradually, building towards an actual, material electron at t_1, the material electron then decays to a massless potential state. The oscillation process is energetic and takes time. The energy is employed during the phase of actualisation which, in principle, cannot be determined until the process has concluded. Precise details of the energy could only be determined for the period of around a zepto second[2], quite reminiscent of Heisenberg's time energy uncertain relations[3].

Each oscillation cycle inherits potential from its immediate antecedent oscillation. The oscillation is analogous to the initial and final states that precede and follow a quantum mechanical evolution and measurement. A single cycle of the oscillatory process is an electron event.

During the course of actualisation, the property of electric charge is actualised. Virtual photons are thus able to deliver potential they have inherited from the charged particle they were created by. Thus there are two sources of potential acquired by the actualising electron event. One is genetic potential inherited from itself, which is important in the maintenance of its identity. And the other is mainly geometric potential which assists in the formation of its geometric relations with other particles, particularily the electron or proton that supplied the virtual photon. Upon completion of the actualisation the electron is briefly a motionless, classical particle, which is the culmination of an electron event. The actual electron then decays instantaneously into a state of pure potential from which the next oscillation cycle begins once again.

---

[2] A zepto second is 1 x 10^{-21} of a second.

[3] The Einstein mass-energy transformation relation is postulated to be involved in the oscillation. The actual state of the electron includes its mass. That state decays to a state of immaterial potential. Energetic actualisation terminates in a classical electron, which has the usual properties, including mass.

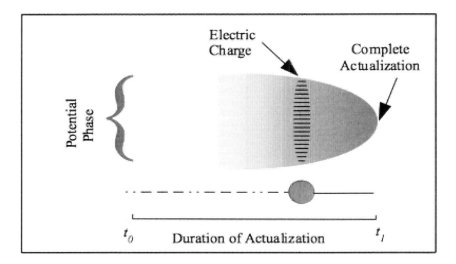

Figure 1. Depicts a single oscillation of the Dirac electron, which is a single electron event. A ball and stick cartoon illustration of the same process is also shown, with electric charge depicted as a filled circle.

Pure potential, being immaterial has no properties, including geometric relations of distance, direction and duration with anything, including its own antecedent electron events. Yet to be actualised pure potential has no space or time relations—it is not somewhere. Space and time are not always everywhere, for elementary particles. The electron is in a vacuum state[4], from which it emerges during the process of actualisation. During actualisation the electron acquires its properties, which include electric charge, mass and geometrical locus.

The process of actualisation of the electron has an organic flavour; it is a development of an immaterial potential system (electron) into a material actual object. The process is quantum mechanical; the end of the process is a fleeting, stationary classical state of the electron. The sum of the quantum process and its classical manifestation or final state constitute the electron event. In quantum mechanical language, the process of actualisation is the evolution of the initial state. The termination of actualisation is the final or stationary state of an oscillation, which also serves as the initial state of the following oscillation. Self actualising potential is always preceded by an actual object, which is its chief source of potential. Potentia always comes from somewhere, in the case of elementary particles that usually means one or more motionless objects.

The frequency of the oscillation is $\sim 2.5 \times 10^{21}$ cycles per second. Its amplitude was shown by Schrödinger to be $\sim 1.9 \times 10^{-13}$ meters. The present theory is extended from the electron to the individual photon and the two atomic nucleons. The oscillation transforms the continuous elementary particles of classical physics into series of actual events which are discrete in space and time. To summarize, the Dirac electron is an oscillation between states of a) zero-energy, immaterial potential and b) energetic actualisation of that potential; the oscillation terminates with a fully actual motionless electron.

---

[4] The vacuum is a state of an individual particle. It's not a space that contain multiple electrons etc. It is the condition of the occurrence of an electron's transformation from potential to actual states.

The Enduring Electron

The ideal one-electron system is comprised of a series of motionless actual events, separate from one other in space and time. Contiguous events of a single electron are separated by immaterial potential, which actualizes to form the next event. The locus of each event, like all electron properties, is influenced by the potential of the event(s) it actualised. Thus, where and when an event occurs is influenced by where and when its immediate antecedent event occurred, and the source of the influence that passes from the earlier event to the later event is its potential. The potential was a serial constituent of both events. This analysis is interpreted as a self-interaction of a one-electron system. Unlike the requirement of quantum field theory, in which an electron acts on itself by the creation of a photon which it subsequently annihilates, here the electron creates its own potential phase and then inherits it, which has the effect of supplying its own endurance. Potential phases bind serial actual phases into an enduring electron. Phases of immaterial potential of the oscillation behave as bosons, while the actual material phases behave as fermions.

The concept of the organic boson first arises with the one-electron system. There, the oscillation gives rise to serial actual events that possess electron properties, including loci in space and time, which constitute part of the classical, physical aspect of the electron. Contiguous events are not physically linked to each other; they are not joined, they are where they are because they were put there by their own action. Where and when an event is located is subject to influence of its own immediate antecedent actual event. This is the bond between contiguous events which confers endurance on the single, ideal electron. An analogy is an electron bubble chamber experiment characterised by the series of *separate* bubbles that form a track; another is seen in cinematic motion pictures which consist of separate stationary images. Nothing moves in the movies

The actualising potential for an oscillating electron plays the rôle of a boson which links pairs of contiguous events in the life of a single electron. The potential phase binds the event that preceded it with the event that follows it. Two contiguous actual electron events are fermions. The potential phase of the oscillation between them is the boson that binds them. The boson is serially a component of each of the two bound fermions. The oscillation continually transforms the electron between bosonic (Bose Einstein or BE) and fermionic (Fermi Dirac or FD) states. Bosonic and fermionic states are serial for a single electron. The enduring electron is an oscillation between motionless fermionic events separated by states of nonlocal bosonic potential.

The oscillating electron shows hints of rudimentary organism; it inherits its immaterial potential from its own preceding electron event and grows it to form a new physically separate material electron-event. The electron grows from zero-mass potential to physical actuality with mass. Each oscillation inherits its generic potential from preceding oscillations; complete actualisation culminates in a material actual electron, which finally decays (perishes in process terminology) to a state of immaterial potential. Reminiscent of the observation that rabbits beget rabbits, electron events beget electron events. The actualisation process is then repeated.

The Photon – Real and Virtual

Light, and especially the speed of light, sometimes presents conceptual difficulties and can cause confusion, because different experiments find that light moves at different speeds[5]. Einstein postulated that the speed of light in a vacuum is constant, and is independent of the speed of its source. And the speed of light has been found to be ~ 3.0 x 10^8 ms^{-1} (usually designated as "c"). The difficulty arises because quantum mechanical experiments have often found that light also moves at much higher speeds, at or close to instantaneously. When analysed from the standpoint of the present event scheme, the different light speeds are reconciled, as follows.

Electric charge is the property of some elementary particles which enables them to exert influence on one another. The agent for the influence is the photon. One of the counterintuitive consequence of the Dirac equation is that the motion of the electron occurs at the speed of light. This consequence of the equation is directly related to the oscillation of the electron. It is hardly surprising that such a proposition is highly controversial.

Photons are discrete quanta of light of zero mass and a wide range of individual energy levels. Famously, photons are both waves and particles. One way to think of the wave/particle duality puzzle is to imagine that a photon can be either a wave or a particle or neither, but only one at the same time. When a photon interacts with a particle detector a particle is detected, when a photon interacts with an antenna a wave is detected. Before either interaction occurs, the photon is neither a wave nor a particle, its the third one. In the present scheme, the photon is immaterial potential to actualise as a particle only if a particle detector participates in the actualisation, and actualises as a wave only if an antenna participates in the process of actualisation. The detector or antenna has to be an active participant. There is no generally accepted theory of quite what photons are before they turn into either a wave or a particle, or how they do it. Photons divide into two groups: real photons and virtual photons.

The *real* photon oscillates between actual and potential states on journeys across a room or across a galaxy. The duration of the journey, at ~ 3 x 10^8 m/s, is explained as due to the sum of the durations *taken* for the real photon to complete the actualisations of its multiple oscillations, during the journey. The real photon changes its position instantaneously between serial contiguous actual photon events, which together sum to the duration of the photon's journey. Among bosons and fermions generally, the photon has the shortest time of actualisation, which coincides with its not experiencing actualisation that produces mass. (see below)

Real photon effects are long range, by comparison with virtual photon effects. They carry heat and light across a room and across the solar system, and beyond; they carry electromagnetic communication, radio, TV, etc. One of the oscillation rules that apply to all the component particles of the atom is: actual events are influenced by the events whose potential they actualise. Real photons do not have a central rôle in the present electron theory.

The *virtual* photon, by contrast, is pure immaterial potential. Like the real photon it also changes its position instantaneously. However, it is created by a charged particle during a potential phase of the particle's oscillation and is immediately annihilated by another charged particle during an actualisation phase of its oscillation. Therefore the virtual photon completes

---

[5] The movement and speed of light in this chapter is used as a familiar shorthand for 'a change of position without motion'. In the present theory electrons and photons are stationary, they do not move. However they do change their position in both space and time. Classical motion of bodies is an emergent phenomenon.

just a single oscillation between its creation and annihilation. Before it has the chance to actualise as its own photon event, it co-actualises with an actualising event of a charged particle (electron or proton) as part of its annihilation. Consequently it co-actualises just once during the actualisation phase of the charged particle event which annihilates it. Being virtual also means it does not experience a space or time interval. Its brief existence is instantaneous.

Virtual photon effects are generally short range. They are the electromagnetic bosons among charged particles, which are the force particles. The forces they mediate include Coulomb attraction and repulsion. They bind atoms together in the formation of chemical molecules. They bind electrons to nuclear protons in the formation of atoms. They drive electrons in the production of electric current. Coordinated virtual photons produce magnetic effects.

Studies of quantum tunnelling have shown that the time taken for a virtual photon to tunnel across an impenetrable barrier 1 mm thick is extremely short. That duration is referred to as the photon dwell time. When the experiment is repeated with the impenetrable barrier thickness increased to 2 mm, the tunnelling time is the same as that for the 1 mm barrier. This is interpreted to mean that quantum tunnelling across impenetrable barriers is an instantaneous process with a small dwell time delay. This phenomenon is the Hartman effect [13] and was discovered in the 1960s. Light changed its position instantaneously, without moving.

Being a theory of *direct inter-particle action* the concept of the field theory of classical electromagnetism of the 1700s, or its recent quantum derivation is not required. Which has the advantage that the electron self-interaction is not mediated by virtual photons, which is consistent with the absence of a theory to explain how an electron emits a photon and then overtakes and annihilates it [14], without a clash with special relativity. The classical field is still with us, it has lasted since Michael Faraday's time in the eighteenth century.

Kinematics

Electron–Electron Interactions

Photon annihilation and creation are the terms used to denote the absorption and emission respectively of real or virtual photons by charged particles. In Figure 2 electron events are depicted by horizontal lines and the property of electric charge by circles on the lines.

As its property of electric charge becomes effective during actualisation, each electron event (Figure 2) annihilates a propitiously related, virtual photon. At the termination of the oscillation the actual electron decays, the result of which is its transformation into a state of pure potential. The potential which is thereby released has two components: a) the *genetic* potential for an actual electron, which will actualise as the electron's next event, and b) the photonic potential which came with the propitiously related virtual photon. This process is assumed to occur among electrons generally. The principle that a charged particle event will be influenced by the source(s) of the potential it actualizes applies generally, not only to electron-electron interactions.

Potential conveyed by a virtual photon derives from the event that created the photon, which is usually a charged particle, (an electron or proton) mediates influence to the actualisation of the electron. Influence transferred as part of the potential is likely to embrace space and time relations among actual events of the immediate region. If a real photon is annihilated by an electron, its transferred potential will have derived from the last photon event in its journey. Its potential could originally have come from the Sun, in which case its energy is conserved but its original geometric relations with other particles is lost among the vast number of serial events the photon experienced since leaving Sun.

Photon annihilation and creation, respectively, are postulated to occur serially during each oscillation of the ideal electron (Figure 2a). When the phases and space separation of two oscillating electrons are suitably related, a virtual photon created by one electron at its potential phase (which occurs at the termination of actualisation) is annihilated by the second electron during its phase of actualisation, when its property of electric charge becomes actual. Figure 2 (b) shows the instantaneous transfer of a virtual photon between two propitiously related electrons e_2 and e_3, which mediates the formation of a bound electron pair. The bond may be of minimal duration, completed just once, or it may be iterated by the same two electrons at subsequent pairs of events.

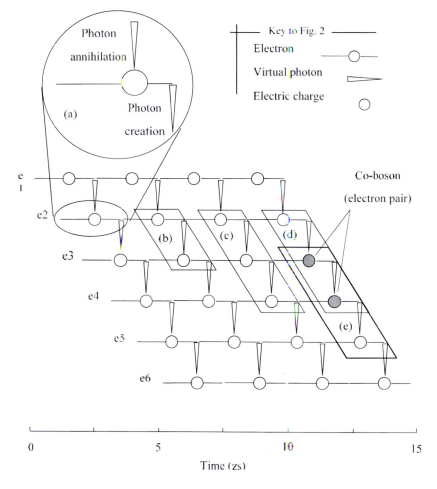

Figure 2. Four events in the life of six oscillating electrons, that form an ideal one-dimensional electron plasma. Photon annihilation and creationare depicted at (a); electron pair formation at (b), treble electron formation at (c), and the treble electron interaction at (d, e).

As seen in Figure 2 (b), the key to the electron pair formation is the single photon which is created by e_2 and annihilated by e_3. The potential of the binding photon derives from e_2 and is co-actualised by e_3. Bond formation occurs in just one direction, there are no reciprocal boson exchanges among bound fermions, in the present scheme. The immediate past supplies the potential for a future actualisation, the major dose is genetic endogenous potential and the other is exogenous geometrical potential. The external transfer between the two electrons augments the genetic potential. Neither geometric nor genetic potential can transfer from the present to the past.

An important consequence of the present scheme is that the bound electron pair is never isolated from other charged particles by way of virtual photons serially annihilated and created. The addition of a third electron to the pair forms an electron treble. Figure 2 (c) shows two photons connecting three electrons in the formation of a treble interaction; the three electrons are not contrary to the Pauli exclusion rule because they form a serial pair of two-particle antisymmetrical ensembles. Two further photons connect the three-electron interaction to an outside source in the past and a sink in the future.

The Pauli Exclusion Principle

Wolfgang Pauli discovered his exclusion principle in the mathematics of the wave function. He knew that it divided wave functions into two classes that differed in terms of their symmetry relations. In addition he also realised that the difference hinged on what he called "classically non-describable, two-valuedness" but initially what the values referred to was not clear; a property had yet to be found. Finally, electrons were discovered to have quantum spin among [15] their properties and Pauli agreed, after considerable persuasion to adopt spin as the classically non-describable property. His principle excluded all but the antisymmetrical two-particle ensemble.

In the present theory, compliance with the Pauli exclusion principle arises naturally. For two quantum mechanical systems to assemble, the mathematics of the two systems (electrons) have to be antisymmetrical. All other combinations are excluded. In the present scheme, one member of the ensemble that constitutes a compound quantum system is in a state of zero energy, being purely potential while the other is in a state of energetic actualisation, which is obviously energetic. A shared virtual photon forms the quantum ensemble, as shown in Figure 2(b). The process can only be anti-symmetrical because, simultaneously, one member of the ensemble is in a zero-energy potential state (Bose Einstein state) of its oscillation, when it creates the virtual photon, and the other member of the ensemble instantly annihilates the photon during its state of energetic actualisation (Fermi Dirac state).

In the discrete scheme, the antisymmetry of the two-electron ensemble arises from the absolute energy difference between the two electrons. This contrasts with the spin antisymmetry of the conventional picture in which angular momentum of the two electrons is geometrically counter-oriented in space. The two energies differ geometrically, because one has spin up and the other has spin down. The principle of the absolute energy difference for the members of an oscillating electron pair extends naturally to larger ensembles of interacting electrons.

The Trebling Imperative

Trebles of charged particles are the focus of attention here because they are a primitive form of structure which results naturally from the oscillation. Each actual electron event, because of its property of charge connects with its close environment in such a way that it both influences and is influenced by other charged particles of that environment, thus forming a rudimentary association. Because of a time constraint the growth cannot go beyond three charged particles (see Figure 3).

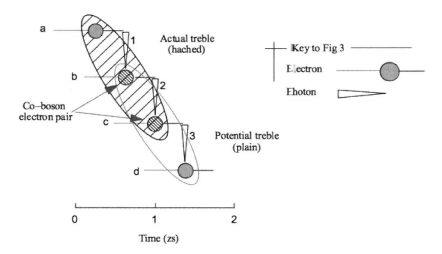

Figure 3. One event in the life of each of four electrons, which form the treble electron (hached) and the treble electron interaction are depicted. The four electron events are serially bound by three numbered photons. One treble is actual (hached oval) and the other is potential (plain oval).

Figure 3 (enlargement of Figures 2 (d) and (e)) depicts the scheme for the electron treble and the electron treble interaction. The treble interaction is the binding interaction between two electron trebles, one of which is actual and the other is potential. This element of the kinematics is particularly important for the theoretical treatment of atomic electrons, in addition to electron-only plasmas.

The electrons depicted in Figure 3 form two trebles; the first treble is actual (hached oval) and the second treble is potential (plain oval). Two electrons, with hached electric charge, are common to both trebles. The actual treble components are electrons a, b, and c. The active principle in the formation of the actual treble is the pair of photons 1, and 2. Approximately half the duration of a zepto second after the creation of photon 1, photon 2 is created, thus completing the actual treble. While photons 1 and 2 are active, photon 3 is yet to be created; the potential treble has not begun to actualize. The creation of photon 3 has four effects: (i) the actual treble decays to a state of immaterial potential, (ii) the potential treble is actualised and (iii) the electron pair co-boson ceases membership of the formerly actual treble, and takes up membership of the new actual treble, all without moving and (iv) the electron pair co-boson thereby forms the bond between the two trebles. The mass of the co-boson when combined with the three photons gives the interaction a good deal more heft than the photons alone.

The kinematics of the treble electron is inevitable, given the development of electric charge as described in an environment of oscillating electrons. For electric charge to actualise as part of the oscillation, it needs to do so well before the overall actualisation has completed. It is only by being ready to annihilate photons in time for photonic potential to be incorporated into the electron event, thus enabling the flow of influence to continue. Only after actualisation can the event emit its own photon. The incoming photon must precede the outgoing photon, if influence (photon potential) is to continue to spread. It is this state of affairs that enables the process of trebling among an electron plasma, which is the precursor to the actualisation of potential in the atomic sector.

The actualisation process has moved up the scale from the single system, then the electron pair to the next step–the electron treble. The two and three electron ensembles are mediated by photons alone. The one, two and three electron systems have not achieved the status of an isolated interaction. The outside source and external sink are still necessary for the actualisation of potential.

Figure 3 shows how Pauli exclusion is obeyed, since two systems do not occupy the same state at the *sametime*. One of the surprises of the Dirac equation was that time was required for a thorough description of the electron. This element of the present scheme is an instance of natural Pauli exclusion.

The Composite Boson and Electron Treble

It is only by being ready to annihilate photons in time for photonic potential to be incorporated into the electron event, thus enabling the flow of influence to continue. In addition, according to the present theory, bosons are serially integral elements of the fermions they bind. The boson-fermion distinction is of some importance in the present theory. Individual particles are readily classified as fermionic or bosonic, whereas grouped particles can present difficulties. Whether a particle is bosonic or fermionic depends upon its wave function, and ultimately its spin statistics. Composite particles are expected to exhibit bosonic quality and components forming the composite are expected to be entangled.

At the nonlocal, potential phase of the one-electron oscillation, being immaterial, the system is not a matter particle and therefore it is not fermionic; it acts as a boson. The bosonic phase of the oscillation, like the virtual photon, behaves nonlocally, it transfers in one direction and is serially integral to both events that it binds. This is the origin of the principle of an actual event being bound to the source of the potential(s) it actualizes. An electron oscillates between BE and FD states.

A composite quantum system which consists of an even number of bound fermions has integer spin and is therefore analysed as a boson. [16] That constraint on composite systems limits the numbers of grouped electrons that can act as bosons. In the present chapter two co-bosons are the subjects of the analysis: (1) a single electron pair and (2) three electron pairs. Being composed of even numbers of electrons, i.e., 2 and 6, and provided they are sufficiently entangled, both are said to have integer spin and thus can occupy the BE state. Importantly, in the atomic sector both co-bosons are part of charged-particle trebles (see below).

The electron treble is depicted in Figure 2(c). The interaction among three bound electrons is virtual-photon mediated. Because individual electrons and odd-numbered bound groups of electrons have half integer spin, the treble electron is a composite *fermion*.

Figure 3 shows one actual event of each of four serially bound electrons that constitute, on the present analysis, one actual treble and one potential treble. The actual treble consists of three of the four electrons, the potential treble is then immaterial; when it subsequently actualizes it consists of the fourth electron plus the electron-pair; this is a natural consequence of the passage of time. The electron pair (e3 and e4) is analysed as a co-boson during its one-way transfer from the first treble to the second treble. The transfer of the electron pair occurs as a fully composite co-boson indistinguishable from a single particle. Both electrons transfer simultaneously; they are fully entangled, and exhibit bosonic character. A potential electron treble consists of just one actual electron since it is deficient of an electron pair. Analogous with the virtual photon, the co-bosonic electron pair is serially a material constituent of the two fermionic trebles between which it forms the bond. Superposed on the co-bosonic bond are the virtual photonic bonds that bind the members of constituent electron pairs.

The issue attaching to the co-boson question is not whether two electrons can be bosonic, Dirac predicted bosonic electrons in 1974. [17] The issue is, can *these* two electrons be treated as co-bosons. The treble electron interaction is analogous with the virtual photon bond of the electron pair. The electron trebles differ only in that one is actual and the other potential, which means that their energies differ absolutely, one being positive and the other zero. Thus, they obey the Pauli exclusion principle by precisely the same analysis as the electron pair. Furthermore, having half-integer spin, both single and treble electrons behave as fermions. The concept of the co-boson has become a topic of increasing interest during recent times.

The Emergence of the Periodic Table

The Atomic Treble and its Interaction

In order to enter the atomic sector, protons are introduced to the narrative. The electron-electron interactions (see Figure 2 above) were all directed to the actualisation of potential, they were not however able to achieve that end without interactions with outside photon sources and sinks. The actual and potential trebles for the interactions were always related spatially and temporally to charged particles that had no direct rôle in the action. These were the outside sources and sinks. Having said that, we cannot describe the individual photon-electron interactions or their timing that make up the one-atom global interactions, they remain nonlocal. Any such description can be nothing more than a likely tale. However, we can detect a principle at work – the actualisation of potential.

The individual atoms of the natural elements each have a global one-atom interaction, among their protons and electrons. Some of these interactions occur without outside photonic sources and sinks and others, indeed most, occur only with the participation of outside photon sources and sinks. These interactions are best visualised in a particle framework rather than fields. At the level of the individual interaction, these are electron–electron only, and proton–electron interactions. The bonds formed by these interactions collectively are its global interaction and give the single atom a structure and a volume, as well as chemical propensity. Chemical propensity arises principally from the interplay of the single atom and its need for external photon sources and sinks.

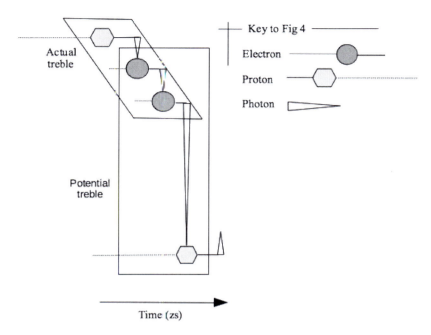

Figure 4. A diagrammatic representation of the components for the BAT, and its interaction. Two atomic trebles are framed. Two protons have their charge depicted as hexagons and two electrons' charges as filled circles, all of which comprise the charged particle ensemble of the basic atomic treble (BAT) interaction. Photons are depicted as vertical arrow blade. The photon annihilated by a proton is shown to be longer than those annihilated by electrons (see below, The Franck – Condon Principle).

The substitution of one proton for one member of an electron treble, binds an electron pair to the proton; which thus forms an actual BAT, bound by the one-way transfer of virtual photons shown in Figure 4. In order to preserve the principle of the conservation of electric charge a second proton is required, which happily completes the complementary potential atomic treble, and, like the potential electron treble, the proton lacks an electron pair. All the requirements for a basic atomic actualisation of potential are now assembled, as depicted in Figure 4. The two-electron two-proton system is the simplest single-atom interaction for the natural elements, beyond hydrogen. However, the interaction depicted in Figure 4 does not occur in isolation.

The two BATs shown in Figure 4 have the two electrons in common. With the passage of time the upper treble loses its proton and thus becomes potential, the electron pair then sends a photon to a second proton, thus actualising the potential treble. The two trebles are bound by the simultaneous one-way transfer of the electron pair, which therefore exhibit bosonic features. The bond occurs between two nucleons in a proton state. The process may be analysed as an electron co-boson provision of a Coulomb contribution to the nuclear bond.

As a consequence of the generalized Pauli principle for a bound atomic nucleus, in the context of quantum mechanics, a nucleon may be found in either a proton state or a neutron state. Consistent with this condition, in the present scheme an individual nucleon at an instant is either a proton or a neutron. For example, the deuteron nucleus is always a proton and a neutron, but each of its two nucleons is not always a proton or a neutron. When an oscillating proton is fully actualised it decays to a potential *neutron state*, without the property of electric charge. The proton decay also creates a virtual photon which, is shown in Figure 4.

One of the consequences of this difference is seen in Figure 4, in which the photon that is annihilated by the proton is depicted to have a long distance between creation and annihilation. Because the nucleon has a mass of more than 1,800 times that of the electron, and it switches between proton and neutron states, its behaviour as a charged particle differs from that of the electron, which needs to be accommodated by the theory.

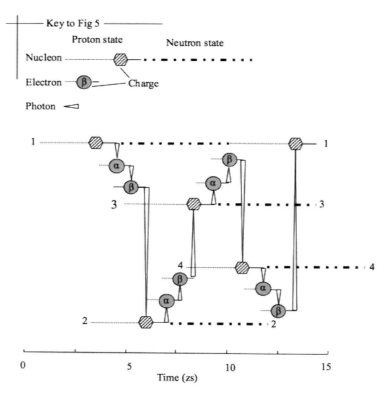

Figure 5 depicts the helium-4 atom electromagnetic interaction. Two electrons (α and β) and four nucleons (1 – 4) are shown. The interaction is equivalent to four BAT interactions. See Figure 4.

The Noble Gases

In order to get a sense of a global interaction for a single atom, an idealised depiction is shown in Figure 5, and a description follows:

A series of single photons go from proton 1 to electron α, then to electron β, then to proton 2, then to electron α, then to electron β, then to proton 3, then to electron α, then to electron β, then to proton 4, then finally to electron α then electron β and the last photon in the interaction goes back to proton 1.

The Franck – Condon Principle [18] has a major rôle in the interaction of electrons with protons. Basically the proton is a good deal slower than the smaller electron. Because elementary particles do not move in order to change their position in space or time, their relative *speeds* apply to the durations of their actualisations, not classical motion. It is the process of actualisation which brings about an apparent motion of the elementary particle.

The duration of photon actualisation is shortest, having zero mass, electrons are slower than photons and nucleons are the slowest of all the atomic components.

As a result of the generalized Pauli principle for a bound atomic nucleus, some nucleons are in a superposition of two states, a proton state and a neutron state. The oscillating nucleon is postulated to alternate between proton and neutron states. The alternating charge state of a proton member of a BAT truncates its participation in the atomic treble as its oscillation takes it from the proton state to the charge-neutral neutron state. The electron pair then interacts with a second oscillating nucleon, which is then in the proton state. Prior to the interaction, the second proton is analysed as a potential atomic treble, because like the potential electron treble the one–proton potential atomic treble is deficient of an electron pair.

At the centre of the BAT interaction is the actualisation of a potential atomic treble, mediated by the electron pair which is analysed as the co-boson of the interaction. The one-way transfer of the electron-pair co-boson between atomic trebles (actual treble to potential treble) is the basis for the bond between the electrons and the nucleus of a single atom. In contrast to the many-electron interaction, the nucleus imposes a cyclic constraint on the one-atom global interaction. This is a direct result of the oscillation and the actualisation of potential. The electron collectivity[6] is shown in Table 1.

Like the one-electron system, the kinematics is cyclical. The collectivity of the actual and potential trebles of the helium–4 interaction are shown in Table I

Table 1. The interaction for the helium-4 atom

Element	Electron collectivity	Co-boson	Orbital configuration[a]
Helium-4	‖ **2**[b] ‖ ‖[c]	| 2 |	$1s^2$

[a]Orbital electron configurations are shown for comparison with electron collectivities
[b]Actual trebles are depicted in bold face and potential trebles and co-bosons in regular face
[c]For clarity, protons are omitted from the tables; only electron components of atomic trebles are shown. Thus, | 2 | means two electrons and one nuclear proton, which represents an actual BAT. A potential BAT is depicted by ‖ ‖, which represents a proton (not shown) and the lack of an electron pair

The Carbon Treble

The charged-particle ensemble of the carbon atom is an atomic treble of trebles, with an electron-pair which forms a low-mass co-boson. Its six electrons form a high-mass co-boson which is three times the mass of the electron pair (see Table 2). Consequently, the carbon atom electron ensemble (not a carbon atom) has a significant rôle in the kinematics of the atomic electromagnetic interaction among the medium- and high-mass natural elements, generally. Therefore its features are outlined here.

Moving up the mass scale of the natural elements from helium, the charged-particle ensemble of the carbon atom is the first element to form an actual global treble of trebles. One electron pair and a proton form a single treble; three electron pairs together with three protons form a treble of BATs. The carbon treble marks the first mass transition of the atomic treble;

---

[6] Electron collectivity is a static representation of charged-particle trebling. The simplest treble in the atomic sector is the BAT. The largest in the present chapter is the palladium treble composed of 46 electrons and 23 protons. Trebles larger than the BAT are composed of smaller trebles, thus forming trebles of trebles etc.

the electron component of the treble has evolved from the single electron pair of helium to the treble electron pair of carbon.

The trebling imperative and the principle of the bond formed between potential and actual trebles is unchanged. Accordingly, a single carbon atom does not form a treble *interaction* in isolation. The actual carbon treble needs a potential carbon treble (two electron pairs) in order to form a treble interaction. The three electron pairs, subject to the interaction imperative, tend to interact with two electron pairs from an outside source, as it does in the formation, for example, of the methane molecule.

Table 2. The electron treble of the carbon atom

Element	Electron collectivity	Co-boson	Orbital configuration
Carbon	‖ 2 | 2 | 2 ‖	| 2 |	$1s^2, 2s^2, 2p^2$

The co-boson for the carbon treble interaction is the same as the co-boson for the BAT interaction; the electron pair. As shown in Table 2, the discrete electron collectivity and the orbital configuration for carbon are the same.

The Neon Interaction

The first one-atom treble interaction beyond the helium atom is neon. It consists of one actual carbon treble and one potential carbon treble. Like the potential BAT, the potential carbon treble lacks an electron pair. Table 3 shows the discrete collectivity of a single neon atom to consist of an actual carbon treble, abbreviated to six electrons, and two BATs which constitute a potential carbon treble.

Table 3. The interaction for the neon atom

Element	Electron collectivity	Co-boson	Orbital configuration
Neon	‖ **6** ‖ 2 | 2 ‖	| 2 |	$1s^2, 2s^2, 2p^6$

Analogous to the pure electron treble interaction, one electron pair of the actual carbon treble forms an actual treble with the two electron pairs of the potential carbon treble. Thereby, that electron pair is the co-boson that binds the two trebles by its one-way transfer between them.

The neon global interaction also forms an actual treble, closely analogous to the BAT (see below) in terms of the mass distribution among its three components. Both consist of one high-mass component and two equal low mass components. Neon differs from the BAT in being a treble of trebles and like single trebles generally, it is not the basis of the one-atom interaction, which arises from its decomposition into the actual and potential carbon trebles. The electron collectivity mirrors the orbital configuration, as shown in Table 3.

The Argon Interaction

Like the evolution of the carbon treble into the neon interaction, the addition of a potential neon treble to an actual neon treble forms a complete global interaction for a single atom, which is the argon interaction. None of the natural elements whose charged-particle numbers lie between those of neon and argon forms a complete one-atom interaction. Like the carbon atom, argon forms a single equal-mass treble of trebles. But unlike carbon it is able to satisfy the imperative of a complete interaction by its decomposition into an actual and a potential treble (in this case, one actual and one potential neon treble), as shown in Table 4.

Table 4. The interaction for the argon atom

Element	Electron collectivity	Co-boson	Orbital configuration
Argon	‖ 6 | 2 | 2 ‖ 6 | 2 ‖	| 2 |	$1s^2, 2s^2, 2p^6, 3s^2, 3p^6$

The argon interaction represents a reversion of the form of the atomic interaction back to its origin, which is the pure electron interaction. Each BAT is analogous to an electron in the interaction of an electron plasma (Figure 2). The constraint on the argon interaction is that, unlike the pure electron interaction, it is cyclical. Whereas the kinematics of the pure electron interaction conforms to the principle of the bond without a cyclical constraint, the interaction among actual and potential neon trebles of the argon interaction is so constrained by one member of each BAT being a bound component of the atomic nucleus. The similarity of the electron orbital configuration and discrete collectivity is seen in Table 4.

The Krypton Interaction

The electron collectivity of the argon interaction cannot be treated as the neon treble by the addition of a potential argon treble, in order to find the next complete one-atom treble interaction. The krypton atom charged-particle ensemble does not form a global treble, and it does not decompose into an actual and potential treble. However, it decomposes into two identical trebles, each of which further decompose into an argon interaction, each of which comprises an actual and a potential neon treble. The two-valuedness of each argon interaction sub-system follows from the oscillation of each of the four neon trebles between an actual and a potential state. The two components of each argon interaction form a superposition of an actual and a potential state.

It can be seen from Table 5 that the mass equality of the two actual neon trebles and the two potential neon trebles enables two serial argon interactions, with the form of a two-particle (two-component) antisymmetric ensemble, each of which individually conforms with the Pauli exclusion principle. The two components of the global interaction are each composite bosons.

The exclusion principle allows compound systems of atomic electrons to form a single quantum system, provided the number of subsystems is restricted to two and each of the two individually complies with the principle. The krypton ensemble of electrons complies with those limitations, rendering it a bosonic ensemble of interacting particles. Therefore the two argon sub-systems are able to interact cyclically as a single quantum system. The two electron-pair co-boson transfers are nonlocal, occur one-way between actual and potential neon trebles and each is a constituent of the trebles they bind.

Table 5. The interaction for the krypton atom

Element	Electron collectivity	Co-boson	Orbital configuration
Krypton	‖ 10 ‖ 6	2 ‖ 10 ‖ 6	2 ‖		2	2		$1s^2, 2s^2, 2p^6, 3s^2, 3p^6, 3d^{10}4s^2, 4p^6$

Like the argon interaction, the krypton global interaction is cyclical. The first neon treble (Table 5) is preceded by the second potential neon treble giving the interaction the form of a closed loop. The global interaction is symmetrical and its two constituents are antisymmetrical, which is consistent with its being a composite boson with integer spin.

The two electron–pair co-bosons are independent of each other and represent a mass-transition from a single electron pair to a double electron pair for a one-atom interaction. Table 5 shows the similarity of the orbital configuration and discrete collectivity.

The Xenon Interaction

The coupling of the ensembles of charged particles of argon and krypton forms the next global treble that decomposes so as to satisfy the interaction imperative and the cyclical constraint. The xenon treble marks the second mass transition of the interaction among the natural elements. It can be seen from Table 6 that the xenon interaction is a high-mass analogue of the argon interaction, and takes the form of an actual and potential zinc treble.

Table 6. The interaction for the xenon atom

Element	Electron collectivity	Co-boson	Orbital configuration
Xenon	‖ 18	6	6 ‖ 18	6 ‖		6		[Kr] $4d^{2,2,6}5s^2, 5p^6$

The binding unit (co-boson) of the xenon interaction is the carbon treble, which represents an evolution from the electron pair for helium, neon and argon, in the low-mass sector, into the intermediate double electron-pair co-bosons of the krypton interaction, and finally the carbon treble of the xenon interaction and the high-mass sector (see below).

The co-boson for the xenon treble interaction (the carbon treble) undergoes a one-way transfer between actual and potential trebles of the neon and argon form (the zinc treble). The argon treble is the high-mass analogue of the carbon treble. The electron collectivity of xenon embraces the underlying BATs and carbon trebles but diverges somewhat from the orbital configuration, as depicted in Table 6.

The Radon Interaction

The electron ensembles of krypton and xenon are respectively double and treble the eighteen electrons of the argon atom. The next multiples of the argon electron ensemble are hafnium and thorium; neither of which form self–contained, one-atom interactions and neither do they comply with the Pauli exclusion principle, being four and five multiples of the argon ensemble, respectively. The exclusion principle restricts composite systems to the antisymmetrical two-particle ensemble.

The addition of a silicon treble to the hafnium ensemble gives the next one-atom ensemble to satisfy the interaction imperative in the high-mass sector; which is the radon interaction. The radon electron ensemble decomposes into the formation of an actual and a potential treble, which differ by a carbon treble (the co-boson of the high-mass sector).

The radon ensemble shows a unique mass relationship among the interacting components of the group 18 elements. Its two trebles, one actual and one potential, are composed of one low-mass treble and two equal high mass trebles, giving the interaction units a more massive form than the lower-mass noble gases. The silicon treble has the same form; one low-mass component and two equal high-mass components. It can be seen in Table 7 that the addition of the silicon treble to the hafnium ensemble transforms the 18 electron trebles into the high-mass form; which is the two argon trebles and one neon treble.

Table 7. The interaction for the radon atom

Element	Electron collectivity	Co-boson	Orbital configuration
Radon	‖ 18 ‖ 18 ‖ 10 ‖ 18 ‖ 18 ‖ 2 ‖ 2 ‖	‖ 6 ‖	[Xe] $4f^{14}$, $5d^{10}$, $6s^2$, $6p^6$

The radon interaction occurs between trebles that differ by a carbon treble. Like the noble-gas interactions generally, the co-boson is shown to derive from a low-mass treble. The form of the kinematics remains unchanged in the transition from the low- to high-mass sectors. Table 7 shows the interaction for the radon atom. Unlike the orbital configuration, the radon ensemble decomposes into an actual and a potential palladium treble. The electron collectivity diverges somewhat from the orbital configuration.

The Mercury Interaction

Mercury is an interesting element because it is not a strict monatomic gas; when ionised it can be diatomic. It has a high first ionisation energy and a low atomic radius, which generally accompany the noble gasses. Its one-atom interaction looks initially as though it might belong to a monatomic gas. However, two possible electron collectivities of mercury each entail a double co-boson. Table 8 shows that unlike the krypton interaction the mercury ensemble of charged particles does not decompose to form two identical interactions which accommodate the two identical co-bosons.

Table 8. Alternative electron collectivities for the mercury atom

Element	Electron collectivity	Co-boson	Orbital configuration
Mercury	‖ 18 ‖ 18 ‖ 6 ‖ 18 ‖ 18 ‖ 2 ‖	‖ 2 ‖ 2 ‖	[Xe] $4f^{14}$, $5d^{10}$, $6s^2$
Mercury	‖ 18 ‖ 18 ‖ 10 ‖ 18 ‖ 10 ‖ 6 ‖	‖ 6 ‖ 6 ‖	

In order for the electrons of an element to form a single-atom, chemically inert global interaction they must be able to form an actual and a potential treble which differ by a boson of a mass suitable for the actual treble. Mercury comes close.

The Helium–3 Interaction

Of the six members of the group 18 elements only the helium atom analysis is restricted to a particular naturally occurring isotope, i.e., helium–4. Helium-3 is the other stable isotope and it is proton–rich. Because helium-3 is unique among the stable bound nuclei of the natural elements, the kinematics of all the other naturally occurring isotopes are accommodated by the theory.

As shown in Figure 5, the oscillations of all four helium-4 nucleons are needed to allow the photon-mediated interaction to proceed without interruption. Because helium–3 consists of three nucleons, of which two occupy the proton state and one the neutron state, unlike He–4, only one proton can oscillate directly into a neutron state and be compensated by a neutron oscillation into a proton state. The other proton oscillation is immediately followed by a proton actualisation. This causes a delay in photon creation for one electron bond. The compensation is required in order to maintain nuclear charge conservation. Consequently, the helium–3 atom cannot behave according to the scheme shown in Figure 5.

Half the nucleon oscillations from the proton state will not occur in the appropriate sequence. The space or time relations among the charged particle components of the helium–3 nucleus will not be as propitious as they are for the helium–4 interaction. This deforms the kinematics by *stretching* the space for the photon transfer. Because of the special relativistic relation between time and space, the two-electron 'cloud' of He–3 will occupy a larger volume than the two electron cloud of the He–4 atom.

Atomic volumes of the noble gases are the smallest in their own region of the periodic table, because their global interactions are complete, and every electron is equally involved.

It is relevant that atomic density measurements have shown that the atomic volume of the He–3 isotope exceeds that of He–4 by 28%. [19] It is also of interest that the volume of the H–1 atom also exceeds that of He–4 by ~ 27%.

The He-3 and H-1 isotopes are therefore close to equal in volume to each other. Because the H-1 isotope nucleus has no neutron, it is not a bound nucleus and therefore the generalised Pauli principle for a bound atomic nucleus does not apply in that case. It is a surprising coincidence that two isotopes whose nuclear masses are so different but each has one electron without a neutron backup in the face of the the Franck – Condon effect.

These well-known facts of He–3 and H–1 are consistent with the present theory. Atomic properties are a consequence not only of atomic electrons, but also of both species of nucleon.

The Periodic Table

The explanation for the periodic table, offered by the theory is that each member of the group 18 elements uniquely have the number of electrons that, together with their nuclear protons, exactly form a pair of homologous trebles, which consist of a potential treble and an actual treble, and includes the full ensemble of electrons. The pair of trebles interact adiabatically with each other by a process that converts an even number of fermionic electrons, at the beginning of the interaction, to the bosonic state. The actual treble decays to a state of immaterial potential and the potential treble actualises; all without a need for photons from an external source or to go to an external sink. This process is unique to the group 18 elements and it is what underlies their inertness. No other natural element has the necessary electron ensemble for the interaction.

The lengths of the periods of Mendeleev's periodic table are due to the number of elements, when arranged in order of atomic number, that fall between successive noble gas elements which are members of the group 18 elements.

Bonding

Molecular formation is also accommodated by the theory. In that respect, the interaction for some diatomic gases is of particular interest. We write the reaction using the number of all the electrons of the molecule. We expect a diatomic gas to form complete trebles of some sort when in the form of a molecule – because it is a gas. Table 9 shows how the electrons are divided so as to form an actual and a potential treble; remembering that the trebles must be the same, differing only by the co-boson.

The CO and NN molecules are known to have respectively, the strongest diatomic bond, and the strongest homonuclear diatomic bond. They are the only two, in Table 9, that are bound by a 6-electron co-bosons. Which in the scheme of group 18 elements is found to be the co-boson for the middle and high mass sectors.[7] Table 9 shows the relationship between bond energy and interaction co-bosons for some diatomic gases.

Table 9. Bonding energy and co-bosons of some diatomic gases

Molecule	Electrons	Actual treble	Potential treble	Co-boson	E (kJ/mol)
HH	2	‖ 2 ‖	‖ ‖[a]	ǀ 2 ǀ	436
CO	14	‖ 6 ǀ 2 ǀ 2 ‖	‖ 2 ǀ 2 ‖	ǀ 6 ǀ	1077
NN	14	‖ 6 ǀ 2 ǀ 2 ‖	‖ 2 ǀ 2 ‖	ǀ 6 ǀ	945
OO	16	‖ 6 ǀ 2 ǀ 2 ‖	‖ 6 ‖	ǀ 2 ǀ 2 ǀ	498
FF	18	‖ 6 ǀ 2 ǀ 2 ‖	‖ 6 ǀ 2 ‖	ǀ 2 ǀ	157
ClCl	34	‖ 6 ǀ 6 ǀ 6 ‖	‖ 6 ǀ 6 ǀ 2 ǀ 2 ‖	ǀ 2 ǀ	239
BrBr	70	Kr (36)	Se (34)	ǀ 2 ǀ	190
II	106	Xe (54)	Te (52)	ǀ 2 ǀ	147

a. See table 1.

Conclusion

We conclude that the organic theory of the electron provides a satisfactory explanation of the unique inertness of the noble gases, and with it the period lengths of Mendeleev's Periodic Table of the Elements.

The theory is pre-atomic, its postulates derive neither from the atomic sector nor from chemical phenomena, and nothing is put in by hand. The theory is derivative of: (1) a fully physical interpretation of the Dirac equation and (2) the Aristotelian doctrine of the actualisation of immaterial potential and (3) Whitehead's doctrine of organic realism. The

---

[7] Stronger bonds, whether of single atoms or molecules, are expected to be associated with higher mass co-bosons.

theoretical element of major importance is the behaviour of the electron. It is not a new principle, but its application to the behaviour of the electron is novel. It is a process that was thoroughly understood well over two thousand years ago. What is the electron doing that makes it turn out the way is does? It is actualising potential.

Physically real waves, shells, orbitals and fields are not required. And electric charge has the comparatively simple rôle of a quantum property of some individual fermions. It is a scheme of discrete, direct, inter-particle action. There are no variable parameters, or nuisance infinities.

The theory is novel, but strongly supported by its congruence with well-known chemical phenomena. From the standpoint of quantum physics, the theory fits well with the Copenhagen interpretation. In essence it is prior to quantum physics, and yet Heisenberg's time-energy uncertainty relation, the Pauli exclusion principle and the absence of properties between initial and final states of a quantum measurement, all fall out of the theory naturally

The answer to Einstein's expectation for the quantum system is perhaps not quite what he expected. But it is a description, a rudimentary organic description, which refers to the individual system. That puts it prior to quantum mechanics, without conflicting with post nineteenth-century physics. Dirac's presumption that ". . . a more drastic revision of our fundamental concepts" is arguably instantiated by this theory. There are distinct features of reproduction, inheritance and growth that culminate in an actual real individual. These features of an elementary particle stand in stark contrast to classical and post classical physics of the last century. It is a major revision of canonical concepts–and *it works*.

The new theory provides a description which is (1) clearly defined, (2) economical of postulates, (3) logical and (4) congruent with several chemical phenomena.

The theory offers alternative natural explanations for several well known facts:

1. The observed unique chemical inertness of the six noble gas elements is explained by an extension of the theory of the single, ideal electron. Chemical inertness of an element is due to the number of electrons and protons of a single atom being exactly divisible into two interacting homologous trebles, one of which is actual and the other is potential. The global single-atom electromagnetic interaction does not include outside sources of or sinks for virtual photons. There are only 6 such natural elements which can comply with these provisions, and they form group 18 of Mendeleev's Periodic Table of the Elements.
2. The lengths of the periods of the periodic system are simply defined by the number of elements that lie between successive members of the group 18 elements, when serially arranged by atomic number. All the natural elements that fall between the six noble gases lack the precise electron numbers that form a pair of homologous trebles without an excess or deficit.
3. Co-boson bonding and like-charge attraction among electrons and protons.
4. The electron orbital configurations of the lower mass noble gases are given an explanation from first principles, unconnected with quantum numbers or chemical data.

Acknowledgments

I am grateful to Vincent Powell for his helpful discussions and to the reviewer for his careful reading of the Ms. and suggestions.

References

[1] Scerri, Eric R. (2007). *The Periodic Table: Its Story and Its Significance.* 203. Oxford: Oxford University Press.

[2] Einstein, A., Podolski, B. & Rosen, N. (1935). "Can Quantum-Mechanical Description of Physical Reality Be Considered Complete?" *Phys. Rev.*, **47**, 777.

[3] Butterfield, J. &Isham, C. J. (2001). "Spacetime and the Philosophical Challenge of Quantum Gravity." In *Physics meets Philosophy at the Planck Scale*, edited by C. Callender, and N. Huggett, 33–89. Cambridge: Cambridge University Press.

[4] 'tHooft, G. (2005). "Determinism Beneath Quantum Mechanics." In *Quo Vadis Quantum Mechanics*, edited by A. Elitzur, S. Dolev, and N. Kolenda, 111. Berlin: Springer.

[5] Home, D. (1997). *Conceptual Foundations of Quantum Physics.*, *17*. New York: Plenum.

[6] Einstein, A. (1955). "Einleitende Bemerkungen über Grund begriffe." In *Louis de Broglieung die Physiker*, 318. Hamburg: Claasen Verlag.

[7] Dirac, P. (1930). *The Principles of Quantum Mechanics*, 255-56. Oxford: The Clarendon Press.

[8] Whitehead, A. N. (1929). *Process and Reality.* corrected edition, edited by David Ray Griffin, and Donald W. Sherburne, 1978. New York: Free Press.

[9] Heisenberg, W. (1958). *Physics and Philosophy: The Revolution in Modern Science*, New York: Harper and Row.

[10] Aristotle, *Physics* Book III $200^b 26 - 201^a 19$, trans. 1999 by R. Waterfield, New York: Oxford University Press., 56–59.

[11] Epperson, M. (2004). *Quantum Mechanics and the Philosophy of Alfred North Whitehead*, New York: Fordham University Press.

[12] Dirac, P. (1965). *Nobel Lectures Physics 1922-1941.* 320–325. Amsterdam: Elsevier.

[13] Wheeler, J. A. & Feynman, R. P. (1949). "Classical Electrodynamics in Terms of Direct Interparticle Action." *Rev. Mod. Phys.*, **21**, 425-433.

[14] Hartman, T. E. (1962). "Tunneling of a Wave Packet." *J. Appl. Phys.*, **33**, 3427.

[15] Goudsmit, S. & Uhlenbeck, G. (1925). "Ersetzung der Hypothese vom unmechanischen Zwang durch eine Forderung bezüglich des inneren Verhaltens jedes einzelnen Elektrons" *Naturwiss.*, **13**, 953.

[16] Ehrenfest, P. & Oppenheimer, J. R. (1931). "Note on the Statistics of Nuclei." *Phys. Rev.*, **37**, 333.
[17] Dirac, P. (1974). *Spinors in Hilbert Space*, *84*. New York: Plenum.
[18] Condon, E. (1928). "Nuclear motions associated with electron transitions in diatomic molecules". *Physical Review.*, **32**, 858–872.
[19] Ifft, D., Edwards, R., Sarwinski, R. &Skertic, M. (1967). "Solubility Curve and Molar Density of Dilute He3-He4 Mixtures." *Phys. Rev. Lett.*, **19**, 831.
[20] Pauli, W. (1964). *Nobel Lectures Physics*, 1922-1941, 27-43. Amsterdam: Elsevier.

In: An Essential Guide to Electrodynamics
Editor: Norma Brewer

ISBN: 978-1-53615-705-5
© 2019 Nova Science Publishers, Inc.

Chapter 2

Maxwell's Electromagnetic Equations, Elementary Introduction

*George J. Spix and V. M. Red'kov**
Institute of Physics, National Academy
of Sciences of Belarus, Minsk, Belarus

Abstract

There is a need to present the detail mathematical steps that are required to prove the equations of Maxwell Text books and course instruction do give students a firm grasp of the equations and their applications. What is often missing is the step by step exposition leading from the basic experiments to the established equations. The following book fills this gap admirably. These are presented for students and erstwhile students who are interested in how the great physicists derived and proved their equations.

All of the math and physics presented here are more than covered in the four year college engineering course. However, students with a passing interest in high school math can easily understand all that is contained in these three papers. For some readers, there will be the revisiting of old friends. Other readers will get the "Ah, Ha!" experience. The examples in using the equations are presented in great detail and are easy to follow.

Good Luck, Any questions?

Introduction

James C. Maxwell (1831-1879) published his electromagnetic field equations in 1864. He brought together previous experimental work and concepts of *Gauss, Ampere and Faraday*, and his own knowledge of mathematics to present his analysis of electromagnetic fields.

With his field equations, Maxwell calculated the speed of electromagnetic propagation. The result showed that the speed of propagation was equal to the speed of light, which

---

*Corresponding Author's E-mail: v.redkov@ifanbel.bas-net.by.

indicated that light is also an electromagnetic field. Maxwell predicted the discovery of the radio waves that Hertz found in 1888. Maxwell's equations are the bases for the theory of electromagnetic fields and waves. They are used in the design of antennas, transmission lines, cavity resonators, fiber optics and solving radiation problems.

This tutorial book derives Maxwell's equations using the methods followed by college texts, and as such must start with the laws and experiments of Gauss, Ampere and Faraday. In the format followed by the book, one of Maxwell's equations, in integral and differential form will be stated at the beginning of each section followed by its derivation. Then one example will be given in the use of the equations.

We will find the velocity of electromagnetic radiation. The calculus required for the derivation is presented in detail to remove the difficulties of the mathematics from being an impediment to the students in making the equations their own. Within Maxwell's equations the Electric and Magnetic vector fields are interdependent. A section presented here disentangles these vectors by using vector analyze and Maxwell's equations. The solutions of these derived equations are sine waves.

Therefore, Maxwell's equations are also the bases that enable scientists to use sine waves to mathematically represent electromagnetic radiation, whether the source is a star, an electronic sine wave generator, or a spark transmitter.

The final section derives Poynting's equation, and demonstrates where electromagnetic radiation theory and circuit theory can be joined.

1. Maxwell's Electromagnetic Field Equation I

The following electrostatic field equations will be developed in this section:
  integral form

$$\oint_{Surface} \mathbf{D} \, d\mathbf{a} = \int_{Volume} \rho \, d\mathbf{v}, \tag{1.1}$$

  differential form

$$\nabla \bullet \mathbf{D} = \rho, \quad \text{or} \quad \text{div } \mathbf{D} = \rho. \tag{1.2}$$

Maxwell's first equation is based on Gauss' law of electrostatics published in 1832, wherein Gauss established the relationship between static electric charges and their accompanying static fields. The above integral equation states that the electric flux through a closed surface area is equal to the total charge enclosed. The differential form of the equation states that the divergence or outward flow of electric flux from a point is equal to the volume charge density at that point.

1.1. Area Integral

We will derive the integral equation by considering the summation of electric flux density on a surface area, and then as a summation of volume containing electric charge. The two integrals are shown to be equal when they are based on the same charge. Two examples using the equations are shown.

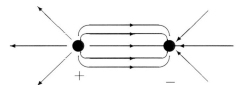

Figure 1.1. Lines of electric flux.

Gauss' Law. Gauss' electrostatics law states that lines of electric flux, **E**, emanate from a positive charge, q, and terminate, if they terminate, on a negative charge. The sketch in Figure 1.1 represents the charges and the three dimensional field. The field is visualized as being made up of lines of flux. For an isolated charge, the lines of flux do not terminate and are considered to extend to infinity.

To obtain the equation relating an electric charge q, and its flux Φ_E, assume that the charge is centered in a sphere of radius r meters. The electric flux density, **D**, is then equal to the electric flux emanating from the charge, q, divided by the area of the sphere

$$D = \frac{\Phi_E}{\text{Area}} \text{ coulombs per square meter},$$

where the area is perpendicular to the lines of flux[1]. The charge enclosed in the sphere is then equal to the electric flux density on its surface times the area enclosing the charge

$$q \text{ (coulombs enclosed)} = D \times 4\pi r^2.$$

The lines of flux contributing to the flux density are those that leave the sphere perpendicular to the surface of the sphere. This leads to the integral statement of this portion of Gauss' law

$$q \text{ (coulombs enclosed)} = \oint_{Surface} \mathbf{D} \bullet d\mathbf{a}. \qquad (1.3)$$

The integral sign indicates the summation of infinitesimal areas, $d\mathbf{a}$, in order to obtain the entire surface area. The circle on the integral sign indicates that the integral or summation of area is taken of a closed continuous surface. Bold face letters indicate that the letter represents a vector, i.e., this quantity has magnitude and direction. Distance, velocity, acceleration and force are common examples of vectors.

The symbol • (said dot) following **D** shows that the vector dot product must be used when multiplying the two vectors, **D** and $d\mathbf{a}$. The dot product indicates that the magnitudes of the two vectors are multiplied together and then that product is multiplied by the cosine of the angle between the two vectors. The dot product here enables us to determine the effective lines of flux flowing through the surface.

A vector dot product application can be illustrated by calculating work in the following physics problem. Recall that

$$\text{work} = (\text{force}) \times (\text{distance}).$$

---

[1] One coulomb is equal to the magnitude of charge of 6.25×10^{18} electrons.

Work is equal to the product of the force, that is in the direction of movement, times the distance the force moves. In the following example, assume a person is pushing a mop across a floor with the mop handle at an angle of 60 degrees to the floor as in Figure 1.2. Arrows are used in this diagram to represent vectors.

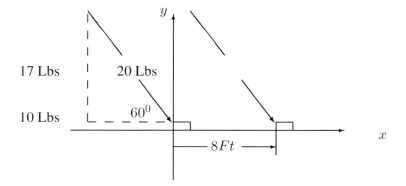

Figure 1.2. Calculation of work.

A force of 20 pounds is applied through the mop handle. As shown in the diagram, only that component of the force in the direction parallel to the floor is used in the calculation of work. We see that the force parallel to the floor is

$$20 \text{ pounds} \times \cos 60^0 = 10 \text{ pounds}.$$

What is the work done when pushing the mop 8 feet across the floor? It is

$$10 \text{ lbs} \times 8 \text{ ft} = 80 \text{ ft} \times \text{lbs}.$$

Using the dot product, the equation for work is

$$\text{Work} = \textbf{Force} \bullet \textbf{distance}.$$

The dot product indicates that work equals the magnitude of the force times the magnitude of the distance moved, times the cosine of the angle between the two vectors. Or, Work = Force on the mop handle, times the distance the force moves, times the cosine of the angle between the force and the floor. Work = 20 pounds times 8 feet times 1/2 = 80 foot pounds. Only that component of the total force in the direction parallel to the floor, as obtained through the dot product, is used in the calculation of work.

This shows that the dot product is defined as the method of vector multiplication in which the vector magnitudes are multiplied together and then that product is multiplied by the cosine of the vector's included angle. Therefore, when the dot product is used in Gauss' law, only the component of flux parallel to the vector representing area will contribute to the total enclosed charge.

Gauss' Law, Area Integral. The method of determining charge by using the dot product of $\textbf{D} \bullet d\textbf{a}$ is similar to finding work as the dot product between applied force and distance. Through this method, only those components of the vector lines of flux in the same direction as the vector representing area will be summed in the calculation of charge.

Or, stated in another way: only those flux lines perpendicular to the surface are incorporated in the result of the dot product to obtain the charge enclosed. A scalar value is always the resultant of a dot product. In this case, the result is a number of coulombs. Examples of other scalar quantities are temperature, mass and power. A scalar quantity, in contrast to a vector, does not have direction.

The differential element of area is $d\mathbf{a}$. A vector representing an area is pointed normal, i.e., perpendicular to that area. Using the dot product between the vector representing area $d\mathbf{a}$ and the flux density $\mathbf{D}$, results in obtaining the effective flux through the area. The summation of the entire area is in square meters. The preliminary equation (Gauss' law) in our procedure to obtain Maxwell's first equation is now

$$\oint_{Surface} \mathbf{D} \bullet d\mathbf{a} = q \,. \tag{1.4}$$

This integral equation states that the amount of electric flux density normal to a surface is caused by a specific amount of charge, q, enclosed by the surface.

Consider the following examples of finding the electric flux density on a spherical surface and on a cylindrical surface.

Flux Density on a Sphere. Assume a charge of one coulomb is centered in a sphere of radius r meters as in Figure 1.3. Calculate the electric flux density $\mathbf{D}$ on the surface on the sphere. The integral or summation of area of the sphere is $4\pi r^2$ square meters.

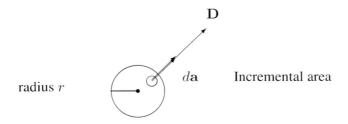

Figure 1.3. Flux Density on a Sphere.

A vector representing an area is directed normal to that area. The vector representing the small area, $d\mathbf{a}$, is then directly in line with a line of electric flux leaving the sphere. $\mathbf{D}$ represents the density of those lines of electric flux leaving the sphere. The angle between the displacement density, $\mathbf{D}$, and the arrow representing the infinitesimal area is zero degrees. The cosine of zero degrees is one. Restating the area integral equation of Gauss' law:

$$q \text{ (coulombs enclosed)} = \oint_{Surface} \mathbf{D} \bullet d\mathbf{a} \,, \quad q = D \times 4\pi r^2 \,.$$

That is

$$D = \frac{q}{4\pi r^2} \quad \text{coulomb per square meter} \tag{1.5}$$

on the surface of the sphere.

Flux Density on a Cylinder. Assume a long line of stationary charges of q coulombs per meter as shown in Figure 1.4. There is a cylinder of length "L" and radius "r" centered on the charges. What is the electric flux density on the surface of the cylinder?

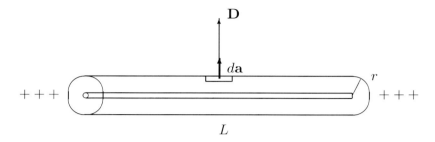

Figure 1.4. Flux density on a cylinder.

Gauss' equation is

$$\oint_{Surface} \mathbf{D} \bullet d\mathbf{a} = Q,$$

where the total charge enclosed

$$Q = q \text{ coulombs per meter} \times L \text{ meter}$$

and the total flux

$$\oint_{Surface} \mathbf{D} \bullet d\mathbf{a} = D\, 2\pi r\, L;$$

that is

$$D = \frac{qL}{2\pi r\, L} = \frac{q}{2\pi r} \qquad (1.6)$$

in coulombs per square meter.

Volume Integral. The total charge enclosed is also written as a volume integral

$$q = \int_{Volume} \rho\, dv. \qquad (1.7)$$

This equation states that the charge enclosed in a volume is equal to the volume charge density ρ summed for the entire volume, where q is the charge enclosed in the volume; ρ is the volume charge density in coulombs per cubic meter; dv is an infinitesimal element of the volume. The entire volume is in cubic meters.

1.2. Integral Form Completed

To obtain the integral form of Maxwell's equation I, assume that an experiment is set up so that the same charge of q coulombs is contained in each of Gauss' law equations. Then the integrals due to the same charge must be equal

$$q = \oint_{Surface} \mathbf{D}\, \mathbf{a}, \quad q = \int_{Volume} \rho\, dv,$$

then

$$\oint_{Surface} \mathbf{D}\, d\mathbf{a} = \int_{Volume} \rho\, dv. \tag{1.8}$$

Thus we have obtained the integral form of Maxwell's equation I. This equation states that the effective electric field through a surface enclosing a volume is equal to the total charge within the volume. The equation shows that the area enclosed by the left hand integral must enclose the volume of the right integral. This is similar to stating that the surface area of a ball or box encloses the volume of the ball or box. The area and volume indicated by the equations need not be observable physical surfaces, often they will be mathematical limits.

1.3. Differential Form

The differential form of Maxwell's equation I is

$$\nabla \bullet \mathbf{D} = \rho, \tag{1.9}$$

where ∇ is a differential operator read "del", $\nabla\bullet$ is read "divergence"; $\mathbf{D}$ is the electric flux density in coulombs per square meter; ρ is the volume charge density in coulombs per cubic meter. The ∇ is the mathematical extension of the ordinary single dimension calculus derivative into three dimensions.

We will begin the discussion by reviewing ordinary derivatives. As an example, a derivative is used as the notation for velocity. Velocity (v) is the increase in distance s, for an increase in time t:

$$\text{Velocity (v)} = \frac{\text{Change in distance}}{\text{Change in time}}.$$

As the change in time is made very small, the differential calculus symbol is used for velocity

$$\text{v} = \frac{ds}{dt}. \tag{1.10}$$

Now consider the ordinary single dimension derivative for acceleration. Recall the equation for obtaining the velocity of an object when it is dropped from a height. The velocity that the object attains is found by

$$\text{Velocity}(v) = g\ (\text{acceleration due to gravity}) \times$$
$$\times t\ (\text{the time during which the object is falling}),$$
$$\text{Velocity} = v = gt, \text{ or } \frac{\text{v}}{t} = g,$$
$$\text{Acceleration} = \frac{\text{change in velocity}}{\text{change in time}} = \frac{dv}{dt} = g.$$

In the integral form of Gauss' law we summed the infinitesimal values of area and volume, da and dv. Here we are using the differential ds, dv and dt to find instantaneous rates of change of distance and velocity with respect to time. Notice that velocity is equal to acceleration $(g) \times$ time(t). So we can take the derivative of velocity in this manner

$$\frac{d\mathrm{v}}{dt} = \frac{d}{dt} gt = g\,.$$

We will now extend that concept of ordinary derivatives to partial derivatives. This will allow us to obtain the rate of change of a volume in three dimensions, which in turn leads to the definition of ∇ (del). To illustrate the rate of change in three dimensions, assume a box is placed at the origin of a rectangular coordinate system, as shown in Figure 1.5

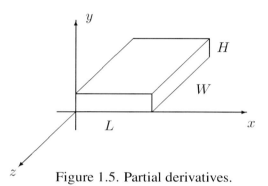

Figure 1.5. Partial derivatives.

$$V(\text{Box volume}) = L(\text{length}) \times W(\text{wifth}) \times H(\text{height})\,, \quad V = LWH\,.$$

What is the rate of change of volume when only the length increases by a small amount, but the width and height remain constant? This is where the symbol for the partial derivative is used.

The symbol for partial derivatives is slightly different than the symbol for ordinary (one dimension) derivatives. The symbol indicates that only one of the independent variables is changing at the moment under consideration. The dependent variable, volume V, changes as determined by changes of the independent variables L, W and H. When only the length changes

$$\frac{\partial}{\partial L} V = \frac{\partial}{\partial L} WHL = WH\,, \qquad (1.11)$$

the partial derivative symbol

$$\frac{\partial}{\partial L} V \qquad (1.12)$$

shows that the change in volume is due to a change in length only. The width and height are being held constant. We also see that the derivative of a variable times a constant is the constant, W times H. It follows, by symmetry, that the rate of change of volume as a function of either width or height, is expressed as a partial derivative

$$\frac{\partial}{\partial W} V = \frac{\partial}{\partial W} WHL = HL, \quad \frac{\partial}{\partial H} V = \frac{\partial}{\partial H} WHL = WL.$$

When the change in volume is due to a simultaneous change in length, width and height, the changes will occur in the x, y and z directions, and the partial derivatives are added to find the resultant rate of change of volume. This is accomplished in vector form by multiplying each partial derivative by unit vectors pointing in the x, y and z directions. Unit vectors are indicated here and discussed further. Using unit vectors and simultaneous changes in three dimensions, the total change in volume is designated by

$$\frac{\partial V}{\partial \text{ (each dimension)}} = \frac{\partial V}{\partial L} \mathbf{x} + \frac{\partial V}{\partial H} \mathbf{y} + \frac{\partial V}{\partial W} \mathbf{z}.$$

For the more general case of a volume, V changing in the x, y and z directions, we have

$$\frac{\partial V}{\partial (x, y, z)} = (\frac{\partial}{\partial x} \mathbf{x} + \frac{\partial}{\partial y} \mathbf{y} + \frac{\partial}{\partial z} \mathbf{z}) V = \nabla V. \tag{1.13}$$

This discussion of ordinary and partial derivatives was aimed at obtaining the group of three partial derivative terms in the above parenthesis, ∇ is shorthand for

$$(\frac{\partial}{\partial x} \mathbf{x} + \frac{\partial}{\partial y} \mathbf{y} + \frac{\partial}{\partial z} \mathbf{z}), \quad \nabla V \text{ means the gradient of } V. \tag{1.14}$$

We must consider the x, y and z components of a vector in rectangular coordinates. In the above, the dot product of two vectors, force and distance, was used to calculate work. We will here calculate the same work using the vector components of force and distance, and employ the dot product.

The vector • (dot) multiplication procedure now is to multiply the vector component magnitudes and the cosine of the angle between them, term by term. This procedure of multiplying vector x, y and z components is followed in performing the product below. The multiplication of vector components, that are always at zero or 90 degrees apart, greatly simplifies vector mathematics. Above we discussed dot product multiplication of vectors using their components. These two concepts are now used to calculate $\nabla \bullet \mathbf{D}$.

1.4. Calculating $\nabla \bullet \mathbf{D}$

The components of vector **D** are its projections on the x, y and z axis. The vector directions of **D** components are designated by the unit vectors **x**, **y** and **z**. In Figure 1.6, the vector **D** starts at the origin, points up and to the right and is indicated as coming out of the paper. The magnitudes of projections of **D** along the axes are D_x, D_y, and D_z. Figure 1.7 shows the unit vectors in the x, y and z directions that give the components of **D** their vector relationship. The same unit vectors are designated in ∇.

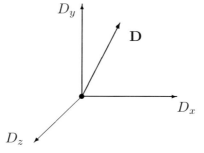

Figure 1.6. Components of a vector.

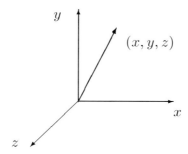

Figure 1.7. Cartesian system.

The equation for vector **D** as projected on the three coordinate axes is

$$\mathbf{D} = D_x \mathbf{x} + D_y \mathbf{y} + D_z \mathbf{z} . \tag{1.15}$$

We will now do the indicated dot product of $\nabla \bullet \mathbf{D}$:

$$\nabla \bullet \mathbf{D} = (\frac{\partial}{\partial x} \mathbf{x} + \frac{\partial}{\partial y} \mathbf{y} + \frac{\partial}{\partial z} \mathbf{z}) \bullet (D_x \mathbf{x} + D_y \mathbf{y} + D_z \mathbf{z}) . \tag{1.16}$$

The dot product indicates that we must multiply the parentheses, term by term, times the cosine of the included angle between each pair of terms. This series of multiplications could result in nine terms, but notice that a unit vector dotted into the same unit vector:

$$\mathbf{x} \bullet \mathbf{x} = 1, \quad \mathbf{y} \bullet \mathbf{y} = 1, \quad \mathbf{z} \bullet \mathbf{z} = 1 .$$

The other six combinations of unit vector dot product multiplications contain the cosine of 90 degrees and they are equal to zero. The final result of the $\nabla \bullet \mathbf{D}$ operation is a scalar of only three terms:

$$\nabla \bullet \mathbf{D} = \frac{\partial D_x}{\partial x} + \frac{\partial D_y}{\partial y} + \frac{\partial D_z}{\partial z} = \rho . \tag{1.17}$$

This equation indicates the sum of a change in electric flux density **D** in each of the three orthogonal directions. This defines the volume charge density.

1.5. Differential Form Completed

By doing the indicated operation $\nabla \bullet \mathbf{D}$ we obtained ρ, the volume charge density. This is the differential statement of Maxwell's equation I:

$$\rho = \nabla \bullet \mathbf{D} = \text{divergence of } \mathbf{D} . \tag{1.18}$$

The equation states that the divergence of the electric flux density at a point is equal to the charge per unit volume at that point.

Divergence Theorem. It is instructive at this point to continue using the integral and differential equations for Maxwell's equation I in order to illustrate a vector identity called, "Gauss' divergence theorem". This identity equates a vector surface integral to a vector volume integral, and will be required later. In the above we had

$$\oint_{Surface} \mathbf{D} \bullet d\mathbf{a} = \int_{Volume} \rho \, dv . \tag{1.19}$$

$$\nabla \bullet \mathbf{D} = \rho . \tag{1.20}$$

By substituting $\nabla \bullet \mathbf{D}$ for ρ in the integral equation, we obtain

$$\oint_{Surface} \mathbf{D} \bullet d\mathbf{a} = \int_{Volume} (\nabla \bullet \mathbf{D}) \, dv . \tag{1.21}$$

This is a typical illustration of Gauss' divergence theorem, using vector $\mathbf{D}$ as the example. The point here is that any time we have a vector surface integral of this type, we can substitute the volume integral. If we have a vector volume integral of the above type, we can substitute the surface integral. The integral of the divergence of a vector summed throughout the volume is equal to the integral of the product of the vector times its effective area summed over the area. The circle on the integral sign indicates that the integral is taken over a continuous area.

If we had simply used Gauss' divergence theorem from a textbook list of vector identities, we could have immediately written down the differential form of Maxwell's equation I from the integral form. This more detailed way of obtaining the identity will be helpful in later derivations.

1.6. Relation of D, E, the Permittivity ϵ

Surrounding an electric charge q, there is an electric field of field strength $\mathbf{E}$. It is the electric field strength $\mathbf{E}$, that will cause an amount of flux density $\mathbf{D}$ depending on the permittivity of the surrounding medium

$$\mathbf{E} = \frac{\mathbf{D}}{\epsilon} . \tag{1.22}$$

$\mathbf{D}$ is in coulombs per square meter; $\mathbf{E}$ is in newtons per coulomb or volts per meter; ϵ is in coulomb^2 per newton meter^2.

Due to this equation, $\mathbf{D}$ is often designated as electric flux displacement density in addition to electric flux density. Also, it will be shown later that a magnetically induced electric field is also designated **E**, with a dimension of volts per meter. That induced electric field is the same field as the static field strength discussed here but it is generated by a changing magnetic field.

The permittivity ϵ is the degree to which the surrounding medium will permit the electric flux density **D** to occur due to a given electric field strength **E**. In the medium of air or free space,

$$\epsilon = 8.85 \times 10^{-12} \frac{\text{coulomb}^2}{\text{newton meter}^2} \,. \tag{1.23}$$

These concepts and definitions will be used below.

1.7. Coulomb's Law

In the above, we found that the electric flux density **D**, due to a charge q located within the sphere is

$$D = \frac{q_1}{4\pi r^2} \;\; \text{coulombs per sqaure meter} \,. \tag{1.24}$$

Then, using **E** and ϵ as defined above, we derive

$$E = \frac{1}{4\pi\epsilon} \frac{q_1}{r^2} \;\; \text{newtons per coulomb} \,. \tag{1.25}$$

When another charge q_2 is placed r meters from q_1, a force is experienced by q_2. The force is E times q_2 newton's:

$$F = \frac{1}{4\pi\epsilon} \frac{q_1 \, q_2}{r^2} \;\; \text{newtons}, \tag{1.26}$$

where q_1 and q_2 designate individual charges. This equation is Coulomb's law.

Coulomb's law states that the force between two charges is proportional to the product of the two charges over the distance between the two charges squared. The equation is the basis for experimentally determining the force between two charges and the permittivity of different mediums. This important electrical law is not included in Maxwell's list, and is not used in these field equations.

This completes the discussion of Maxwell's equation I.

2. Maxwell's Electromagnetic Equation II

The following magnetic field equations will be developed in this section:

integral form

$$\oint_{Surface} \mathbf{B} \bullet d\mathbf{a} = 0, \tag{2.1}$$

differential form

$$\nabla \bullet \mathbf{B} = 0, \;\; \text{div } \mathbf{B} = 0 \,. \tag{2.2}$$

These equations have the same form as Gauss' electrostatic equations of Chapter 1. It is well to compare all these equations now. The integral form of this magnetic field equation states that the summation of the effective lines of magnetic flux density, **B**, entering and leaving a closed surface is zero. The differential form of the magnetic field equation states that the divergence of the magnetic field is zero. This is another way of saying that lines of magnetic flux do not emanate from nor converge upon a single point.

2.1. Gauss' Equations Applied to Magnetic Flux Lines

On reviewing Gauss' electrostatic field equations, we see that the summation of flux lines emanating from a charge, when the charge is included within the closed surface, will always have some value. That value is the charge enclosed. The divergence of the electric flux density always has the value of the charge density. However, when we apply Gauss' equation to lines of magnetic flux we will find that their summation is always zero. An identical finding of zero is determined when we find the divergence of magnetic flux lines. Maxwell's equation II is an application of Gauss' equations to magnetic flux lines. Magnetic flux equations are based on the experimental fact that a single isolated magnetic pole does not exist. All experiments on a bar magnet show that the lines of flux leave the designated north pole, flow outside the bar, enter the south pole and continue through the bar. This continuity of magnetic flux lines is universal for all magnetic sources, whether due to physical magnets or due to electric current flow or displacement current. Magnetic flux lines are always continuous and close upon themselves. The magnetic flux lines do not emanate from or converge to a point. This is in sharp contrast to Gauss' electrostatic laws which state that single charges exist, and electric flux lines can originate at one charge and terminate on another.

Assume that an area in one plane is placed perpendicular to the lines of flux of a bar magnet, and the total magnetic flux entering the area is Φ_M. Then

$$\frac{\Phi_M}{\text{Area}} = B . \tag{2.3}$$

B is the magnetic flux density in lines per square meter. The effective area is in square meters at a right angle to the lines of flux.

2.2. Divergence of Magnetic Flux Density

Experiments with magnetic fields show that isolated magnetic poles do not exist and that lines of magnetic flux are continuous. Therefore a group of magnetic lines (or one flux line) will follow a continuous path and close upon themselves. This experimental data provides an analytical method for describing magnetic field by the equation

$$\nabla \bullet \mathbf{B} = 0 . \tag{2.4}$$

Place a bar magnet in an orthogonal coordinate system, and construct the vector representing flux density **B**, flowing approximately through a continuous circle including the bar magnet as in Figure 2.1 (where C is some constant)

$$\mathbf{B} = C \left(-y \, \mathbf{x} + x \, \mathbf{y} \right) .$$

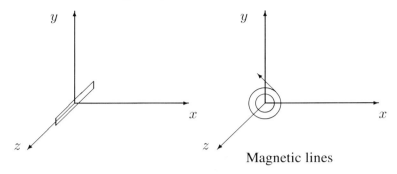

Figure 2.1. Bar magnet and magnetic field.

Calculate $\nabla \bullet \mathbf{B}$:

$$\nabla \bullet \mathbf{B} = \left(\frac{\partial}{\partial x}\mathbf{x} + \frac{\partial}{\partial y}\mathbf{y} + \frac{\partial}{\partial z}\mathbf{z}\right) C\left(-y\,\mathbf{x} + x\,\mathbf{y}\right) = 0\,.$$

That is

$$\nabla \bullet \mathbf{B} = 0\,.$$

The divergence of $\mathbf{B}$ is zero. This is what we wanted to show.

2.3. Maxwell's Equation II, Differential and Integral Forms

Based on the above development, Maxwell's magnetic field equation in differential form is

$$\nabla \bullet \mathbf{B} = 0\,. \tag{2.5}$$

This equation states that the divergence of magnetic flux is zero. Wherein: ∇, read *del*, is the partial differential operator; $\nabla\bullet$, read *divergence*, and abbreviated div. This is the differential form of Maxwell's equation II. It is a mathematical proof that an isolated magnetic pole does not exist.

The integral form of the magnetic flux equation, presented as Maxwell's equation II:

$$\oint_{Surface} \mathbf{B} \bullet d\mathbf{a} = 0\,. \tag{2.6}$$

Consider an imaginary sphere surrounding the end of a bar magnet and encompassing a volume of space surrounding it. All the lines of flux will enter the sphere flowing through the bar magnet. Then all the lines will flow through the volume and then exit the sphere. The number of lines that reenter the sphere will exactly equal the effective lines exiting the sphere. So the algebraic sum of all the lines entering the sphere plus all the lines exiting the sphere is zero. This is different than electrostatic theory where the lines of electric flux due to an enclosed charge simply exit a sphere. The analytical method for developing this integral equation is to use the vector identity "Gauss' divergence theorem" to equate a vector volume integral to its surface integral. We will use vector $\mathbf{B}$ in place of vector $\mathbf{D}$:

$$\int_{Volume} (\nabla \bullet \mathbf{B})\, dv = \oint_{Surface} \mathbf{B} \bullet d\mathbf{a} = 0\,. \tag{2.7}$$

2.4. Magnetic Flux Density, Field Strength, Permeability

The most common way of generating magnetic flux is with current flow in a wire. Many experiments have shown that lines of magnetic flux form concentric circles around the length of the wire. For example, assume there is a wire carrying a steady current in line with the z axis as in Figure 2.2.

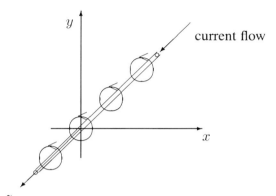

Figure 2.2. Magnetic field of the current in a wire.

The z axis is represented perpendicular to the paper, and all the concentric circles of magnetic flux density **B** are parallel to the $x - y$ plane. These circles of flux density are thought of as being placed together with no distance between them and then visualized as concentric tubes of flux surrounding the length of the wire. When we take the divergence of **B** completely around the circumference of any circle, the divergence is zero.

Ampere's law shows that the flow of current in the wire causes a stress in the surrounding space. This stress is termed the magnetic field strength **H**. The magnetic field strength in turn, causes the magnetic flux density **B**. The lines of magnetic flux in any pictorial can represent either **B** or **H**.

The amount of **B** that appears is dependent upon the permeability, μ (mu), of the medium surrounding the wire

$$\mathbf{B} \text{ (the flux density)} = \mu \text{ (permeability)} \times \mathbf{H} \text{ (field strength)},$$

or

$$\mathbf{B} = \mu \mathbf{H} . \tag{2.8}$$

The constant μ, in air or in space, is assigned the value of $4\pi \times 10^{-7}$ magnetic lines per square meter over amperes per meter.

The first two of Maxwell's equations describe the characteristics of stationary, non-varying electric and magnetic fields. The following sections develop Maxwell's third and fourth equations that concern varying electric and magnetic fields.

3. Maxwell's Electromagnetic Equation III

The following equations will be developed in this section:

integral form

$$\oint_{Closed\ path} \mathbf{H} \bullet d\mathbf{s} = \oint_{Area} \left(\frac{\partial \mathbf{D}}{\partial t} + \mathbf{J} \right) \bullet d\mathbf{a} . \qquad (III)$$

differential form

$$\nabla \times \mathbf{H} = \frac{\partial \mathbf{D}}{\partial t} + \mathbf{J} , \ or \ \text{curl } \mathbf{H} = \frac{\partial \mathbf{D}}{\partial t} + \mathbf{J} . \qquad (III')$$

Maxwell's equation III is based upon Ampere's electrical experiments. The concept of generated fields, in contrast to simply action at a distance, was placed into Ampere's laws by Maxwell. The first part of this derivation will develop Ampere's law for steady magnetic fields. We will then follow Maxwell in adding a time varying electric displacement density field to Ampere's equation.

This addition implements Maxwell's current continuity principle, and shows that every time-varying electric field is accompanied by a magnetic field. The resulting equation is one of Maxwell's important contributions to electromagnetic field theory.

3.1. Ampere's Law of Steady Magnetic Fields, Integral Form

Ampere published the results of his experiments in 1826. His equation for steady magnetic fields, in present day notation and units, is Ampere's law

$$\text{magnetomotive force } (mmf) = \oint_{Closed\ path} \mathbf{H} \bullet d\mathbf{s} = I \text{ amperes} . \qquad (3.1)$$

We will now discuss each term in Ampere's equation, and then give several examples in using the equation. The indicated closed path integral of $\mathbf{H}$ defines *mmf*. The unit of *mmf* is ampere. An ampere is defined as a coulomb of charge passing a point per second. The circle on the integral sign means that, when taking the integral, a complete enclosing path must be followed around the current source that is causing the $\mathbf{H}$. The magnetic field strength in amperes per meter; I is the current in amperes; $d\mathbf{s}$ is a small distance on the complete path indicated by the circle on the integral sign. The unit of distance along the path is meter; The dot product $\bullet$ indicates that only the $\mathbf{H}$ tangent to the path of integration is to be summed to obtain *mmf*.

Below are several examples of increasing complexity to show the applications of Ampere's integral equation.

3.2. Magnetic Field Due to Current in a Wire, Examples

Consider the current carrying wire in Figure 3.1.

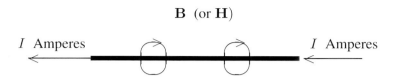

Figure 3.1. Magnetic field due to current in a wire.

Let us calculate the magnetic field strength **H**, and magnetic flux density **B**, around a long straight wire carrying current of I amperes. Ampere's law will be used to determine the magnetic field strength at every point around the wire. The field strength will depend upon the current and distance from the wire. This distance is the radius r of the circle that defines the path of **H**.

$$\oint_{Closed\ path} \mathbf{H} \bullet d\mathbf{s} = I \text{ amperes}.$$

The summation of the small distances $d\mathbf{s}$ around the circumference of a circle of **H** is $2\pi r$. **H** is exactly in line with this circular path. The angle between the path traversed and **H** is zero degrees. The cosine of zero degrees is one. Therefore

$$2\pi r\, H \cos 0^0 = I, \text{ or } H = \frac{I}{2\pi r} \frac{\text{amperes}}{\text{meter}}. \tag{3.2}$$

This equation shows that magnetic field strength H increases with current and decreases with distance from the wire. Since $\mathbf{B} = \mu \mathbf{H}$,

$$B = 4\pi \times 10^{-7} \frac{\text{lines per meter}^2}{\text{amperes per meter}} \frac{I}{2\pi r} \text{ amperes per meter},$$

and

$$B = \frac{2 \times 10^{-7}}{r} \frac{\text{magnetic flux lines}}{\text{square meter}} \tag{3.3}$$

at the distance r meters from the center of the wire. This is the steady state flux density surrounding a current carrying wire.

The convention used to find the direction of vector **H**, is to grasp the wire with the right hand with the thumb in the direction of current flow. The curve of the fingers circled around the wire shows the direction of **H**. The vector **H** is tangent to the circle at every point on the circumference of the circle. Every length of the wire is surrounded by concentric circles of **H**. The planes of the circles are perpendicular to the wire. A sensitive compass needle moved carefully around a current carrying wire will continuously align itself tangent to a circle. The wire would be at the center of the circle.

Let us calculate magnetic field around wire group. What is **H** in the vicinity of the wires, when several wires are bundled together, each carrying the same current? Assume there are three (or any number) of wires, as in Figure 3.2.

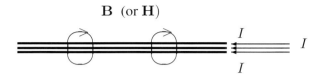

Figure 3.2. Magnetic field around wire group.

Ampere's magnetics law for steady current is

$$\oint_{Closed\ path} \mathbf{H} \bullet d\mathbf{s} = I \text{ amperes enclosed}.$$

The concentric circles of **H** are generated by the total current in the wire bundle. Each small distance $d\mathbf{s}$, along each circumference, is exactly in line with **H**

$$H\ 2\pi r = 3\ I \text{ amperes}, \quad H = \frac{3\ I}{2\pi r}. \tag{3.3}$$

Let us calculate **H** in toroid coil. Consider a toroid of N closely wound turns as shown in Figure 3.3. All the magnetic flux is essentially contained within the toroid form. The flux from each turn add and reinforce inside the toroid coil and cancel outside the coil. Lines of **H** form circles throughout in the entire inside area of the coil. Current I amperes flows in the wire. What is the **H** at a distance r meters from the center of the circle to the middle of the toroid?

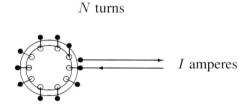

Figure 3.3. Magnetic field in toroid coil.

$$\oint_{Closed\ path} \mathbf{H} \bullet d\mathbf{s} = I.$$

The path is a circle of radius r. The path followed by the lines of flux is tangent to the path, so the angle between the small distances $d\mathbf{s}$ and **H** is zero degrees

$$\oint_{Closed\ path} \mathbf{H} \bullet d\mathbf{s} = H\ 2\pi r = NI, \quad H = N\ \frac{I}{2\pi r}\ \frac{\text{amperes}}{\text{meter}}. \tag{3.4}$$

Let us calculate **H** in a Solenoid. Consider the long solenoid whose cross section is sketched in Figure 3.4

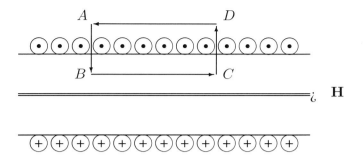

Figure 3.4. Magnetic field of a solenoid.

The solenoid has n turns per meter of closely wound wire carrying a current of I amperes. The dots are arrowhead points showing current coming out of the wire turn. The plus signs are arrow feathers showing current going into the turn. As indicated in the figure there is essentially no magnetic field outside of the solenoid. The magnetic field is inside the coil only. Calculate the **H** within the solenoid. Ampere's law is

$$\oint_{Closed\ path} \mathbf{H} \bullet d\mathbf{s} = I \text{ (total enclosed) }.$$

Sum the **H** • d**s** around the rectangular path: A to B, B to C, C to D, and D back to A

$$\int_{A-B} \mathbf{H} \bullet d\mathbf{s} + \int_{B-C} \mathbf{H} \bullet d\mathbf{s} + \int_{C-D} \mathbf{H} \bullet d\mathbf{s} + \int_{D-A} \mathbf{H} \bullet d\mathbf{s} = I. \qquad (3.5)$$

The summation of the closed path around the current will be calculated in piecewise steps. The path from A to B is at a right angle to the lines of **H**. Recall that the dot product means that the cosine of the included angle between the two vectors is used multiply the product of the vector magnitudes. As the cosine of 90 degrees is zero, the contribution to the integral along this part of the path is zero. The path from B to C is in line with the **H** vector so that the angle between the path traversed and the **H** vector is zero degrees, with a cosine of one. The distance from B to C is "s" meters. The summation of **H** dotted into **s** is **H** • **s**. According to Ampere's law, this summation must be equal to the current to be enclosed. The current enclosed is equal to the current in the coil wire times the number of turns included by the distance "s" meters.

The current in the coil wire is I, there are n turns of wire per meter and the distance is "s" meters. The current to be enclosed is snI. Proceed along the path from c to d. The integral along the path from c to d is zero because the path is traversed at right angles to the **H** vector. The integral along the path from d to a is zero, because there are essentially no lines of magnetic flux outside a closely wound solenoid.

Therefore, Ampere's law applied to a solenoid is

$$\oint_{Closed\ path} \mathbf{H} \bullet d\mathbf{s} = I \ , \quad \text{or} \quad HS = nSI \ .$$

The S cancels and the result is that
$$H = n\,I \qquad (3.5b)$$

The magnetic field strength, H, in amperes per meter of width within a solenoid, is the number of turns enclosed times the current in the coil.

3.3. Derivation of Differential Form of Ampere's Law

We will follow the eight steps to obtain the Stokes' theorem.

1. Recall the integral form of Ampere's law:

$$\oint_{Closed\ path} \mathbf{H} \bullet d\mathbf{s} = \int_{Surface} \mathbf{J} \bullet d\mathbf{a}.$$

2. Place a small mathematics cube at the origin of the three dimensional coordinate system, as shown in Figure 3.5.

3. Assume that current density $\mathbf{J}$ is flowing at some angle. $\mathbf{J}$ is composed of three orthogonal vectors; J_x, J_y and J_z. These current densities generate $\mathbf{H}$ in the three orthogonal planes perpendicular to the current densities.

4. Using Ampere's integral equation, sum the $\mathbf{H} \bullet d\mathbf{s}$ around the three cube faces.

5. Show that $J = J_x\mathbf{x} + J_y\mathbf{y} + J_z\mathbf{z}$ results from step 4.

6. Perform a typical vector cross product $\mathbf{A} \times \mathbf{B}$ operation.

7. Calculate the cross product of $\nabla \times \mathbf{H}$.

8. Show that $\nabla \times \mathbf{H}$ has the same constituent terms as

$$\oint_{Closed\ path} \mathbf{H} \bullet d\mathbf{s}$$

of step 4. Therefore $\nabla \times \mathbf{H}$ also equals $\mathbf{J}$. This proves the differential form of Ampere's law.

9. By substituting $\nabla \times \mathbf{H}$ for $\mathbf{J}$ in Ampere's integral equation we show an example of Stokes' theorem.

10. Define curl.

Figure 3.5 a, b, c shows the small cube with dimensions Δx, Δy, Δz, (delta x, delta y, delta z,) placed at the origin of the coordinate system.

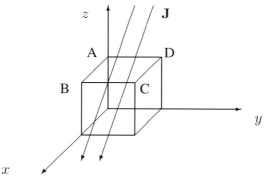

Figure 3.5a. Differential form of Ampere's Law, I.

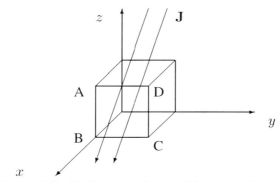

Figure 3.5b. Differential form of Ampere's Law, II.

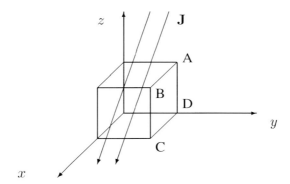

Figure 3.5c. Differential form of Ampere's Law, III.

Assume that conduction current sheets of J amperes per square meter are passing through the three dimensional space. **J** is indicated by parallel lines placed at any angle with respect to the cube. **J** goes in through three sides of the cube and out through the other three sides of the cube, representing the most general case of current flow. Recall that no matter how a cube is turned, the maximum number of sides that can be directly seen, is three. So the maximum number of sides through which **J** can enter the cube is three. Ampere's law in integral form, as shown above is

$$\oint_{Closed\ path} \mathbf{H} \bullet d\mathbf{s} = \int_{Surface} \mathbf{J} \bullet d\mathbf{a} . \tag{3.6}$$

We will use this integral equation to show that

$$\nabla \times \mathbf{H} = \mathbf{J} . \tag{3.7}$$

The procedure is to sum $\mathbf{H} \bullet d\mathbf{s}$ on the four edges of the three sides of the cube. Remember that the path followed to calculate **H** must encompass all the **J** that is causing the **H**. This sum of $\mathbf{H} \bullet d\mathbf{s}$ is equal to the amount of current that is passing through the cube face normally, i.e., at right angles to the face.

The vector **J** is composed of $J_x\mathbf{x} + J_y\mathbf{y} + J_z\mathbf{z}$, as projected on the x, y, z axes of the orthogonal coordinate system. The amount of current that passes through the face of the cube is the component of **J** normal to the face times the area of that face. The calculated current must be equal to the sum of $\mathbf{H} \bullet d\mathbf{s}$ that we will calculate along the four edges of the cube face. Only that component of **J** at right angles or normal to the cube face will cause the $\mathbf{H} \bullet d\mathbf{s}$ summation of that face. This is why we must use dot products in the integrals. This follows from the previous examples of Ampere's law, where the path followed when calculating the $\mathbf{H} \bullet d\mathbf{s}$ was either at right angles to **H** or parallel to **H**.

Now, we calculate $\mathbf{H} \bullet d\mathbf{s}$ and current through cube.

The first cube face to be considered is the face $ABCD$ parallel to the $(x-y)$ plane, shown in Figure 3.5a. What is the current entering that surface? Current density J_z normal to the face parallel to the $(x-y)$ plane. Only current due to the J_z component of **J** can be responsible for the $\mathbf{H} \bullet d\mathbf{s}$ calculated for that face. Ampere's integral equation applied to that face of the cube is

$$\oint_{Path\ around\ face} \mathbf{H} \bullet d\mathbf{s} = \int_{Face\ area} \mathbf{J} \bullet d\mathbf{a}.$$

Designate the average **J** in the area $ABCD$ as

$$\mathbf{J}^{\mathrm{av}} = J_x^{\mathrm{av}}\,\mathbf{x} + J_y^{\mathrm{av}}\,\mathbf{y} + J_z^{\mathrm{av}}\,\mathbf{z}, \tag{3.8}$$

and the area $d\mathbf{a}$ $d\mathbf{a} = \Delta x \Delta y\,\mathbf{z}$. Then the integral over the face area is

$$\int_{Face\ area} \mathbf{J} \bullet d\mathbf{a} = J_z\,\Delta x \Delta y. \tag{3.9}$$

Now continue by determining the path integral by multiplying the **H** of each edge by the distance along the edge. Then sum the four products. This sum is equal to $J_z\,\Delta x \Delta y$ amperes, the current that is causing the **H**. The edges along which to calculate the path integral are

$$A\ \text{to}\ B, \quad B\ \text{to}\ C, \quad C\ \text{to}\ D, \quad \text{and}\ D\ \text{back to}\ A.$$

The path A to B is $d\mathbf{s} = \Delta x\,\mathbf{x}$. Designate the average **H** in the $x-z$ plane over the distance x as

$$\mathbf{H}^{\mathrm{av}} = H_x^{\mathrm{av}}\,\mathbf{x} + H_y^{\mathrm{av}}\,\mathbf{y} + H_z^{\mathrm{av}}\,\mathbf{z}. \tag{3.10}$$

Then the path integral from A to B is

$$\int_{Path\ A\ to\ B} \mathbf{H} \bullet d\mathbf{s} = H_x^{\mathrm{av}}\,\Delta x. \tag{3.11}$$

Calculate the integral from B to C. It consists of two parts. The first part is $H_y^{\mathrm{av}}\,\Delta y$. The second part is the correction required because H_y, for edge B to C, is at distance x from the $y-z$ plane.

Question: How does H_y change as we move, in the **x** direction, for a distance x, away from the $y-z$ plane? Or, what is the value of H_y when we are at a distance x from the $y-z$ plane? Answer: We must use a partial derivative expression to denote that change:

$$\frac{\partial H_y}{\partial x}.$$

The partial derivative with respect to x is used to show that the only movement is in the x direction. What is the change in H_y at a distance x from the x axis? The change is

$$\frac{\partial H_y}{\partial x} \Delta x .$$

What contribution is made to the *mmf* by moving along from B to C? That contribution is

$$\left(\frac{\partial H_y}{\partial x} \Delta x \right) \Delta y .$$

So the total integral from B to C consists of the two parts:

$$\int_{Path\ B\ to\ C} \mathbf{H} \bullet d\mathbf{s} = H_y^{av} \Delta y + \frac{\partial H_y}{\partial x} \Delta x\, \Delta y .$$

Factoring gives

$$\int_{Path\ B\ to\ C} \mathbf{H} \bullet d\mathbf{s} = \left(H_y^{av} + \frac{\partial H_y}{\partial x} \Delta x \right) \Delta y . \tag{3.12}$$

Next, calculate the integral from C to D. When traversing the edge of the cube from A to B, the H generated was in the same direction as the movement from A to B. Then the angle between the movement and the generate H was zero degrees. The cosine of zero degrees is positive one. But now when we traverse the cube from C to D, the angle between the generated H and the movement is 180 degrees. The cosine of 180 degrees is negative one. The integral from C to D is

$$\int_{Path\ C\ to\ D} \mathbf{H} \bullet d\mathbf{s} = -H_x^{av} \Delta y + \text{correction} . \tag{3.13}$$

The correction, determined by using the partial derivative method as

$$-\left(\frac{\partial H_x}{\partial y} \Delta y \right) \Delta x .$$

Then

$$\int_{Path\ C\ to\ D} \mathbf{H} \bullet d\mathbf{s} = -\left(H_x^{av} + \frac{\partial H_x}{\partial y} \Delta y \right) \Delta x . \tag{3.14}$$

Calculate the integral for the path D back to A. The movement is directly above the y axis, so there is no correction to be calculated. But the traversing direction is opposite to the direction of the generated **H**. The angle between the direction of motion and **H** is 180 degrees and the cosine is negative one:

$$\int_{Path\ D\ to\ A} \mathbf{H}\, d\mathbf{s} = -H_y^{av} \Delta y . \tag{3.15}$$

Summing the integrals around the four edges of one face of the cube, or integrating around the closed path:

$$\int_{A-B} \mathbf{H} \bullet d\mathbf{s} + \int_{B-C} \mathbf{H} \bullet d\mathbf{s} + \int_{C-D} \mathbf{H} \bullet d\mathbf{s} + \int_{D-A} \mathbf{H} \bullet d\mathbf{s} = \text{current}.$$

Replacing the integrals with their values found above:

$$+H_x^{av} \Delta x + (H_y^{av} + \frac{\partial H_y}{\partial x} \Delta x) \Delta y - (H_x^{av} + \frac{\partial H_x}{\partial y} \Delta y) \Delta x - H_y^{av} \Delta y.$$

Simplifying,

$$\int_{ABCDA} \mathbf{H} \bullet d\mathbf{s} = \left(\frac{\partial H_y}{\partial x} - \frac{\partial H_x}{\partial y} \right) \Delta x \Delta y = I. \tag{3.16}$$

Question: What current is generating the magnetic flux density that we just calculated? Answer: Component J_z of current density **J** flows through face ABCD of Figure 3.5a. The current flowing through this face is the current density component times the area of the face. This is $J_z \Delta x \Delta y$.

The vector representing the area is normal to the area. The unit vector representing area $\Delta x \Delta y$ is **z**. Therefore,

$$\left(\frac{\partial H_y}{\partial x} - \frac{\partial H_x}{\partial y} \right) \mathbf{z} \, \Delta x \Delta y = J_z \, \mathbf{z} \, \Delta x \Delta y. \tag{3.17}$$

Canceling $\Delta x \Delta y$ from both sides:

$$\left(\frac{\partial H_y}{\partial x} - \frac{\partial H_x}{\partial y} \right) \mathbf{z} = J_z \, \mathbf{z}. \tag{3.18}$$

The second cube face to be considered is the face parallel to the $y - z$ plane. We will sum the path integrals around the edges of that face caused by the **J** component through that face, J_x times the area of the face, $\Delta y \Delta z$. Figure 3.5b indicates the face of the cube parallel to the $y - z$ plane. Summing the integral around ABCDA of this second face, following the same procedure as with the first face above, we obtain,

$$\left(\frac{\partial H_z}{\partial y} - \frac{\partial H_y}{\partial z} \right) \Delta y \Delta z = I_x \, \Delta y \Delta z. \tag{3.19}$$

he unit vector representing this area, $\Delta y \Delta z$, is **x**. Component J_x of current density **J** flows through his face. The current flowing through this second face then is $J_x \Delta y \Delta z$. Therefore,

$$\left(\frac{\partial H_z}{\partial y} - \frac{\partial H_y}{\partial z} \right) \mathbf{x} \, \Delta y \Delta z = J_x \, \mathbf{x} \, \Delta y \Delta z. \tag{3.20}$$

Canceling $\Delta x \Delta y$ from both sides:

$$\left(\frac{\partial H_z}{\partial y} - \frac{\partial H_y}{\partial z} \right) \mathbf{x} = J_x \, \mathbf{x}. \tag{3.21}$$

Figure 3.5c shows the next path to be traversed, and the third face to be encompassed. These are the edges of the side of the cube parallel to the $x - z$ plane. The current density component of **J** flowing through this face is J_y. The unit vector that represents the area, $\Delta z \Delta x$, of this face is **y**.

Following the same summation procedure as with the first face and second face of the cube, we obtain

$$\left(\frac{\partial H_x}{\partial z} - \frac{\partial H_z}{\partial x}\right)\mathbf{y} = I_y\,\mathbf{y}. \tag{3.22}$$

Let us study the components of current density **J**. In Figure 3.6, the current density **J** completely fills the three dimensional space in the direction indicated. We can represent that whole sheet of vectors with one vector arrow **J** as in the orthogonal coordinate system of Figure 3.8. In that figure, **J** starts at the origin, is directed out of the paper, up and to the right. The components of **J** are projected on the x, y and z axie as J_x, J_y and J_z. Unit vectors **x**, **y** and **z** are also shown in Figure 3.6.

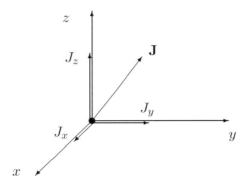

Figure 3.6. Current density **J**.

Figure 3.6 shows that components $J_x\,\mathbf{x} + J_y\,\mathbf{y} + J_z\,\mathbf{z} = \mathbf{J}$. These components of **J** flowing through the cube in their orthogonal coordinates, generated the three expressions with the components of **H**: J_z generates the function of H_y and H_x; J_x generates the function of H_z and H_y; J_y generates the function of H_x and H_z. Therefore, since both factors are vectors, they must be equal in magnitude and direction

$$J_x = \frac{\partial}{\partial y}H_z - \frac{\partial}{\partial z}H_y\,,\; J_y = \frac{\partial}{\partial z}H_x - \frac{\partial}{\partial x}H_z\,,\; J_z = \frac{\partial}{\partial x}H_y - \frac{\partial}{\partial y}H_x. \tag{3.23}$$

Replace J_x, J_y and J_z with their H terms and the vector sum is equal to **J** as shown in Figure 3.6:

$$(\frac{\partial H_z}{\partial y} - \frac{\partial H_y}{\partial z})\mathbf{x} + (\frac{\partial H_x}{\partial z} - \frac{\partial H_z}{\partial x})\mathbf{y} + (\frac{\partial H_y}{\partial x} - \frac{\partial H_x}{\partial y})\mathbf{z} = \mathbf{J}. \tag{3.24}$$

This completes step 5 to derive the differential form of Ampere's law.

We now have to show that is also equal to **J**, by showing that $\nabla \times \mathbf{H}$ has the same constituent terms as does **J**. But, before we can do that, we will demonstrate the typical × (cross) product of two vectors, $\mathbf{A} \times \mathbf{B}$, per step 6 of outline.

We have previously used the vector • (dot) product that always results in a scalar value. We will now discuss the vector × (cross) product that results in obtaining another vector. Consider two vectors **A** and **B**, shown in Figure 3.7 with their projections on the orthogonal axes. The projections of **A** are denoted A_x, A_y and A_z. The projections of **B** are B_x, B_y and B_z. The projections are multiplied by the unit vectors, **x**, **y** and **z** to give the projections their vector orientation.

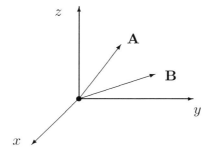

Figure 3.7. Scalar product.

Then,

$$\text{vector } \mathbf{A} = A_x\,\mathbf{x} + A_y\,\mathbf{y} + A_z\,\mathbf{z}, \quad \text{vector } \mathbf{B} = B_x\,\mathbf{x} + B_y\,\mathbf{y} + B_z\,\mathbf{z}.$$

We will discuss the cross product of two electrical vectors by using the "right hand rule". The rule is taught by demonstration. Spread out the thumb, forefinger and middle finger of the right hand. Whenever two digits are folded into each other, that is "cross" multiplied, the vector resulting is represented (direction pointed) by the third digit. For example, if we cross the thumb into the forefinger, Then the middle finger points in the direction of the resulting vector. If we cross the forefinger into the middle finger, the thumb points the direction of the resulting vector. Note also, that the resulting vector is always perpendicular to the plane determined by the two original vectors. When two given vectors are not at right angle to each other, the right hand rule still applies. Two of the right hand digits, representing the vectors, are folded into each other in a counterclockwise direction and the third digit will show the direction of the resulting vector. The definition of the vector × (cross) product, keeping in mind the direction of the resulting vector, is

$$\mathbf{C} = \mathbf{A} \times \mathbf{B} = \mathbf{C_0}\ |A|\,|B|\ \sin\,(\text{of the included angle}), \qquad (3.25)$$

where $\mathbf{C_0}$ is a unit vector perpendicular to the plane determined by the two original vectors. The vertical bars mean "the magnitude of". In this case they show the magnitude of **A** multiplied by the magnitude of **B**.

Another method of obtaining the vector cross product is to multiply the vector projections of each vector on the **x**, **y** and **z** axes, and then using vector × (cross) multiplication. The unit vector times the amplitudes of the vectors, **A** and **B**, projected on the axes, result in the constituent vectors. To do this × (cross) multiplication, each term of the first vector must be multiplied by each term of the second vector. A resulting result will have nine terms:

$$\begin{aligned}
\mathbf{A} \times \mathbf{B} &= (A_x\mathbf{x} + A_y\mathbf{y} + A_z\mathbf{z}) \times (B_x\mathbf{x} + B_y\mathbf{y} + B_z\mathbf{z}) \\
&= A_xB_x\,(\mathbf{x}\times\mathbf{x}) + A_xB_y\,(\mathbf{x}\times\mathbf{y}) + A_xB_z\,(\mathbf{x}\times\mathbf{z}) \\
&\quad + A_yB_x\,(\mathbf{y}\times\mathbf{x}) + A_yB_y\,(\mathbf{y}\times\mathbf{y}) + A_yB_z\,(\mathbf{y}\times\mathbf{z}) \\
&\quad + A_zB_x\,(\mathbf{z}\times\mathbf{x}) + A_zB_y\,(\mathbf{z}\times\mathbf{y}) + A_{zx}B_z\,(\mathbf{z}\times\mathbf{z}).
\end{aligned} \qquad (3.26)$$

Maxwell's Electromagnetic Equations, Elementary Introduction

The vector cross product symbol is shown between each pair of unit vectors. The vector cross product definition and the right hand rule are used to obtain the following unit vector cross multiplication results

$$\begin{aligned} \mathbf{x} \times \mathbf{x} &= 0, & \mathbf{x} \times \mathbf{y} &= +\mathbf{y}, & \mathbf{x} \times \mathbf{z} &= -\mathbf{y}, \\ \mathbf{y} \times \mathbf{x} &= -\mathbf{z}, & \mathbf{y} \times \mathbf{y} &= 0, & \mathbf{y} \times \mathbf{z} &= +\mathbf{x}, \\ \mathbf{z} \times \mathbf{x} &= +\mathbf{y}, & \mathbf{z} \times \mathbf{y} &= -\mathbf{x}, & \mathbf{z} \times \mathbf{z} &= 0. \end{aligned} \qquad (3.27)$$

By noting these results of unit cross product multiplications, we get simplifies to

$$\mathbf{A} \times \mathbf{B} = (A_y B_z - A_z B_y)\mathbf{x} + (A_z B_x - A_x B_z)\mathbf{y} + (A_x B_y - A_y B_x)\mathbf{z}. \qquad (3.28)$$

This completes step 6. Whenever it is required to form the cross product of two vectors that are in orthogonal form, the results can be written down directly by copying the results arrived at above, but using the given two vectors.

We are now have enough basic equations to obtain the differential form of Ampere's law: Using the above example of $\times$ product, let

$$\mathbf{A} = \nabla, \text{ and } \mathbf{B} = \mathbf{H}.$$

The vector representing magnetic field strength, $\mathbf{H}$ is made of constituent orthogonal vectors exactly as vector $\mathbf{J}$ in Figure 3.6.

$$\begin{aligned} \nabla &= \frac{\partial}{\partial x}\mathbf{x} + \frac{\partial}{\partial y}\mathbf{y} + \frac{\partial}{\partial z}\mathbf{z}, \\ \mathbf{H} &= H_x\,\mathbf{x} + H_y\,\mathbf{y} + H_z\,\mathbf{z}. \end{aligned} \qquad (3.29)$$

Follow the results of $\mathbf{A} \times \mathbf{B}$ above and write $\nabla \times \mathbf{H}$ as

$$\nabla \times \mathbf{H} = \left(\frac{\partial H_z}{\partial y} - \frac{\partial H_y}{\partial z}\right)\mathbf{x} + \left(\frac{\partial H_x}{\partial z} - \frac{\partial H_z}{\partial x}\right)\mathbf{y} + \left(\frac{\partial H_y}{\partial x} - \frac{\partial H_x}{\partial y}\right)\mathbf{z}. \qquad (3.30)$$

In Step 5, Ampere's integral equation was used to obtain these same three terms of $\mathbf{H}$ with the same unit vectors. The sum of these terms were shown to be equal to $\mathbf{J}$. Therefore,

$$\nabla \times \mathbf{H} = \mathbf{J}. \qquad (3.31)$$

We have just derived Ampere's law in differential form. This completes Step 8 of the outline.

This is what we had set out to prove with these eight steps; that the differential form of Ampere's law is derivable from the integral form. Also, having obtained this equivalent form for $\mathbf{J}$, we can now demonstrate Stokes' theorem.

We will use the above derivation of the differential form of Ampere's law to illustrate the vector identity of Stokes' theorem. The differential form of Ampere's equation was proven above to be

$$\nabla \times \mathbf{H} = \mathbf{J}. \qquad (3.32)$$

The integral form of Ampere's law is

$$\oint_{Closed\ path} \mathbf{H} \bullet d\mathbf{s} = \int_{Surface} \mathbf{J} \bullet d\mathbf{a}. \qquad (3.33)$$

Substituting $\nabla \times \mathbf{H}$ for $\mathbf{J}$ in the surface integral we obtain

$$\oint_{Closed\ path} \mathbf{H} \bullet d\mathbf{s} = \int_{Surface} (\nabla \times \mathbf{H})\, d\mathbf{a}. \qquad (3.34)$$

This vector identity is true for all field vectors. The curl (the $\nabla \times$ operation) of a field vector summed over a specific area is equal to the same vector summed over a continuous path enclosing that area.

Below we will use Stokes' theorem and substitute the surface integral of field vector $\mathbf{E}$ for its path integral as we develop the differential form of Maxwell's equation IV. The differential forms of Maxwell's equations can all be derived from the integral forms (and vice versa) by using vector identities and operations.

We have now completed the development of Ampere's law for steady magnetic fields. As determined above, the calculus forms are:

$$\text{integral form,} \quad \oint_{Closed\ path} \mathbf{H} \bullet d\mathbf{s} = \int_{Surface} \mathbf{J} \bullet d\mathbf{a}, \qquad (3.35)$$

$$\text{differential form,} \quad \nabla \times \mathbf{H} = \mathbf{J}, \quad \text{or} \quad \text{curl } \mathbf{H} = \mathbf{J}. \qquad (3.36)$$

3.4. Maxwell's Equation III

We proceed now to consider the term that Maxwell added to Ampere's equations to account for non-steady circuit currents and magnetic fields generated by varying electric fields. The previous Section was resented the mathematical basis for Ampere's equations. These equations completely describe the magnetic field strength generated by steady current in a continuous circuit. The question then arises: how should the equations be changed to account for a non-steady flow of current or to account for a circuit interruption? Maxwell accounted for the change in the magnetic field due to those current changes by adding one term to Ampere's equation. Figure 3.8 is a sketch of a theoretical experiment that will illustrate the concepts involved in reaching Maxwell's conclusion.

The circuit consists or the battery and switch wired to two disks separated by a tiny gap. When the switch is closed, the current starts from some value and quickly falls to zero. From Ampere's equations we know that circles of magnetic field strength, $\mathbf{H}$, will surround the wire while any amount of current is flowing. The magnetic field will begin at some maximum value and fall to zero as the current falls to zero.

What is happening in the gap between the two disks during the time that the current is changing? Maxwell postulated that circles of $\mathbf{H}$ would be generated in and around the gap similar to the circles of $\mathbf{H}$ that appear around the wires. After the switch is closed, electrons will be accumulating on one disk and leaving the other, and thus changing the electric flux displacement density, $\mathbf{D}$, of the gap. The phenomena that is causing the circles of $\mathbf{H}$ field to appear around and in the gap is this changing electric displacement density, $\mathbf{D}$, in the gap. The displacement density is changing because of the changing charge on the disks. No charges are moving across the gap.

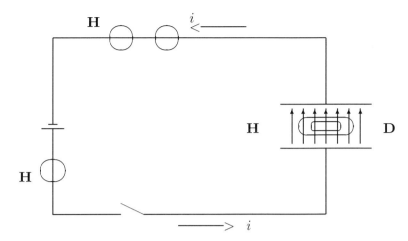

Figure 3.8. Maxwell equation III.

The greatest change in the electric displacement density occurs at the instant the switch is closed and the current is falling from its maximum. When the conduction current decays to zero, the decreasing displacement density and its generated magnetic field fall to zero, and the constant battery voltage is then across the disks.

Recall that current, i, is a certain quantity of charge passing a point per unit time. The definition of one ampere is

$$\frac{\text{one coulomb of charge changing at a point}}{\text{one second}} = \frac{dq}{dt} = i. \qquad (3.37)$$

The decrease of current in the circuit is a decrease in the change of charge per unit time. Less charges are passing a point per unit time at a changing rate. When a moving car is decreasing its velocity and slowing down, we say that the car is decelerating. A similar statement applies here. This deceleration of charges is responsible for the creation of the **H** field by first changing the electric displacement density. When there is no more change in displacement density, that is, when the voltage across the plates is equal to the battery voltage, the displacement density is constant and the generated **H** between the plates is zero. Of course, when there is no current flow, there is no generated **H** in the circuit. The electric displacement density **D** will occur whether the current is increasing or decreasing in the wire[2].

Since Maxwell postulated a current continuity principle and stated that the changing electric field, i.e., changing electric displacement density, would cause a magnetic field, Maxwell concluded that a new term must be included in Ampere's magnetic field equation.

The differential form of Ampere's equation was changed to be Maxwell's electromagnetic equation III:

$$\nabla \times \mathbf{H} = \mathbf{J} + \frac{\partial \mathbf{D}}{\partial t}. \qquad (3.38)$$

---

[2] Changing currents are considered further with Faraday's Experiments.

The addition of this new term

$$\frac{\partial \mathbf{D}}{\partial t} \qquad (3.39)$$

was one of Maxwell's important contributions to electromagnetic field theory. Other reasons for Maxwell's addition of this term are discussed below.

For the experimental set-up, the displacement density term applies to the gap and not to the rest of the circuit. The time varying **D** field does generate the **H** field in the surrounding space. The conduction current term, **J**, applies only to the rest of the circuit and not to the gap. The conduction current creates the magnetic field **H** around the wire. The change in displacement density creates the magnetic field, **H**, in and around the disks.

The partial derivative symbol is used in Maxwell's term to indicate that there may be other independent variables affecting **D**. Here we are concerned with a change in time only, and the space variables are held constant. The dimension of $\frac{\partial \mathbf{D}}{\partial t}$ is coulombs of charge, per square meter-second. **D** is density of charge, and charge density being displaced in some time has the same units as **J**. Amperes are coulombs per second, and **J** is given in amperes per square meter, Since the dimensions are the same, the two terms are shown to be added (as vectors) at every instant of time. This is why $\frac{\partial \mathbf{D}}{\partial t}$ is called displacement current, and cited as part of Maxwell's continuity principle as stated above. But it must be remembered that no charges flow in the air gap or in space. The use of the term "electric displacement density" for **D** now seems reasonable, since it is the change in charge density with time that must have the same units as **J**.

The amount of **H** developed by changing **D** is not measurable by simple testing. The experimental proof that a changing electric field caused a changing magnetic field had to wait for Hertz' confirming radiation experiments.

The sketch in Figure 3.8 shows the planes of the circles of generated **H** are perpendicular to the lines of changing displacement density **D**, just like the circles of **H** are perpendicular to current carrying wires. When radiation takes place, these fields will stay perpendicular to each other as they appear here, but then the electric fields will also be closed loops surrounding the magnetic fields that in turn generate electric fields.

Maxwell had four reasons to add his displacement current term to Ampere's equation:

1. Implementation of his current continuity principle as discussed before.

2. Natural symmetry of magnetic and electric fields: Maxwell knew all about Faraday's experiments, where a changing magnetic field in time caused an electric field to be generated in space. To Maxwell, natural symmetry of electric and magnetic fields demanded that a changing electric field would also generate a magnetic field.

3. Ampere's equation fails the divergence test as shown below.

4. If radiation were to occur, a magnetic field must be part of the radiation. Since there are no conduction currents in space, there must be another source of "current" available to generate a magnetic field. This source is the change of displacement density with time.

3.5. Divergence Test of Ampere's Equation

To solve some vector equations, it is common practice to take the divergence of both sides of the equation. This operation is similar to taking the logarithm of both sides of an algebraic

Maxwell's Electromagnetic Equations, Elementary Introduction

equation to aid in its solution. When the divergence $\nabla \bullet$ of both sides of a correct vector equation is taken, an equality will result. But, when the divergence of both sides of the differential form of Ampere's equation

$$\nabla \times \mathbf{H} = \mathbf{J}$$

is taken, an inequality results. Perform $\nabla \bullet (\nabla \times \mathbf{H})$:
the first term
$$\frac{\partial}{\partial x}\left(\frac{\partial H_z}{\partial y} - \frac{\partial H_y}{\partial z}\right),$$
the second term
$$\frac{\partial}{\partial y}\left(\frac{\partial H_x}{\partial z} - \frac{\partial H_z}{\partial x}\right),$$
the third term
$$\frac{\partial}{\partial z}\left(\frac{\partial H_y}{\partial x} - \frac{\partial H_x}{\partial y}\right).$$

Summing three terms and simplifying we get

$$\nabla \bullet (\nabla \times \mathbf{H}) = 0. \tag{3.40}$$

So the divergence of the left side of Ampere's differential equation is zero. Take the divergence of $\mathbf{J}$:

$$\nabla \bullet \mathbf{J} = \nabla \bullet (J_x \mathbf{x} + J_y \mathbf{y} + J_z \mathbf{z}) = \frac{\partial J_x}{\partial x} + \frac{\partial J_y}{\partial y} + \frac{\partial J_z}{\partial z}. \tag{3.41}$$

The right side of the equal sign will have a value for every given current density. So Maxwell added a term that can mathematically make the right side of the equation zero when the divergence is taken, (only when the current is changing):

$$\nabla \bullet (\nabla \times \mathbf{H}) = \nabla \bullet (\mathbf{J} + \frac{\partial \mathbf{D}}{\partial t}) = 0. \tag{3.42}$$

When the current $\mathbf{J}$ is changing with time in the experimental circuit

$$\frac{\partial \mathbf{D}}{\partial t}$$

will also be changing and the vector sum at each instant will be zero. When $\mathbf{J}$ is zero, in the experimental circuit, the displacement current is zero. Maxwell's addition to Ampere's equation, does not invalidate Ampere's equation. When

$$\frac{\partial \mathbf{D}}{\partial t} = 0,$$

Ampere's original equation is obtained.

3.6. Displacement Current Term Meets Radiation Requirement

The change of electric displacement density with time gives rise to a three dimensional magnetic field in space, as indicated by the magnetic field between the disks. This was also indicated analytically by Maxwell in the differential form of his third equation. In the equation, the time derivative term of displacement density is on one side of the equal sign and a space derivative operator and the magnetic field on the other side. Since there are no conducting currents in space, only the displacement density term can be the source of the magnetic field in space.

Since $\mathbf{D} = \epsilon\, \mathbf{E}$, electric field strength, is also displaced from $\mathbf{H}$, magnetic field strength, by 90 spatial degrees. This 90 degree spatial separation of the $\mathbf{H}$ and $\mathbf{E}$ fields always exists and if radiation occurs, the 90 degree spatial separation continues as the fields move through space. As mentioned previously, the amount of $\mathbf{H}$ developed by changing $\mathbf{D}$ is not measurable by simple testing. The experimental proof that a changing electric field caused a changing magnetic field had to wait for Hertz' confirming radiation experiments. Maxwell proceeded to use his equations to find the speed of electromagnetic propagation. The speed was determined to be the same as the measured speed of light. This result was the confirming proof of Maxwell's equations up to the time that Hertz proved them by his experiments.

3.7. Integral Form of Maxwell's Equation III

Maxwell's current continuity principle allows us to add the displacement current to the conduction current in Ampere's integral equation. Ampere's integral equation is

$$\oint_{Closed\ path} \mathbf{H} \bullet d\mathbf{s} = \int_{Surface} \mathbf{J} \bullet d\mathbf{a}. \tag{3.43}$$

With Maxwell's current continuity term added, the equation becomes

$$\oint_{Closed\ path} \mathbf{H}\bullet; d\mathbf{s} = \int_{Surface} (\mathbf{J} + \frac{\partial \mathbf{D}}{\partial t}) \bullet d\mathbf{a}. \tag{3.44}$$

Either

$$\mathbf{J} \quad \text{or} \quad \frac{\partial \mathbf{D}}{\partial t}$$

may be zero or both may have a value and the equation will still be true. If

$$\frac{\partial \mathbf{D}}{\partial t}$$

is zero, Ampere's equation results. When $\mathbf{J}$ is zero,

$$\oint_{Closed\ path} \mathbf{H} \bullet d\mathbf{s} = \int_{Surface} \frac{\partial \mathbf{D}}{\partial t} \bullet d\mathbf{a} \tag{3.45}$$

is the result. This last equation is an important equation in electromagnetic device design and wave propagation. We will use it below to find the speed of electromagnetic propagation.

3.8. Maxwell's Field Equation III, Conclusion

Ampere's steady current magnetic field equations:
integral form

$$\oint_{Closed\ path} \mathbf{H} \bullet d\mathbf{s} = \int_{Surface} \mathbf{J}\ d\mathbf{a}, \qquad (3.46)$$

differential form

$$\nabla \times \mathbf{H} = \mathbf{J}, \quad \text{or} \quad \text{curl } \mathbf{H} = \mathbf{J}. \qquad (3.47)$$

They were modified by Maxwell to account for the generation of magnetic fields by time varying electric fields:
integral form

$$\oint_{Closed\ path} \mathbf{H}\ d\mathbf{s} = \int_{Surface} (\mathbf{J} + \frac{\partial \mathbf{D}}{\partial t})\ d\mathbf{a}, \qquad (3.48)$$

differential form

$$\nabla \times \mathbf{H} = \mathbf{J} + \frac{\partial \mathbf{D}}{\partial t}, \quad \text{or} \quad \text{curl } \mathbf{H} = \mathbf{J} + \frac{\partial \mathbf{D}}{\partial t}. \qquad (3.49)$$

As is seen clearly in the differential form, Maxwell presented the space derivatives of the magnetic field as a function of the time derivative of an electric field. In so doing he showed conclusively that an electric field changing in time could produce a magnetic field in space. **D** is density of electric charge and that charge density being displaced in some time has the same units as current density, **J**. This completes the development of Maxwell's equation III. We will use Maxwell's equation III, together with equation IV, to obtain the velocity of electromagnetic radiation in free space.

4. Maxwell's Electromagnetic Equation IV

The following equations will be developed in this chapter.
integral form

$$\oint_{Closed\ path} \mathbf{E} \bullet d\mathbf{s} = -\frac{d}{dt} \int_{Surface} \mathbf{B} \bullet d\mathbf{a}, \qquad (4.1)$$

differential form

$$\nabla \times \mathbf{E} = -\frac{\partial}{\partial t}\mathbf{B}, \quad \text{or} \quad \text{curl } \mathbf{E} = -\frac{\partial}{\partial t}\mathbf{B}. \qquad (4.2)$$

Maxwell's equation IV is based on the results of Faraday's experiments published in 1831. Faraday investigated the generation of electrical voltage and current due to a changing magnetic field. The idea of magnetic fields originated with Faraday. Maxwell formulated the above equations to describe Faraday's findings in the way in which they could be applied to electromagnetic radiation

4.1. Faraday's Experiments

Figure 4.1 is a theoretical sketch of Faraday's experimental equipment. One coil of wire is connected to a battery, switch and a variable resistor for controlling the current. A second coil (pick-up coil) is placed parallel to the first and connected to a zero-at-center-scale voltmeter.

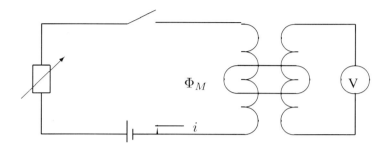

Figure 4.1. Faraday Experiment.

When the switch is closed, current in the first coil starts from zero, increases and reaches a constant amount. The amount of current depends upon the total circuit resistance and the battery voltage. While the current is rising, the meter needle deflects in one direction due to the build up of a magnetic flux, Φ_M, and then settles back to zero at center scale as the current in the first coil reaches its steady value.

When the switch is opened, the magnetic flux lines that had previously expanded to a certain magnitude will collapse. The meter needle deflects in the opposite direction and then settles back to center scale zero.

The voltage that appears between the output terminals of the pick-up coil is due to the change in the number lines of magnetic flux that link the pick-up coil. The strength of the magnetic field and therefore the density of magnetic flux lines is dependent upon the current as described in Maxwell's equation III[3].

To observe the output of the pick-up coil, when the output is due to an increase or decrease in the number of flux linkages with time only, the following test can be made:

Assume that the resistor of Figure 4.1 is set so that a current of one ampere is flowing through the first coil. Therefore, there is a fixed number of flux lines passing through the area inside the pick-up coil. Since the flux linkages are not changing, there is no voltage output. Then, for the next five seconds, we reduce the resistance smoothly. For this reduction in resistance, assume the current increases smoothly to six amperes and we cause the increase in current to be directly proportional to the elapsed time. During the first second, the current increases smoothly and continuously to two amperes. At the instant that the next second had passed, the current would be at three amperes, and so on.

---

[3] A change of flux lines through the pick-up coil can also be made by other means, such as moving either coil or moving a magnet through either coil. This kind of spatial change of flux is the concern of motor-generator theory. Maxwell was here interested in the basis of electromagnetic radiation. So he looked at Faraday's tests for voltage output due to the change in magnetic flux with time only.

4.2. Results of Faraday's Experiments

The results of the tests are these:

1. The meter needle shows a constant output voltage all the time that the flux lines are increasing directly proportional to time.

2. When the flux lines stop changing, the output is zero.

3. When we increase the resistance in the same smooth way, and cause the current and flux to decrease smoothly, the same voltage output is observed but the meter needle points in the opposite direction.

4. When we make the same current change in one-half the time, the output voltage doubles, and is steady all of the time the current is changing proportionately with time. These tests show the correlation between the output voltage and the rate of change of flux linkages.

5. When the output voltage is directly proportional to a constant rate of flux change, as in 1. to 4. above, the flux must be equal to some constant, K times time. Or, $\Phi_M = K\,t$. The differential calculus method of showing the rate of change of magnetic flux with time is

$$\frac{d}{dt}\Phi_M.$$

Then we get

$$\frac{d}{dt}\Phi_M = \frac{d}{dt}Kt = K \text{ (constant output in volt)}.$$

6. Experiments also show that when the plane of pick-up coil is placed at some random angle with respect to the plane of coil one, less output voltage is induced. When the pick-up coil is made of a few turns and very thin, and placed at a right angle to the plane of the first coil, there are no flux linkages and therefore no output voltage is induced in the pick-up coil.

7. The rate of change of the circuit current, and therefore the change of magnetic flux, most often varies with time. However, the output voltage will always be proportional to the derivative of the flux term.

As example, if the flux was changing as a sine wave, the voltage output would change as the derivative of the sine wave.

8. Much faster, but variable, rates of change of current and the generated flux linkages can be obtained by opening and closing the switch. The meter needle would flick to the full scale stops indicating large instantaneous voltages are generated.

The tests results noted above lead to the following equation for the voltage output of the pick-up coil

$$V_{\text{output}} = -N\,\frac{d\Phi_M}{dt}. \tag{4.3}$$

This is Faraday's Law of magnetic induction. Wherein, N is the number of turns in the pick-up coil, and

$$\frac{d}{dt}\Phi_M$$

is the rate of change of magnetic flux.

Faraday's Law shows that:

1. The output voltage is constant for constant change of flux linkages per unit time.
2. If the flux, M is varying at a non-constant rate, (perhaps as the sine wave used later), the output would have to vary as the derivative of the flux change.
3. More flux linkages per unit time generate larger output voltage.
4. Larger number of turns provide greater output.
5. Rising number of flux linkages per unit time provide the opposite output polarity than that polarity due to falling number of flux linkages per unit time.
6. Experiments have shown that an increase or decrease of 104 lines of flux in a square centimeter per second through a one turn loop will generate one volt.

4.3. Faraday's Law Applied to Electromagnetic Radiation

Maxwell postulated several principles that are indicated by Faraday's Law and experiments:

1. Only one loop would be needed to obtain a voltage output if sensitive test equipment was used. Therefore, N can be one in Faraday's law.
2. The voltage output was dependent upon the generated electric field strength **E**.
3. The voltage developed by Faraday's pick-up coil was equal to the electric field strength **E**, summed for each small length of wire, or distance around a loop.
4. The **E** field strength would be generated by the change of magnetic flux, whether a sensing wire loop was at that location or not. In the absence of a pick-up loop, the electric displacement **D**, that results from that field strength, would be dependent on the surrounding medium.
5. The **E** field that is being generated by the change of the magnetic flux, is at right angles to the flux. Therefore, to calculate the developed **E** field at other than a right angle to the flux, we will have to use a dot product to calculate the effective **E** field in that plane.
6. The magnetic flux **H** generating **E**, can be represented by single or multiple closed loops. Also the **E** can be represented by single or multiple closed loops. This shows a natural symmetry in the generating and generated fields. The equations representing the fields must be of the same form.

4.4. Maxwell's Equation IV, Integral Form

The total voltage generated by the change of the magnetic lines of flux is the integral or summation of the electric field strength, **E**, in volts per meter, dot multiplied with the distance along the loop. The dot product is required because only that **E** directly in line with each small distance, $d\mathbf{s}$, is effective in the summation of the generated voltage. The summation of the $\mathbf{E} \bullet d\mathbf{s}$ will be equal to the negative of the rate of flux linkage change as stated by Faraday's Law.

These considerations lead to the first part of Maxwell's fourth equation

$$\oint_{Closed\ path} \mathbf{E} \bullet d\mathbf{s} = -\frac{d}{dt} \Phi_M . \qquad (4.4)$$

In the above equation, **E** is the electric field strength in volts per meter.

Maxwell's Electromagnetic Equations, Elementary Introduction   63

The symbol • (dot) directs that the dot product be formed between **E** and d **s** so that only the **E** tangent to the path is included in its summation. The effective length of the path is exactly in line with the generated **E**.

The quantity d**s** is an incremental length of wire in the pick up loop or an incremental increase in the mathematical path indicated by the circle on the integral sign. This closed path must enclose all the magnetic flux that is causing the **E**. The vector **E** is induced in each incremental length of the path or wire loop.

The quantity Φ_M is the magnetic flux linking the pick-up loop or linking the mathematical path of the generated **E** and $\frac{d}{dt}\Phi_M$ is the rate of change of magnetic flux with time.

The surface integral (for the right side of Maxwell's equation below) of the magnetic flux is obtained through the following considerations: The total flux, Φ_M is equal to the flux density **B**, in lines per square meter times the effective area through which the flux lines flow. The effective area is only that amount of area that is at a right angle to the flow of flux. That means we must use the dot product, of the area and **B**, summed across the surface area to obtain the total effective flux, Φ_M.

The above considerations lead to the complete Integral form of Maxwell's equation IV:

$$\oint_{Closed\ path} \mathbf{E} \bullet d\mathbf{s} = -\frac{d}{dt}\int_{Surface} \mathbf{B} \bullet d\mathbf{a}. \qquad (4.5)$$

This equation states that the electric field strength, **E**, generated along a closed path is equal to the change of flux density, **B**, with time, times the enclosed area perpendicular to **B**. Or, stated differently, when the effective magnetic flux through an area bounded by a closed mathematical path changes with time, electric field strength, **E**, will be generated along that path. The closed path must enclose all the area that contained the magnetic flux. The dot product that is indicated between the **E** vector and the distance d **s** assures that only **E** tangent to the distance along the path is calculated.

Similarly, the dot product that is indicated between flux density **B**, and the area d **a**, determines that only the lines of flux at right angles to the to the computed area are effective in producing **E**. The negative sign indicates that the **E** field will have one polarity for a growing magnetic field and the opposite polarity for a decreasing magnetic field. The partial derivative symbol $\frac{\partial}{\partial t}$ denotes that all other independent variables are held constant and only t (time) is allowed to change. The **B** will be increasing or decreasing with time to cause **E** to appear.

Maxwell knew that magnetic lines of flux were continuous, and to have the mathematics show that electric lines of flux must also be continuous, was a natural symmetric phenomenon. Then, when electromagnetic radiation was produced, magnetic and electric field vectors would manifest themselves as complete loops.

4.5.   Maxwell's Equation IV, Differential Form

We will apply Stokes' theorem to the integral form of Maxwell's fourth equation to obtain its differential form. The integral form copied from above is

$$\oint_{Closed\ path} \mathbf{E} \bullet d\mathbf{s} = -\frac{d}{dt}\int_{Surface} \mathbf{B} \bullet d\mathbf{a}. \qquad (4.6)$$

Through the use of Stokes' identity, we can simply write down the surface integral that is equal to the closed path integral of the same vector. Stokes identity indicates that the closed path integral must include all the area that is designated by the area (surface) integral.

$$\oint_{Closed\ path} \mathbf{E} \bullet d\mathbf{s} = \int_{Surface} (\nabla \times \mathbf{E}) \bullet d\mathbf{a}. \tag{4.7}$$

Therefore,

$$\int_{Surface} (\nabla \times \mathbf{E}) \bullet d\mathbf{a} = -\frac{d}{dt} \int_{Surface} \mathbf{B} \bullet d\mathbf{a}. \tag{4.8}$$

When the integrals are evaluated over the same area, the result is

$$\nabla \times \mathbf{E} = -\frac{\partial \mathbf{B}}{\partial t}. \tag{4.9}$$

This is the differential form of Maxwell's equation IV. The equation shows a changing magnetic field in time generating an electric field in space. Symmetry in nature would indicate that a changing electric field in time would cause a magnetic field to be generated in space as Maxwell shows in his third equation incorporating **D** and **H**. In the above equation $\nabla \times \mathbf{E}$ a three dimensional space derivative of **E** is shown to be equal to the time rate of change of magnetic flux. When the curl of E is taken a new vector is formed that is a function of **B**. In Maxwell's equation III, we see that the curl of **H** will generate a new vector that is a function of **D**. This symmetry helped Maxwell form his equations.

4.6. Conclusion to Maxwell's Equation IV

Maxwell's equation IV describes the generation of an electric field in space due to the effect of a changing, (with time only), magnetic field. The generation of an E field due to changing with time H field, was demonstrated easily by Faraday by using a pick-up coil and measuring the output voltage. So the formulation of Maxwell's equation IV is far more obvious than the formulation of his equation III. In the latter formulation, the discovery of the existence of an H field due to a changing electric field (the displacement density changing in time) depended upon Maxwell's intuition. There is no sensor for picking up the H field that is generated by the changing D field.

When radiation occurs, the two generated fields are interdependent. Maxwell was able to prove their existence and method of generation with one demonstration. Maxwell demonstrated their existence by calculating the speed of electromagnetic radiation using his equations. The speed turned out to be the same as the speed of light, proving light to be an electromagnetic radiation. About 20 years later Hertz experimentally proved the existence and method of generation of both fields by demonstrating electromagnetic radiation.

5. Electromagnetic Field Propagating

5.1. The Concept of Field

Faraday's electric fields, generated by changing-in-time magnetic field strength, and Maxwell's magnetic fields, generated by changing-in-time electric field strength, are sketched together as an electromagnetic field in Figure 5.1. The electromagnetic field is produced by the radio transmitting antenna powered by an alternating current source.

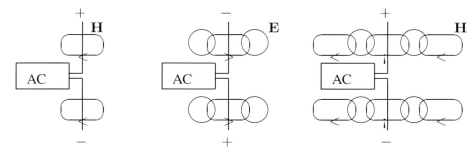

Figure 5.1. Electromagnetic field propagation.

As the fields disengage from the antenna, they propagate through space in self sustaining fields as sketched in Figures 5.1. In the figure, lines of **E** changing in time, stretch from end to end of the antenna and generate an **H** field in space perpendicular to the **E** field, as described by Maxwell's equation III. At the same moment, the changing in time H fields are permeated, at right angles, by the rising in space of **E** fields as described by Maxwell equation IV.

The basis for the propagation of the electromagnetic wave rests on the ability of the magnetic field varying with time, to generate an electric field in space, and the ability of the electric displacement density field varying with time, to generate a magnetic field in space.

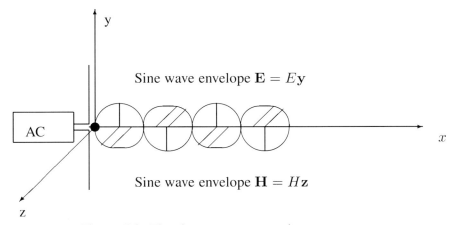

Figure 5.2. The sine wave propagation.

Figure 5.2 illustrates the **E** and **H** fields continuously leaving the antenna at the sine wave frequency determined by the AC source. The sine wave envelopes conceptualize the relative length of the **E** and **H** vectors at every instant of time. For the axes noted, the **E** vectors are placed parallel to the y axis and perpendicular to the x axis. The **H** vectors must be placed at right angles to the **E** vectors so they are placed parallel to the z axis. If the **H** vectors are placed first, the positions of the **E** vectors are defined, and vice versa. The vectors are, of course, defined by Maxwell's equations III and IV. We do not need these representations, but they help to form a mental picture of the mathematics. The vectors of one field stay parallel to themselves and perpendicular to the other generating field vector during propagation. The waves remain 90 degrees apart in space as they propagate in the x direction.

The vector relationship that describes the method whereby these fields leave the antenna is $\mathbf{E} \times \mathbf{H}$. To apply the right hand rule: let the index finger (pointing in the direction of the y axis) represent the $\mathbf{E}$ vector. Let the middle finger (pointing in the direction of the z axis) represent the $\mathbf{H}$ vector. When the $\mathbf{E}$ vector is crossed (folded) into the $\mathbf{H}$ vector (move index finger toward the middle finger), the thumb points in the direction of the resultant motion. This is the direction of propagation along the x axis.

Electromagnetic fields will propagate from the antenna in the directions of the x axes. Electromagnetic fields carry energy, momentum, show 90 degree polarization, and travel through space at a definite speed. These fields, called radio waves, will radiate a full 360 degrees around the y axis (of Figure 5.2).

Since the radio waves fill the space around the antenna, we can look at the $\mathbf{E}$ and $\mathbf{H}$ vectors along any x axis or parallel to an x axis. In the Figure 5.3 view, the $\mathbf{E}$ and $\mathbf{H}$ fields are sine waves looked at on edge. The $\mathbf{E}$ and $\mathbf{H}$ fields are represented as arrows that grow in height (each in its own direction), reach a maximum, decrease to zero and then grow in height in the opposite direction. As the waves leave the antenna they spread out as ever expanding spheres. At some distance from the antenna the spheres are so large that a part of the sphere's surface would approximate a flat plane. This radiation is termed a plane wave. Figure 5.3 is the sketch of the edge view of the sine waves on that flat plane. The edge view of $\mathbf{E}$ and $\mathbf{H}$ are presented as vectors.

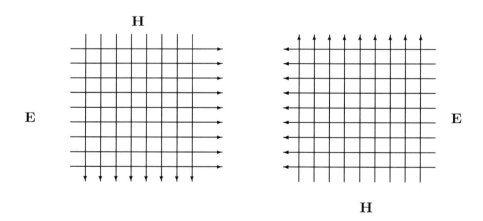

Figure 5.3. Electromagnetic $\mathbf{E}$, $\mathbf{H}$-wave.

5.2. Equation of Stationary Sine Wave

The radio antenna is transmitting electromagnetic fields represented by vectors whose amplitudes vary as sine waves. At what speed do these radio waves travel, and is the speed constant? Maxwell used only his equations to find the answer to these questions.

Maxwell's Electromagnetic Equations, Elementary Introduction

We will begin the problem of finding the speed of electromagnetic propagation by considering the equation of a stationary sine wave, and then proceed to the equation of traveling sine waves of **E** and **H**. We will then insert the traveling sine wave terms into the integral forms of Maxwell's equations III and IV. We will obtain two ratios that enable us to solve for the speed of propagation. Figure 5.4 shows a stationary sine wave, formed by plotting the A axis amplitude of a counterclockwise rotating radius vector.

Let the radius vector be designated **A**, and let it rotate counterclockwise at some constant revolutions per second. Figure 5.4 shows the plot of the sine wave generation and the relationship of the angles generated by the rotating vector as its A axis amplitude is plotted.

$$\mathbf{A} = A_{max} \cos \omega t \; \mathbf{x} + A_{max} \sin \omega t \; \mathbf{y} ,$$

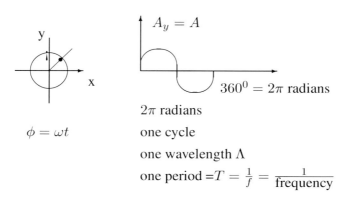

Figure 5.4. Stationary sine wave.

The revolutions per second of the radius vector is equal to the frequency generated in cycles per second. The distance along the x axis can be plotted in length, degrees, radians and time. The period of the sine wave is designated T. It is the time in seconds required for one complete cycle of alternation. Therefore

$$T^{-1} = f \tag{5.1}$$

is the number of cycles per second. The following equations are indicated by Figure 5.4.

$$A = A_{max} \sin (\text{angle } \phi \text{ in radians}) ,$$

or

$$A(t) = A_{max} \sin \omega t . \tag{5.2}$$

Defining period T by

$$\omega T = 2\pi , \quad \omega = \frac{2\pi}{T} = 2\pi f , \tag{5.3}$$

the above statements lead to the equation of a stationary sine wave

$$A(t) = A_{max} \sin \left(2\pi \frac{t}{T}\right) = \sin(2\pi f t) . \tag{5.4}$$

5.3. Equation of Traveling Sine Wave

The equation of the stationary sine wave will now have a term added to obtain the equation of a traveling sine wave. Figure 5.5 shows the sine wave moving in the $+x$ direction at velocity v:

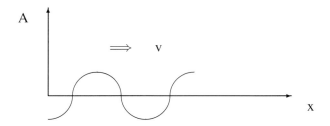

Figure 5.5. Traveling sine wave.

The maximum instantaneous amplitude remains A_{max}, whether the sine wave is moving or not, but the instantaneous value is now a function of time, t, and the distance, x. In mathematical terms

$$A(x,t) = A_{max} \sin(\text{angle}), \quad \text{angle that depends upon } x \text{ and } t.$$

Note in Figure 5.5, that the instantaneous value of A at the origin, goes negative if the stationary sine wave is simply moved along the x axis in the $+x$ direction. So the equation to give the value of A for a wave traveling in the $+x$ direction, must be of the form

$$A(x,t) = A(x=0, t - \frac{x}{\text{v}}), \tag{5.5}$$

or

$$A(x,t) = A_{max} \sin \omega (t - \frac{x}{\text{v}}). \tag{5.6}$$

The wavelength in meters must be equal to the velocity of transmission in meters per second times the period in seconds:

$$\text{wavelength} = \text{velocity} \times \text{period}, \quad \Lambda = \text{v}\, T, \tag{5.7}$$

or

$$\Lambda = \text{v}\, \frac{2\pi}{\omega}. \tag{5.8}$$

In the discussions that follow, A and A_{max} will be replaced by the E, E_{max} and H, H_{max} of the propagating field. The alternating current source feeding the antenna will maintain the emitted frequency constant. The velocity of the transmission is constant because the electric characteristics of the medium (space) through which the transmission occurs are constant. So the wavelength will also be constant.

5.4. Electric and Magnetic Fields as Traveling Sine Waves

The **E** and **H** fields will be leaving the antenna as traveling sine waves, so we need only replace A in the general traveling wave equation with E and H to get the expressions for the traveling fields.

The general equation for traveling sine wave is

$$A(x,t) = A_{max} \sin \omega (t - \frac{x}{V}). \tag{5.9}$$

The envelope of the **E** vectors leaving the antenna is a sine wave. Therefore the equation of the propagating E field is

$$E(x,t) = E_{max} \sin \omega (t - \frac{x}{V}). \tag{5.10}$$

The magnetic **H** field is symmetric to the **E** field and 90 space degrees displaced from it. The equation for the H field is the same form as with the E field

$$H(x,t) = H_{max} \sin \omega (t - \frac{x}{V}). \tag{5.11}$$

The independent variables of both equations are x, t. It is necessary to use these two equations with modified Maxwell's equations III and IV to find v, the velocity of electromagnetic propagation.

5.5. Sketch of E and H Waves for Velocity Determination

Figures 5.6 and 5.7 are separate views of the electromagnetic sine waves that were shown in Figure 5.2. In addition, the figures indicate the meshing of the **E** and **H** fields as they propagate through space. The meshing shows the 90^0 space separation of the fields.

Figure 5.6 is a view of the $x - z$ plane, showing the H sine wave envelope of the **H** vector arrows. Figure 5.7 is a view of the $x - y$ plane showing the **E** sine wave envelope and the **E** vector arrows.

$$\mathbf{E} = E_z(x,t)\,\mathbf{z} = E_{max} \sin \omega (t - \frac{x}{V})\,\mathbf{z}. \tag{5.12}$$

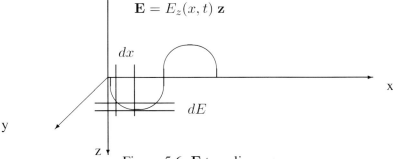

Figure 5.6. **E** traveling wave.

$$\mathbf{H} = H_y(x,t)\,\mathbf{y} = H_{max}\sin\omega(t - \frac{x}{\mathrm{V}})\,\mathbf{y} \qquad (6.9)$$

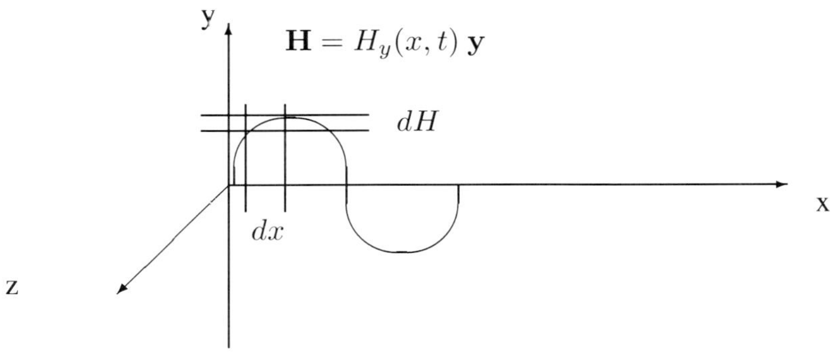

Figure 5.7. **H** traveling wave.

5.6. Maxwell's Equation III

The integral form of equation III is

$$\oint_{Closed\ path} \mathbf{H} \bullet d\mathbf{s} = \int_{Area} \left(\frac{\partial \mathbf{D}}{\partial t} + \mathbf{J}\right) \bullet d\mathbf{a}. \qquad (5.13)$$

Conduction current does not exist in free space, therefore **J** is zero. Here we will use the essence of Maxwell's contribution to electromagnetic theory. A change with time of electric displacement density causes the generation of a magnetic field in space. The velocity of radiation (and many other electromagnetic principles) depend on this concept. Recall that **D** is electric displacement density, and as **D** changes with time (t) is equivalent to current. This is the current that is generating the **H** on the left side of the equation (5.13).

As we will see later, it is necessary to incorporate the permittivity of free space into this problem of finding velocity of electromagnetic radiation. As indicated above,

$$\mathbf{D} = \epsilon\,\mathbf{E}, \qquad (5.14)$$

showing that **D** is dependent both on the electric field strength and the permittivity of the surrounding medium. The permittivity ϵ is constant in space and can be taken through the integral sign. These changes modify equation III to be

$$\oint_{Closed\ path} \mathbf{H} \bullet d\mathbf{s} = \epsilon \int_{Area} \frac{\partial \mathbf{E}}{\partial t} \bullet d\mathbf{a}. \qquad (5.15)$$

To evaluate the left hand integral we need a closed path around the **H** field. The closed path can be designated analytically or geometrically. Here we choose a geometrical representation. In the Figure 5.7 trace a rectangle around the **H** vectors. In the Figure 5.7 trace a rectangle around the **H** vectors. Designate two amplitudes H and $H + dH$. We also designate a directed distance h down from the x axis as the distance we must travel to reach the distance of H plus dH. The distance h is a convenient amplitude designation that will cancel later. The rectangle is completed with a distance dx along the x axis. This closed path enables us to calculate H of the left hand integral and the rectangular area enclosed by the path enables us to calculate the E of the right hand integral.

5.7. Evaluate Magnetic Field Strength Closed Path Integral

Evaluate

$$\oint_{Rectangular\ path} \mathbf{H} \bullet d\mathbf{s} . \qquad (5.16)$$

We have to integrate, or sum, the dot product of **H** times the directed distance we are moving along the path of the rectangle. Using the rectangle of Figure 5.7, the closed line integral of $\mathbf{H} \bullet d\mathbf{s}$ will now be obtained. The path of travel is counterclockwise around the rectangle. Start at the upper right hand corner of the rectangle: Move to the left along the x axis a distance dx. The arrowheads of **H**, the end points of which make up the sine wave envelope, are pointing in the $+z$ direction. Since we are moving at right angles to the **H** vectors, the cosine of the angle between the direction of motion and the **H** vectors is zero

$$\oint_{Distance\ dx} \mathbf{H} \bullet d\mathbf{s} = H\,dx \cos 90^0 = 0 .$$

Traverse down the left side in the $+z$ direction. This movement is in the same direction as the **H** vector

$$\oint_{Distance\ h} \mathbf{H} \bullet d\mathbf{s} = Hh \cos 0 = Hh .$$

Then traverse the bottom of the rectangle from left to right

$$\oint_{Distance\ dx} \mathbf{H} \bullet d\mathbf{s} = H\,dx \cos 90^0 = 0 .$$

The final step is to traverse up the right side of the rectangle to the beginning point

$$\oint_{Distance\ h} \mathbf{H} \bullet d\mathbf{s} = (H + dH)h \cos 180^0 = -hH - h\,dH .$$

Then

$$\oint_{Rectangular\ path} \mathbf{H} \bullet d\mathbf{s} = -h\,dH . \qquad (5.17)$$

The solution of this closed path integral of the modified Maxwell's III equation is $-h\,dH$.

The modified equation III shows that, to obtain the magnetic field strength **H**, the closed path of the left integral must enclose all the displacement current in the right hand integral that is causing the **H**. The displacement current, (the change of **D** over a change of time) that is giving rise to **H**, is contained in the area dx by h of Figure 5.6. We can know that the closed path of the left integral surrounds the area of the right integral by using the same rectangle for both integrals.

5.8. Evaluate Electric Field Strength Area Integral

The displacement current is indicated by the right hand integral

$$\epsilon \frac{d}{dt} \int_{Area} \mathbf{E}\,d\mathbf{a} . \qquad (5.18)$$

This is the amount of **E**, changing with time, that is causing the H that was calculated in the left integral. When we dot multiply the change of **E** with time by the area of the rectangle, and by ϵ, the result must be the displacement current causing the **H** of the left integral. The summation of the area of the rectangle is of course $h\ dx$. The vector representing area is normal to the area, or in line with the **E** vectors. The angle indicated by the dot product is zero degrees with a cosine of one. So the right hand integral simplifies to

$$\epsilon \frac{\partial E}{\partial t}\ \text{area} = \epsilon \frac{\partial E}{\partial t}\ h\ dx\ . \tag{5.19}$$

With the integrations completed, the modified equation III is this differential equation

$$-h\ dH = \epsilon \frac{\partial E}{\partial t}\ h\ dx\ .$$

Cancel h, divide by dx and obtain

$$\frac{\partial H}{\partial x} = \epsilon \frac{\partial E}{\partial t}\ . \tag{5.20}$$

This is the first of the two differential equations required for computing the velocity of propagation. Notice that the mathematics show that the change of **E** in time only, is causing **H** to be generated in space. The second differential equation required to compute the velocity of electromagnetic propagation will now be obtained from Maxwell's equation IV.

5.9. Maxwell's Equation IV

The integral form of Maxwell's equation IV is

$$\oint_{Closed\ path} \mathbf{E} \bullet d\mathbf{s} = -\frac{d}{dt} \int_{Area} \mathbf{B} \bullet d\mathbf{a}\ . \tag{5.21}$$

As Maxwell had postulated, a change with time of the magnetic field **B** will cause the generation of the electric field **E** in space.

Recall that the magnetic flux density **B** is equal to the permeability of the surrounding medium μ times the magnetic field strength **H**

$$\mathbf{B} = \mu\,\mathbf{H}\ . \tag{5.22}$$

The quantity μ is a constant in free space and is required in this problem to find the velocity of radiation. This constant can be taken through the integral sign. With these changes, the equation is modified to be

$$\oint_{Closed\ path} \mathbf{E} \bullet d\mathbf{s} = -\mu \frac{d}{dt} \int_{Area} \mathbf{H} \bullet d\mathbf{a}\ . \tag{5.23}$$

5.10. Evaluate Electric Field Strength, Closed Path Integral

The integral

$$\oint_{Closed\ path} \mathbf{E} \bullet d\mathbf{s} \quad (5.24)$$

must now be evaluated. The left hand integral of Maxwell's equation IV tells us to sum the **E** that is tangent to a path that we select. This path must be continuous and surround the area of the changing **H** that is generating the **E** along the path that we have selected.

Since the **E** vector amplitude arrowhead is creating a sine wave envelope, we need only designate a random point on the x axis and traverse a distance dx from that random point. And then state that the two vertical sides of the imagined rectangle are the amplitudes of **E** vectors. The amplitudes are **E** and $\mathbf{E} + d\mathbf{E}$ on either side of the distance dx, and are shown in the rectangle of Figure 5.7. The integral

$$\oint_{Closed\ path} \mathbf{E} \bullet d\mathbf{s}$$

indicates the summation of vector **E** dot multiplied with the incremental distance along the rectangular path. The integration is performed in piecewise continuous steps, going counterclockwise, starting at the upper right corner.

Traversing the top of the rectangle:

$$\oint_{Distance\ dx} \mathbf{E} \bullet d\mathbf{s} \approx \mathbf{E} \bullet d\mathbf{x} = E\ dx \cos 90^0 = 0 .$$

Let h be a distance equal to $\mathbf{E} + d\mathbf{E}$ directed up from the x axis. Then going down the left side of the rectangle:

$$\oint_{Distance\ dh} \mathbf{E} \bullet d\mathbf{s} \approx \mathbf{E} \bullet \mathbf{h} = E\ h\ \cos 180^0 = -E\ h .$$

Going across the bottom of the rectangle along the x axis:

$$\oint_{Distance\ dx} \mathbf{E} \bullet d\mathbf{s} \approx \mathbf{E} \bullet d\mathbf{x} = E\ (-dx)\ \cos 90^0 = 0 .$$

Going up the right side of the rectangle:

$$\oint_{Distance\ dh} \mathbf{E} \bullet d\mathbf{s} \approx (\mathbf{E} + d\mathbf{E}) \bullet \mathbf{h} = E\ h\ \cos^0 = +(E + dE)\ h .$$

Summing the distances around the rectangle:

$$\oint_{Rectangle} \mathbf{E} \bullet d\mathbf{s} = +h\ dE . \quad (5.25)$$

Now we must evaluate the right hand integral of modified Maxwell's equation IV:

$$-\mu \frac{d}{dt} \int_{Area} \mathbf{H} \bullet d\mathbf{a} . \quad (5.26)$$

The changing H that is effective in generating the E is at right angles to the E. The vector that represents this area is perpendicular to the area dx by h and is in the $+z$ direction. The angle between the **H** vectors and the vector representing the area is zero degrees, with a cosine of one. The integral, or summation of area, is the area of the rectangle, dx times h.

$$-\mu \frac{d}{dt} \int_{Rectangle} \mathbf{H} \bullet \mathbf{da} = -\mu \frac{\partial H}{\partial t} h \, dx \cos 0 = -\mu \frac{\partial H}{\partial t} h \, dx. \quad (5.27)$$

Using the results of the procedures above, the modified equation IV is

$$+ h \, dE = -\mu \frac{\partial H}{\partial t} h \, dx. \quad (5.28)$$

Cancel h, divide by dx and obtain

$$\frac{\partial E}{\partial x} = -\mu \frac{\partial H}{\partial t}. \quad (5.29)$$

This equation illustrates the concept where a change of **H** with time causes a change of **E** in space. Also this is the second of the two differential equations required to be used with the traveling sine wave equation in order to obtain the velocity of propagation.

5.11. Obtain the First E/H Ratio Using Equation III

When the electromagnetic field is propagating though space, exactly the same E and exactly the same H are designated in both equations III and IV. Therefore, we may equate E/H ratios from both equations, and use the ratios in simultaneous equations. But first we must go back to the above two differential equations, and solve those equations for ratios of E and H. The first differential equation is based on Maxwell's equation III.

$$-\frac{\partial H}{\partial x} = +\epsilon \frac{\partial E}{\partial t}. \quad (5.30)$$

Insert the traveling sine wave equations for E and H found above

$$-\frac{\partial}{\partial x}\left(H_{max} \sin \omega(t - \frac{x}{t})\right) = +\epsilon \frac{\partial}{\partial t}\left(E_{max} \sin \omega(t - \frac{x}{t})\right). \quad (5.31)$$

The derivative (or rate of change) of a sine wave is a cosine wave. This can be seen by plotting the cosine of the angle generated by the radius vector in Figure 5.4 in addition to the sine wave. When the sine wave is going through its maximum rate of change, as it passes through the x axis, the cosine wave is maximum. When the sine wave is going through its point no change at its maximum or minimum, the cosine wave is passing through zero on the x axis.

The calculus rule then for differentiating a general sine function like

$$\sin \omega(t - \frac{x}{v}), \quad (5.32)$$

when x is the variable is

$$\frac{\partial}{\partial x} \sin \omega(t - \frac{x}{v}) = -\frac{\omega}{v} \cos \omega(t - \frac{x}{v}). \quad (5.33)$$

Maxwell's Electromagnetic Equations, Elementary Introduction

When t is the variable, the rule for differentiating is

$$\frac{\partial}{\partial t} \sin \omega(t - \frac{x}{v}) = +\omega \cos \omega(t - \frac{x}{v}). \tag{5.34}$$

When we differentiate the above wave equation (5.31) containing the sine wave functions, according to these rules, we obtain

$$H_{max} \left(\frac{\omega}{v}\right) \cos \omega(t - \frac{x}{v}) = E_{max} (\epsilon \omega) \cos \omega(t - \frac{x}{v}),$$

or

$$\frac{H_{max}}{E_{max}} = \epsilon v. \tag{5.35}$$

This is the first of the two H/E ratios needed to find the velocity.

5.12. Obtain the Second E/H Ratio Using Equation IV

The second H/E ratio, with v again constant, is obtained from the differential equation based on Maxwell's fourth equation. The simplified equation from above is

$$\frac{\partial E}{\partial x} = -\mu \frac{\partial H}{\partial t}. \tag{5.36}$$

The traveling sine wave equations for E and H can now be inserted into the equation and differentiated

$$+\frac{\partial}{\partial x}\left(E_{max} \sin \omega(t - \frac{x}{t})\right) = -\mu \frac{\partial}{\partial t}\left(H_{max} \sin \omega(t - \frac{x}{t})\right). \tag{5.37}$$

Applying the differentiation rules to the above equation we obtain

$$E_{max} \left(\frac{\omega}{v}\right) \cos \omega(t - \frac{x}{v}) = H_{max} (\mu \omega) \cos \omega(t - \frac{x}{v}),$$

or

$$\frac{E_{max}}{H_{max}} = \mu v. \tag{5.38}$$

This result is the second E/H ratio. These two ratios will enable us to find the velocity of electromagnetic propagation.

5.13. Solving for Velocity of Electromagnetic Propagation

The ratios obtained in previous sections are

$$\frac{H_{max}}{E_{max}} = \epsilon v, \quad \frac{E_{max}}{H_{max}} = \mu v. \tag{5.39}$$

Multiplying two relations we obtain for velocity v

$$1 = v^2 \epsilon \mu, \quad v = \frac{1}{\sqrt{\epsilon \mu}}. \tag{5.40}$$

The value of the permeability μ is approximately 1.26×10^{-6} and permittivity ϵ is 8.8×10^{-12}. Therefore, the speed of electromagnetic radiation is

$$\text{v} = 3 \times 10^8 \text{ meters per second}. \tag{5.41}$$

This is what Maxwell discovered. The speed is not relative to anything. The answer is not that the radiation went some distance from the emitting antenna in one second. The equations state, that in one second, one of the wave alternations will be 3×10^8 meters from wherever it is now. This important idea was one of the clues that Einstein had for his relativity theory. More important to us, these equations of Maxwell are the bases for modern communications.

5.14. Differential Maxwell's Equations and Electromagnetic Waves

The all four Maxwell's equations in the empty space may be written as follows

$$\nabla \bullet \mathbf{E} = 0, \tag{5.42}$$

$$\nabla \bullet \mathbf{B} = 0, \tag{5.43}$$

$$\nabla \times \mathbf{E} = -\frac{\partial \mathbf{B}}{\partial t}, \tag{5.44}$$

$$\frac{1}{\epsilon\mu}\nabla \times \mathbf{B} = +\frac{\partial \mathbf{E}}{\partial t}. \tag{5.45}$$

Now we are to show that the four Maxwell's equations have a wave solution of the form

$$\mathbf{E} = \mathbf{x}\, E_{max} \sin \omega(t - \frac{z}{\text{v}}), \quad \mathbf{B} = \mathbf{y}\, B_{max} \sin \omega(t - \frac{z}{\text{v}}). \tag{5.46}$$

Let us substitute (5.46) into (5.42) and (5.43):

$$\nabla \bullet \mathbf{E} = \frac{\partial}{\partial x}E_x + \frac{\partial}{\partial y}0 + \frac{\partial}{\partial z}0 = \frac{\partial}{\partial x}E_{max}\sin\omega(t - \frac{z}{\text{v}}) = 0,$$

$$\nabla \bullet \mathbf{B} = \frac{\partial}{\partial y}0 + \frac{\partial}{\partial y}B_y + \frac{\partial}{\partial z}0 = \frac{\partial}{\partial y}B_{max}\sin\omega(t - \frac{z}{\text{v}}) = 0.$$

Substituting (5.46) into the left hand side of eq. (5.44):

$$\nabla \times \mathbf{E} = (\frac{\partial E_z}{\partial y} - \frac{\partial E_y}{\partial z})\mathbf{x} + (\frac{\partial E_x}{\partial z} - \frac{\partial E_z}{\partial x})\mathbf{y} + (\frac{\partial E_y}{\partial x} - \frac{\partial E_x}{\partial y})\mathbf{z}$$

$$= \frac{\partial E_x}{\partial z}\mathbf{y} = -\frac{\omega}{\text{v}}E_{max}\cos\omega(t - \frac{z}{\text{v}})\mathbf{y}.$$

The right hand side of eq. (5.44) gives

$$-\frac{\partial \mathbf{B}}{\partial t} = -\frac{\partial}{\partial t}B_{max}\sin\omega(t - \frac{z}{\text{v}})\mathbf{y} = -B_{max}\omega\cos\omega(t - \frac{z}{\text{v}})\mathbf{y}.$$

From the equality

$$-\frac{\omega}{\text{v}}E_{max}\cos\omega(t - \frac{z}{\text{v}})\mathbf{y} = -B_{max}\,\omega\cos\omega(t - \frac{z}{\text{v}})\mathbf{y}$$

it follows
$$\frac{E_{max}}{B_{max}} = \text{v}. \tag{5.47}$$

Now, let us substitute eq. (5.46) into the Maxwell's equation (5.45); the left hand side of eq. (5.45) gives

$$\frac{1}{\epsilon\mu}\nabla \times \mathbf{B} = \frac{1}{\epsilon\mu}\left[(\frac{\partial B_z}{\partial y} - \frac{\partial B_y}{\partial z})\mathbf{x} + (\frac{\partial B_x}{\partial z} - \frac{\partial B_z}{\partial x})\mathbf{y} + (\frac{\partial B_y}{\partial x} - \frac{\partial B_x}{\partial y})\mathbf{z}\right]$$

$$= -\frac{1}{\epsilon\mu}\frac{\partial B_y}{\partial z}\mathbf{x} = \frac{1}{\epsilon\mu}\frac{\omega}{\text{v}}B_{max}\cos\omega(t-\frac{z}{\text{v}})\mathbf{x}.$$

The right hand side of eq. (5.44) gives

$$\frac{\partial \mathbf{E}}{\partial t} = \frac{\partial}{\partial t}E_{max}\sin\omega(t-\frac{z}{\text{v}})\mathbf{x} = E_{max}\omega\cos\omega(t-\frac{z}{\text{v}})\mathbf{x}.$$

From the equality

$$\frac{1}{\epsilon\mu}\frac{\omega}{\text{v}}B_{max}\cos\omega(t-\frac{z}{\text{v}})\mathbf{x} = E_{max}\omega\cos\omega(t-\frac{z}{\text{v}})\mathbf{x}$$

it follows

$$\frac{B_{max}}{E_{max}} = \text{v}\,\epsilon\mu. \tag{5.48}$$

Two relations, (5.47) and (5.48), will enable us to find the velocity of electromagnetic propagation. It suffices to multiply (5.47) by (5.48):

$$1 = \text{v}^2\,\epsilon\mu.$$

Therefore, the speed of electromagnetic wave is

$$\text{v} = \frac{1}{\sqrt{\epsilon\mu}}. \tag{5.49}$$

This is again what Maxwell discovered.

6. Electromagnetic Field Represented by Propagating Waves

The derivation of the sine wave equations that represent electric field **E** and magnetic field **H** is presented in this section. These are the same sine wave equations that were assumed for **E** and **H** before, which will now be discussed in detail. A group of four Maxwell's equations is the starting point for this derivation. One vector identity is then used to obtain independent vector differential equations of **E** and **H**. These independent equations are then solved to obtain the sine wave equations of **E** and **H**. The space and electric-magnetic phase relations of **E** and **H** are then determined.

6.1. Vectors E and H Are Entangled in Maxwell's Equations

Maxwell's Electromagnetic equations for free space are:

$$\nabla \bullet \mathbf{E} = 0, \tag{6.1}$$

$$\nabla \bullet \mathbf{B} = 0, \tag{6.2}$$

$$\nabla \times \mathbf{H} = \epsilon \frac{\partial \mathbf{E}}{\partial t}, \tag{6.3}$$

$$\nabla \times \mathbf{E} = -\mu \frac{\partial \mathbf{H}}{\partial t}. \tag{6.4}$$

The last two equations, (6.3) and (6.4), seem to indicate some circular reasoning. According to the third equation the derivative of **E** must be known in order to obtain **H**, and according to the fourth equation the derivative of **H** must be known to obtain **E**. Therefore equations three and four must be disentangled in order to obtain the independent equations of **E** and **H**. The separation of **E** and **H** is accomplished through the use of a specific vector identity discussed below.

6.2. Selection and Proof of the Required Vector Identity

This section shows the procedure for calculating the vector cross product $\mathbf{A} \times \mathbf{B}$. That product is shown to be equal to the vector sum of six terms. Now use that same procedure to calculate a second cross product $\mathbf{C} \times (\mathbf{A} \times \mathbf{B})$, where **C** is a another general vector

$$\mathbf{C} = C_x \mathbf{x} + C_y \mathbf{y} + C_z \mathbf{z}.$$

The result of taking this second cross product is a vector sum of twelve terms. That is

$$\mathbf{C} \times (\mathbf{A} \times \mathbf{B}) = (C_x \mathbf{x} + C_y \mathbf{y} + C_z \mathbf{z})$$

$$\times [\, (A_y B_z - A_z B_y)\, \mathbf{x} + (A_z B_x - A_x B_z)\, \mathbf{y} + (A_x B_y - A_y B_x)\, \mathbf{z} \,]$$

$$= C_x\, (A_z B_x - A_x B_z)\, \mathbf{z} - C_x (A_x B_y - A_y B_x)\, \mathbf{y} +$$
$$- C_y (A_y B_z - A_z B_y)\, \mathbf{z} + C_y (A_x B_y - A_y B_x)\, \mathbf{x}$$
$$+ C_z\, (A_y B_z - A_z B_y)\, \mathbf{y} - C_z (A_z B_x - A_x B_z)\, \mathbf{x}\,. \tag{6.5}$$

We must now select a vector identity that contains certain cross and dot products and show that its calculated result is the same twelve terms. The following identity is listed among a group of similar identities in vector mathematics texts

$$\mathbf{C} \times (\mathbf{A} \times \mathbf{B}) = -(\mathbf{C} \bullet \mathbf{A})\mathbf{B} + \mathbf{A}\,(\mathbf{C} \bullet \mathbf{B}). \tag{6.6}$$

The reason for calculating the second cross product above is to show that this identity is true and meets our present derivation requirement. This identity is proven by doing the indicated dot products and multiplications on the right side and then obtaining the same twelve terms listed above.

Two of the general vectors, **A** and **C**, can be replaced by ∇ in the vector identity and still maintain the identity. Thus,

$$\nabla \times (\nabla \times \mathbf{B}) = -(\nabla \bullet \nabla)\mathbf{B} + \nabla(\nabla \bullet \mathbf{B}),$$

which is

$$\nabla \times (\nabla \times \mathbf{B}) = -\nabla^2 \mathbf{B} + \nabla(\nabla \bullet \mathbf{B}). \tag{6.7}$$

This is vector identity required to disentangle **E** and **H**. Note that

$$\nabla^2 \mathbf{B} = \left(\frac{\partial^2}{\partial x^2} + \frac{\partial^2}{\partial y^2} + \frac{\partial^2}{\partial z^2}\right)\mathbf{B}, \tag{6.8}$$

then in component form

$$\nabla^2 \mathbf{B} = \left(\frac{\partial^2}{\partial x^2} + \frac{\partial^2}{\partial y^2} + \frac{\partial^2}{\partial z^2}\right)B_x\,\mathbf{x}$$
$$+ \left(\frac{\partial^2}{\partial x^2} + \frac{\partial^2}{\partial y^2} + \frac{\partial^2}{\partial z^2}\right)B_y\,\mathbf{y} + \left(\frac{\partial^2}{\partial x^2} + \frac{\partial^2}{\partial y^2} + \frac{\partial^2}{\partial z^2}\right)B_z\,\mathbf{z}. \tag{6.9}$$

We now have obtained the required vector identity that is key to untangling **E** and **H** in Maxwell's equations. It is used below to derive the vector differential equations that lead to ordinary scalar differential equations. The solutions of the scalar differential equations are the sine wave equations of **E** and **H** that represent the propagating waves.

6.3. Independent Equations of E and H

We will now proceed to use the required vector identity together with Maxwell's equations to disentangle the **E** and **H** equations. Let us derive the Differential equation of **E**. We start with equation

$$\nabla \times \mathbf{H} = \epsilon \frac{\partial \mathbf{E}}{\partial t}, \tag{6.10}$$

$$\nabla \times \mathbf{E} = -\mu \frac{\partial \mathbf{H}}{\partial t}. \tag{6.11}$$

From eq. (6.10) it follows

$$\frac{\partial}{\partial t}\nabla \times \mathbf{H} = \epsilon \frac{\partial}{\partial t}\frac{\partial \mathbf{E}}{\partial t},$$

or

$$\nabla \times \frac{\partial \mathbf{H}}{\partial t} = \epsilon \frac{\partial^2 \mathbf{E}}{\partial t^2}. \tag{6.12}$$

With the use of eq. (6.11), eq. (6.13) takes the form

$$-\nabla \times (\nabla \times \mathbf{E}) = \epsilon\mu \frac{\partial^2 \mathbf{E}}{\partial t^2}. \tag{6.13}$$

Now it is time to use the formula (see (6.7))

$$-\nabla \times (\nabla \times \mathbf{E}) = +\nabla^2 \mathbf{E} - \nabla(\nabla \bullet \mathbf{E}), \qquad (6.14)$$

then eq. (6.13) will look

$$\nabla^2 \mathbf{E} - \nabla(\nabla \bullet \mathbf{E}) = \epsilon\mu \frac{\partial^2 \mathbf{E}}{\partial t^2},$$

or with eq. (6.1) it becomes

$$\nabla^2 \mathbf{E} = \epsilon\mu \frac{\partial^2}{\partial t^2} \mathbf{E}. \qquad (6.15)$$

Thus we have obtained a vector differential equation for vector $\mathbf{E}$ only.

Now, to obtain the equation for $\mathbf{H}$ follow the same procedure that was used above in obtaining the independent equation of $\mathbf{E}$. Again let us start with

$$\nabla \times \mathbf{H} = \epsilon \frac{\partial \mathbf{E}}{\partial t}, \quad \nabla \times \mathbf{E} = -\mu \frac{\partial \mathbf{H}}{\partial t}. \qquad (6.16)$$

From (6.16) it follows

$$\nabla \times \frac{\partial \mathbf{E}}{\partial t} = -\mu \frac{\partial^2 \mathbf{H}}{\partial t^2}, \qquad (6.17)$$

which with the use of (6.16) becomes

$$\nabla \times (\nabla \times \mathbf{H}) = -\epsilon\mu \frac{\partial^2 \mathbf{H}}{\partial t^2}. \qquad (6.18)$$

Now again it is time to use the formula (see (6.7))

$$\nabla \times (\nabla \times \mathbf{H}) = -\nabla^2 \mathbf{H} + \nabla(\nabla \bullet \mathbf{H}),$$

then

$$-\nabla^2 \mathbf{H} + \nabla(\nabla \bullet \mathbf{H}) = -\epsilon\mu \frac{\partial^2 \mathbf{H}}{\partial t^2}. \qquad (6.19)$$

Taking in mind $\nabla \bullet \mathbf{H} = 0$, eq. (6.19) takes the form

$$\nabla^2 \mathbf{H} = \epsilon\mu \frac{\partial^2}{\partial t^2} \mathbf{H}. \qquad (6.20)$$

This differential equation of for vector $\mathbf{H}$ only.

We have obtained two vector differential equations of $\mathbf{E}$ and $\mathbf{H}$ that are now independent of each other:

$$\nabla^2 \mathbf{E} = \epsilon\mu \frac{\partial^2}{\partial t^2} \mathbf{E}, \qquad (6.21)$$

$$\nabla^2 \mathbf{H} = \epsilon\mu \frac{\partial^2}{\partial t^2} \mathbf{H}. \qquad (6.22)$$

These equations have the form of the classic wave equation that was solved by Jean d'Alembert in 1747. We will be following some of his steps as we arrive at the solution to the equations. A wave equation is a function of time and distance. These wave equations are also found in the mathematics of the physics of vibrating taut strings, speaker cones, diaphragms, and water waves. So **E** and **H** will also behave with the same wave motion. The solution of these wave equations must be twice differentiable with respect to space and time as is shown in the equations. When taking the partial derivatives of the **E** and **H** equations, distance is changing on the left side of the equation and time is held constant. On the right side, distance is held constant and the derivatives are taken with respect to time.

Through the above vector mathematics procedure we have obtained independent equations of **E** and **H**. This does not mean that the fields can propagate independently during electromagnetic radiation. The fields are interdependent and remain so just as described by Maxwell's equations.

Also the above independent equations do not show the electric-magnetic and space phase relationships of **E** and **H**. To recover those phase relationships we must put the wave equation solution back into Maxwell's equations. As mentioned above the independent differential electromagnetic equations have the form of the classic wave equations so now we must discuss wave motion.

6.4. Wave Motion

The term wave is used to describe a pattern that propagates or moves along. The term wave is applied to the water pattern on a pond, vibrations on taut string and electromagnetic radiation. The term plane wave is also applied to electromagnetic radiation. The plane wave is described with a water wave analogy. When a stone is dropped into the center of a pond the water waves spread out in concentric circles. The water moves up and down but does not move outward with the waves. A cork on the surface would just bob up and down. (The water wave analogy does not apply completely to electromagnetic waves as electromagnetic waves do not need a carrier). To continue the analogy we have to establish the x, y, z axes of the orthogonal coordinate system. Imagine a pole is placed vertically in the center of the pond as the y axis. A radius drawn in any direction on the surface of the pond is the x axis. The z axis is drawn from the intersection of the x and y axes and perpendicular to the $x - y$ plane. At some distance from the center of the pond the circumference of the wave circle becomes large. The circumference of the wave circle can be thought of as being made up of many small dashes drawn tangent to the circle in the $x - z$ plane. So a plane containing the tangent dash, perpendicular to the x axis and parallel to the y axis can be drawn. Imagine that this plane is attached to any point of the water wave and proceeds out with it. The plane does not bob up and down because any point on the wave just moves out from the origin. This wave is termed a plane wave.

Also if we look from the origin along the x axis we can plot the amplitude of the wave on the y axis as the wave moves up and down. Therefore, we can say that we are plotting the wave amplitude sub y. The amount that the wave moves up and down is independent of the y axis and the scale of the y axis. There is no movement of the plane wave in the z direction, so the wave amplitude and direction is also independent of z.

6.5. Scalar Differential Equations of E and H

The three dimensional vector differential equations of the independent **E** and **H** fields are

$$\nabla^2 \mathbf{E} = \epsilon\mu \frac{\partial^2}{\partial t^2} \mathbf{E}, \tag{6.23}$$

$$\nabla^2 \mathbf{H} = \epsilon\mu \frac{\partial^2}{\partial t^2} \mathbf{H}. \tag{6.24}$$

The component form of the equation of **E** using the x, y, z axes rectangular coordinate system is

$$(\frac{\partial^2}{\partial x^2} + \frac{\partial^2}{\partial y^2} + \frac{\partial^2}{\partial z^2})(E_x \mathbf{x} + E_y \mathbf{y} + E_z \mathbf{z})$$
$$= \epsilon\mu \frac{\partial^2}{\partial t^2}(E_x \mathbf{x} + E_y \mathbf{y} + E_z \mathbf{z}). \tag{6.25}$$

The component form of **E** is also written as three scalar equations:

$$(\frac{\partial^2}{\partial x^2} + \frac{\partial^2}{\partial y^2} + \frac{\partial^2}{\partial z^2}) E_i = \epsilon\mu \frac{\partial^2}{\partial t^2} E_i. \tag{6.26}$$

where i stands for x, y and z. The component equations for **H** has the same form

$$(\frac{\partial^2}{\partial x^2} + \frac{\partial^2}{\partial y^2} + \frac{\partial^2}{\partial z^2}) H_i = \epsilon\mu \frac{\partial^2}{\partial t^2} H_i. \tag{6.27}$$

We will find below that a solution of these component **E** and **H** differential equations is the desired sine wave equation.

6.6. Sine Wave Equation for E and H

Consider the component scalar differential equation for E:

$$\frac{\partial^2}{\partial x^2} E = \epsilon\mu \frac{\partial^2}{\partial t^2} E, \tag{6.28}$$

we will designate this E as E_y. This is a classic d'Alembert wave equation in one dimension since the y and z directions are not included within the equation. The solution is the wave equation for E and is twice differentiable both with respect to space and time. The equation states that the solution must be differentiated twice with respect to distance x, and then be equal to the same solution differentiated twice with respect to time, times $\epsilon\mu$. In our case the wave movement is plotted along the x axis and the wave amplitude plotted on the y axis.

Since the equation of E is differentiated with respect to distance and time, the equation must be a function of distance x and time t. Each term in the parenthesis however, must have the same dimension, either distance or Assume that the electric wave

$$E = F(x - ct). \tag{6.29}$$

Here distance is the common dimension of each term. Then x and t will vary and the parenthesis term will be differentiable. We have not yet determined the function F, but we can place $F(x - ct)$ into the scalar differential equation of E and determine that an equality results. This will show that we have obtained a part of the solution.

For simplicity of notation, let w $= (x - ct)$. Then, does

$$\frac{\partial^2}{\partial x^2} F(w) = \epsilon\mu \frac{\partial^2}{\partial t^2} F(w) \,? \tag{6.30}$$

In order to answer that question, we will first apply the differential calculus rules to the left side of the equation:

$$\frac{\partial^2}{\partial x^2} F(w) = \frac{\partial}{\partial x}\frac{\partial}{\partial x} F(w) = \frac{\partial}{\partial x}\left(\frac{dF}{dw}\frac{\partial w}{\partial x}\right).$$

Since

$$\frac{\partial w}{\partial x} = \frac{\partial(x-ct)}{\partial x} = 1,$$

the result for the left side of the equation (6.30) is

$$\frac{\partial^2}{\partial x^2} F(w) = \frac{\partial}{\partial x}\left(\frac{dF}{dw}\right) = \frac{d^2 F}{dw^2}. \tag{6.31}$$

Now apply the differential calculus rules to the right side of the equation

$$\epsilon\mu \frac{\partial^2}{\partial t^2} F(w) = \epsilon\mu \frac{\partial}{\partial t}\left(\frac{dF}{dw}\frac{\partial w}{\partial t}\right).$$

Since

$$\frac{\partial w}{\partial t} = \frac{\partial}{\partial t}(x - ct) = -c,$$

the result for the right side of the equation (6.30) is

$$\epsilon\mu \frac{\partial^2}{\partial t^2} F(w) = \epsilon\mu\, c^2 \frac{d^2 F}{dw^2}. \tag{6.32}$$

Note that as shown before

$$\epsilon\mu\, c^2 = 1.$$

The next step in obtaining the complete solution is to find out just what function F represents. The function is to represent a wave of E. This means that E must fluctuate periodically in distance and with time similarly to a water wave. There are several functions that repeat with differentiation and will solve the equation. These functions include sine, cosine and e (logarithm base) to an algebraic power. Jean d' Alembert's solutions of the classic wave equation contain these same functions.

The appearance of water waves and waves created on a taut string suggest using a sine wave function for F. Then F must be a function of the sine of an angle that contains within the angle the distance $(x - ct)$. As shown above, the angle in radians will depend upon the frequency and velocity of the wave. To convert frequency to radians the factor of 2π radians per cycle is included. We here consider only one frequency at a time as even the most complicated wave form is made up of individual sine waves.

The equation of E with this assumed function for F has the form

$$E = E_{max} \sin[\omega(t - \frac{x}{c})] = E_{max} \sin[\frac{\omega}{c}(ct - x)]. \qquad (6.33)$$

Now we have to repeat the operation with the differential equation that we did above. This time the whole sine wave equation is placed into the differential equation to determine again that the differential equation is still true.

First substitute this proposed sine wave solution into the scalar differential equation for E

$$\frac{\partial^2}{\partial x^2} E = \epsilon \mu \frac{\partial^2}{\partial t^2} E. \qquad (6.34)$$

Does

$$\frac{\partial^2}{\partial x^2} E_{max} \sin[\omega(t - \frac{x}{c})] = \epsilon \mu \frac{\partial^2}{\partial t^2} E_{max} \sin[\omega(t - \frac{x}{c})] ? \qquad (6.35)$$

To answer this question, first follow the rules of differentiation as applied to the left side of the equation:

$$E_{max} \frac{\partial}{\partial x} \frac{\partial}{\partial x} \sin[\omega(t - \frac{x}{c})] = E_{max}(-\frac{\omega}{c}) \frac{\partial}{\partial x} \cos[\omega(t - \frac{x}{c})]$$

$$= E_{max}(-\frac{\omega}{c})(+\frac{\omega}{c}) \sin[\omega(t - \frac{x}{c})] = -\frac{\omega^2}{c^2} E_{max} \sin[\omega(t - \frac{x}{c})]. \qquad (6.36)$$

Follow the rules of differentiation for the right side of the equation and obtain

$$\epsilon \mu E_{max} \frac{\partial}{\partial t} \frac{\partial}{\partial t} \sin[\omega(t - \frac{x}{c})]$$

$$= \epsilon \mu E_{max} \omega) \frac{\partial}{\partial t} \cos[\omega(t - \frac{x}{c})]$$

$$= -\epsilon \mu \omega^2 E_{max} \sin[\omega(t - \frac{x}{c})]. \qquad (6.37)$$

Recall that $\epsilon \mu = 1/c^2$. The results obtained for the left and right side of the equation are identical. Therefore, the assumed function for E

$$E = E_{max} \sin[\omega(t - \frac{x}{c})] \qquad (6.38)$$

is a solution to the differential equation of E and is the equation that represents the propagating wave of E. The same mathematics can be repeated to determine an identical sine wave equation for H. Then the differential equation is solved when

$$H = H_{max} \sin[\omega(t - \frac{x}{c})]. \qquad (6.39)$$

We must yet determine the space and electric-magnetic phase relationship of the sine wave equations of E relative to H. That is, we must determine the correct subscripts, x, y, or z for E and H as propagating waves.

6.7. Determine Phase Relation of E and H Sine Waves

Two of Maxwell's equations are used to determine the electric-magnetic and space phase relationship of the **E** and **H** sine waves. We will use the divergence and curl of **E**. The divergence and curl of **H** could be used just as well.

$$\nabla \bullet \mathbf{E} = 0, \tag{6.40}$$

$$\nabla \times \mathbf{E} = -\mu \frac{\partial \mathbf{H}}{\partial t}. \tag{6.41}$$

Eq. (6.40), expressed in component form, is

$$\frac{\partial E_x}{\partial x} + \frac{\partial E_y}{\partial y} + \frac{\partial E_z}{\partial z} = 0.$$

As in the water wave analogy, this wave of $\mathbf{E} = (0, E_y, 0)$ is proceeding in the x direction independent of the y or z directions. Therefore,

$$\frac{\partial E_y}{\partial y} = 0, \qquad \frac{\partial E_z}{\partial z} = 0. \tag{6.42}$$

Since these two terms are zero, then

$$\frac{\partial E_x}{\partial x} = 0. \tag{6.43}$$

Recall that the derivative of the function with respect to x is zero, if the function is zero or a constant. A constant will not be part of propagation and can be discarded. This shows mathematically that there is no component of the E vectors in the direction of propagation. In vector mathematics $\nabla \bullet \mathbf{E} = 0$ is the definition of a plane wave. If we had used the divergence of **H**, this step would also show that there is no component of **H** in the x direction.

Now consider

$$\nabla \times \mathbf{E} = -\mu \frac{\partial \mathbf{H}}{\partial t}. \tag{6.44}$$

Equating the x vector components:

$$\frac{\partial E_z}{\partial y} - \frac{\partial E_y}{\partial z} = -\mu \frac{\partial H_x}{\partial t}. \tag{6.45}$$

The wave amplitude and direction is independent of the y and z directions, so the left side of the equation is zero. This again shows that the right side is also zero and confirms that there are no components of **H** vectors in the x direction. Of course, if H_x was constant, it would not propagate. Equating the y vector components:

$$\frac{\partial E_x}{\partial z} - \frac{\partial E_z}{\partial x} = -\mu \frac{\partial H_y}{\partial t} \implies 0 = 0. \tag{6.46}$$

Here again the left an the right sides are equal to zero.

So now we want to obtain the phase relationship between E_y and its correct propagating H- counterpart. We find this by equating the z vector components.

$$\frac{\partial E_y}{\partial x} - \frac{\partial E_x}{\partial y} = -\mu \frac{\partial H_z}{\partial t}. \tag{6.47}$$

There is no component of **E** in the x direction, (the direction of propagation), so

$$\frac{\partial E_x}{\partial y} = 0.$$

Then

$$\frac{\partial E_y}{\partial x} = -\mu \frac{\partial H_z}{\partial t}. \tag{6.48}$$

This differential equation containing E_y and H_z will establish the electric-magnetic and space phase relation between E_y and H_z. Place the sine wave equation of E_y into the above differential equation:

$$\frac{\partial E_y}{\partial x} = \frac{\partial}{\partial x} E_{max} \sin[\omega(t - \frac{x}{c})] = E_{max}(-\frac{\omega}{c}) \cos[\omega(t - \frac{x}{c})].$$

So, eq. (6.48) becomes

$$E_{max}(-\frac{\omega}{c}) \cos[\omega(t - \frac{x}{c})] = -\mu \frac{\partial H_z}{\partial t}. \tag{6.49}$$

Follow integral calculus rules and integrate both sides of the equation:

$$\int E_{max} \frac{\omega}{c} \cos[\omega(t - \frac{x}{c})] \, dt = \mu \int dH_z.$$

So that

$$H_z = \frac{1}{\mu c} E_{max} \sin[\omega(t - \frac{x}{c})]. \tag{6.50}$$

This equation shows that H_z is in the exact electrical phase as E_y and 90 degrees clockwise in space from E_y. This is true since the amplitude of H_z is plotted on the z axis and its movement on the x axis. Note that the amplitudes of E and H are proportional to each other in their sine wave representations.

We have also shown that the common engineering practice of representing electromagnetic fields as sine waves is based upon Maxwell's equations. This completes the discussion of the sine wave representations of the electric and magnetic fields, **E** and **H**.

7. Electromagnetic Fields and Energy

The topic that is often addressed after completing the study of Maxwell's equations is the concept of electromagnetic energy. Discussion of this concept will conclude our study of Maxwell's electromagnetic equations. The vector **S** = **E** × **H** is used to establish the direction of electromagnetic propagation. The vector **S**. is named for Poynting who first published its characteristics in 1865, (about a year after Maxwell had published his equations).

7.1. Poynting Equation

Since electrical energy is required to establish the **E** and **H** fields it follows that energy is stored and transmitted by the fields. Poynting's equation is used to indicate the amount of energy and its distribution in an electromagnetic field. The two Poynting equations will be derived, The first one is

$$\oint_{Surface} (\mathbf{E} \times \mathbf{H}) \bullet d\mathbf{a} = -\frac{d}{dt} \int_{Volume} \left(\frac{\epsilon E^2}{2} + \frac{\mu H^2}{2} \right) dv, \tag{7.1}$$

this equation is used for electromagnetic propagation energy problems where there are no electric charges within the volume occupied by the fields. All the energy leaving the volume of the right-hand integral will pass through the surface of the left-hand integral since that surface envelops the volume of the field.

And second one is

$$\oint_{Surface} (\mathbf{E} \times \mathbf{H}) \bullet d\mathbf{a} = -\frac{d}{dt} \int_{Volume} \left(\mathbf{E} \bullet \mathbf{J} + \frac{\epsilon E^2}{2} + \frac{\mu H^2}{2} \right) dv. \tag{7.2}$$

This equation can be used in those applications where electric charges are within the volume of the fields.

The right-hand integral shows that the energy in the volume is proportional to the sum of the energies of the E^2 and H^2 terms and the charged particles (if any) in the volume.

7.2. Derivation of Poynting Equation

The electromagnetic field begins to spread out or diverge as it leaves an antenna. So taking the divergence of $\mathbf{E} \times \mathbf{H}$, that is $\nabla \bullet (\mathbf{E} \times \mathbf{H})$, will lead to more information concerning the propagating fields. But first we must locate a vector identity that will assist in expanding $\nabla \bullet (\mathbf{E} \times \mathbf{H})$. This required identity is found in a list of identities in a vector analysis text. In this section the required identity is again presented as an equation of three general vectors $\mathbf{A}, \mathbf{B}$ and $\mathbf{C}$:

$$\mathbf{C} \bullet (\mathbf{A} \times \mathbf{B}) = \frac{1}{2} \left[\mathbf{B} \bullet (\mathbf{C} \times \mathbf{A}) - \mathbf{A} \bullet (\mathbf{C} \times \mathbf{B}) \right]. \tag{7.3}$$

This identity can be verified by writing down the vector equations $\mathbf{A}, \mathbf{B}$ and $\mathbf{C}$. The left side is

$$\mathbf{C} \bullet (\mathbf{A} \times \mathbf{B}) = C_x(A_y B_z - A_z B_y) + C_y(A_z B_x - A_x B_z) + C_z(A_x B_y - A_y B_x).$$

The two terms from the right are

$$\mathbf{B} \bullet (\mathbf{C} \times \mathbf{A}) = B_x(C_y A_z - C_z A_y) + B_y(C_z A_x - C_x A_z) + B_z(C_x A_y - C_y A_x),$$

and

$$-\mathbf{A} \bullet (\mathbf{C} \times \mathbf{B}) = -A_x(C_y B_z - C_z B_y) - A_y(C_z B_x - C_x B_z) - A_z(C_x B_y - C_y B_x).$$

So identity (7.3) is proven. In the above identity (7.3), now replace **A**, **B** and **C** with **E**, **H** and ∇. The resultant identity reads[4]

$$\nabla \bullet (\mathbf{E} \times \mathbf{H}) = [\mathbf{H} \bullet (\nabla \times \mathbf{E}) - \mathbf{E} \bullet (\nabla \times \mathbf{H})]. \tag{7.4}$$

This equation shows the divergence of $\nabla \bullet (\mathbf{E} \times \mathbf{H})$ in terms that are contained in Maxwell's equations. By substituting those terms we can solve the vector equation.

The resultant vector equation is

$$\nabla \bullet (\mathbf{E} \times \mathbf{H}) = [\mathbf{H} \bullet (\nabla \times \mathbf{E}) - \mathbf{E} \bullet (\nabla \times \mathbf{H})]. \tag{7.5}$$

Instead of Maxwell's equation III

$$\nabla \times \mathbf{H} = \mathbf{J} + \epsilon \frac{\partial \mathbf{E}}{\partial t}, \tag{7.6}$$

to simplify this development, we will not use **J** for the moment and use

$$\nabla \times \mathbf{H} = \epsilon \frac{\partial \mathbf{E}}{\partial t}. \tag{7.7}$$

Later, we will include **J** to obtain the complete Poynting equation. From Maxwell's equation IV we have

$$\nabla \times \mathbf{E} = -\mu \frac{\partial \mathbf{H}}{\partial t}. \tag{7.8}$$

Now insert these two terms into (7.5):

$$\nabla \bullet (\mathbf{E} \times \mathbf{H}) = -\left(\mu \mathbf{H} \bullet \frac{\partial \mathbf{H}}{\partial t} + \epsilon \mathbf{E} \bullet \frac{\partial \mathbf{E}}{\partial t} \right). \tag{7.9}$$

Simplifying, we get

$$\nabla \bullet (\mathbf{E} \times \mathbf{H}) = -\frac{\partial}{\partial t} \left(\mu \frac{\mathbf{H} \bullet \mathbf{H}}{2} + \epsilon \frac{\mathbf{E} \bullet \mathbf{E}}{2} \right). \tag{7.10}$$

We can integrate both sides of the equation over the same volume and still maintain an equality. Thus,

$$\int_{Volume} \nabla \bullet (\mathbf{E} \times \mathbf{H}) \, dv = -\frac{d}{dt} \int_{Volume} \left(\mu \frac{\mathbf{H} \bullet \mathbf{H}}{2} + \epsilon \frac{\mathbf{E} \bullet \mathbf{E}}{2} \right). \tag{7.11}$$

Equation states that the energy leaving the right side volume integral per second is entering the left integral volume. Now apply the divergence theorem

$$\oint_{Volume} \nabla \bullet (\mathbf{E} \times \mathbf{H}) \, dv = \oint_{Surface} (\mathbf{E} \times \mathbf{H}) \bullet d\mathbf{a}. \tag{7.12}$$

Substituting the area integral for the volume integral in the above equation (7.11), we obtain

$$\oint_{Surface} (\mathbf{E} \times \mathbf{H}) \bullet d\mathbf{a} = -\frac{d}{dt} \int_{Volume} \left(\epsilon \frac{\mathbf{E} \bullet \mathbf{E}}{2} + \mu \frac{\mathbf{H} \bullet \mathbf{H}}{2} \right). \tag{7.13}$$

---

[4] Take notice: in comparison with (7.3) here the factor 1/2 must be changed to 1

This is Poynting's equation for energy in the electromagnetic field. The equation assumes that there are no electric charges within the volume. This condition is generally assumed in air or space. The equation states that all of the energy leaving the volume will go through the surface area (left integral) of that volume. This equation is another mathematical model, as are Maxwell's equations, that describe how electromagnetic fields exist and propagate.

Consider the terms in the equation (7.13): $\mathbf{E} \times \mathbf{H}$ is the energy area density of the surface of the volume in watts per square meter. The integral quantity

$$\oint_{Surface} (\mathbf{E} \times \mathbf{H}) \bullet d\mathbf{a} \tag{7.14}$$

is the energy in watts leaving the surface the volume.

$$\epsilon \frac{\mathbf{E} \bullet \mathbf{E}}{2} + \mu \frac{\mathbf{H} \bullet \mathbf{H}}{2} \tag{7.15}$$

is the electric and magnetic fields volume energy density in joules per cubic meter. In expression

$$-\frac{d}{dt} \int_{Volume} \left(\epsilon \frac{\mathbf{E} \bullet \mathbf{E}}{2} + \mu \frac{\mathbf{H} \bullet \mathbf{H}}{2} \right),$$

the negative sign indicates a decrease in the energy that is stored or being carried by the fields within the volume.

7.3. Electromagnetic Energy, Examples

We now prove that

$$\epsilon \frac{\mathbf{E} \bullet \mathbf{E}}{2}, \quad \text{and} \quad \mu \frac{\mathbf{H} \bullet \mathbf{H}}{2}$$

are the electric and magnetic volume energy densities of the fields. We prove this by calculating the volume energy densities of the fields in a capacitor and a toroid.

Determine the volume energy density in a capacitor electric field. In Figure 7.1 a capacitor and a resistor are placed across the battery by closing the switch.

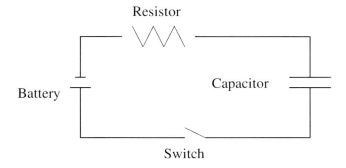

Figure 7.1. Energy density in a capacitor electric field.

The capacitor is made of two parallel metal plates, each of area, A, separated by a small air gap, d. The capacitance, C, of a capacitor can be increased by increasing the area of the plates and by decreasing the distance between them. Capacitance is directly dependent upon the permittivity, of the medium (air) between the plates. That is

$$\text{Capacitance} = \epsilon \frac{\text{Area of the plates}}{\text{Distance between the plates}}, \text{ or } C = \epsilon \frac{A}{d}. \tag{7.16}$$

When the switch is closed the voltage across the capacitor increases gradually until the capacitor voltage equals the battery voltage. This causes the current to stop flowing. Some amount of electrons have been removed from one capacitor plate and the same number has been placed on the other by the battery. That is, a charge Q has been placed on the capacitor plates causing an electric field, $\mathbf{E}$, to appear across the capacitor air gap d. No energy is required to maintain the $\mathbf{E}$ field once it is in place. Later, when the switch is opened the electric field decreases to zero and its energy is dissipated.

The amount of final accumulated charge, Q, is equal to the voltage across the capacitor times its capacitance. That is

$$Q = C\,U. \tag{7.17}$$

The unit of capacitance is coulomb per volt or farad.

There are several calculations now to be made to find the energy in the electric field of the capacitor. During the charging of the capacitor a small increase in the voltage across the capacitor, dU, is accompanied by a small increase in electric charge dq, depending upon the capacitance C of the capacitor. That is

$$dQ = C\,dU.$$

To find the energy transferred from the battery we must sum the work done in transferring the total charge Q, from zero capacitor voltage until the full battery voltage is established across the capacitor plates. That is

$$W_E = \text{Energy} = \int \text{Work done} = \int_0^U CU\,dU.$$

By following the rules for integration we obtain

$$W_E = \text{Energy} = \int_0^U CU\,dU = \frac{CU^2}{2}. \tag{7.18}$$

The unit of electric field strength E is in volts per meter. So the electric field voltage within the air gap is the voltage across the gap divided by the gap distance or

$$E = \frac{U}{d}, \quad \text{and} \quad U = E\,d. \tag{7.19}$$

We have found that the energy placed into the capacitor electric field is

$$W_E = \text{Energy in the field} = \frac{CU^2}{2}. \tag{7.20}$$

Substituting $E\,d$ for U we obtain

$$W_E = \text{Energy in the field} = C\,\frac{E^2 d^2}{2}.$$

We know that (see (7.16)) $C = \epsilon\,(A/d)$. Then, substituting for C,

$$W_E = \text{Energy in the field} = \epsilon\,\frac{A}{d}\,\frac{E^2 d^2}{2} = \frac{\epsilon E^2}{2}\,Ad. \qquad (7.21)$$

Since distance in the gap times the area of the plates is equal to the volume $V = d\,A$ of the air gap we obtain

$$w_E = \frac{\text{Energy in the field}}{\text{Volume of the field}} = \frac{\epsilon E^2}{2}. \qquad (7.22)$$

Which means that the volume energy density in an E field is

$$w_E = \frac{W_E}{V} = \frac{\epsilon E^2}{2}. \qquad (7.23)$$

We see that the volume energy density in the static electric field is exactly the same as the propagating volume electric energy density in Poynting's equation

We must now find the similar term for the magnetic energy volume density in a magnetic field. Figure 6.2 shows the circuit of a battery, switch, resistor and an air core toroid inductor.

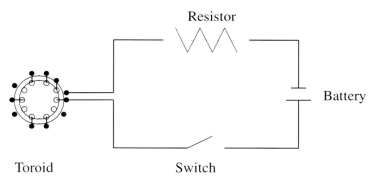

Figure 7.2. Volume density in a magnetic field.

When the switch is closed the current in the circuit starts from zero and increases with time until it reaches a fixed steady-state current. The calculus term for a small increase in that current per a small increase in time is

$$\frac{di}{dt}.$$

The factor that controls this increase of current with time is the inductance, L, of the toroid. The instantaneous voltage e appearing across the toroid is

$$e = L\,\frac{di}{dt}. \qquad (7.24)$$

The inductance of a toroid is determined by the radius, R, from its center, the number of turns, N, the area A of the core and the permeability of the medium μ (air) in the core. That is

$$L = \mu \, \frac{A \, N^2}{2\pi R} \, . \qquad (7.25)$$

The denominator is the average circular length of the toroid. N is squared because if you double the turns you double the magnetic flux lines passing through each turn. So inductance increases with the square of the turns. Inductance is measured in henries.

When the current stops increasing the steady-state current is reached and the voltage across the inductance decreases to zero. The energy in the magnetic field stays constant after steady-state current is reached. Although that amount of current continues to flow in the wires no more energy is placed into the magnetic field. Later when the switch is opened the energy in the field is dissipated in the circuit.

We must now determine the energy removed from the battery and placed into the magnetic field of the toroid. The instantaneous power put into the toroid field is equal to the instantaneous voltage, e, across the toroid times the instantaneous current, i, through the toroid. The instantaneous power in the magnetic field is

$$\text{Instantaneous power} = e \, i \, . \qquad (7.26)$$

As an equation

$$\text{Instantaneous power} = \left(L \, \frac{di}{dt} \right) i \, . \qquad (7.27)$$

The total power placed into the field is the sum of all the instantaneous powers from the instant of switch closing to the instant, T, when steady-state current is reached. Energy input to the field is the total power input for that time period, T. So the summation integral is again required[5]. Total energy placed into the magnetic field following integration rules is

$$W_H = \int \left(L \, \frac{di}{dt} \, i \right) dt = L \int i \, di = \frac{L i^2}{2} + K \, , \qquad (K = 0) \, . \qquad (7.28)$$

The constant K is zero since time and current start from zero. The total energy in the magnetic field divided by the volume of the air core

$$V = A \, 2\pi R$$

is the volume energy density w_H of the field:

$$w_H = \frac{W_H}{V} = \frac{1}{A \, 2\pi R} \, \frac{i^2 L}{2} \, . \qquad (7.29)$$

We know that the inductance L of the toroid is a function of its geometrical constants and μ (see (7.25)). Therefore,

$$w_H = \frac{1}{A \, 2\pi R} \, \frac{i^2}{2} \, \mu \, \frac{A N^2}{2\pi R} \, , \quad \text{or} \quad w_H = \frac{\mu}{2} \left(\frac{N \, i}{2\pi R} \right)^2 \, . \qquad (7.30)$$

---

[5]The summation integral was used above to calculate the energy placed into the capacitor E field.

We also know that H is a function of the current, i, in a toroid:

$$H = \frac{N\,i}{2\pi R},$$

therefore eq. (7.30) reads

$$w_H = \frac{\mu\,H^2}{2}, \tag{7.31}$$

which is the same term in Poynting's equation. Before, we determined that the volume energy density of the E field is

$$w_E = \frac{\epsilon\,E^2}{2}. \tag{7.32}$$

Both results are independent of the boundaries imposed on the field volumes. Also, it is to be noted that the volume energy density in E and H fields generated under static conditions is exactly the same for E and H fields generated under the dynamic conditions indicated by the Maxwell and Pointing equations.

7.4. Energy Carried by Poynting Vector Fields

The Poynting equation is

$$\oint_{Surface} (\mathbf{E} \times \mathbf{H}) \bullet d\mathbf{a} = -\frac{d}{dt}\int_{Volume}\left(\epsilon\,\frac{\mathbf{E}\bullet\mathbf{E}}{2} + \mu\,\frac{\mathbf{H}\bullet\mathbf{H}}{2}\right). \tag{7.33}$$

We have proven that

$$w_E = \frac{\epsilon\,E^2}{2}, \quad \text{and} \quad w_H = \frac{\mu\,H^2}{2}$$

are the volume energy densities of the E and H fields. So the above right-hand integral is the total energy leaving that volume per second. As the energy leaves the volume it must leave through the surface of that volume. This means that the left integral shows the energy leaving the surface of the volume containing the fields. The left integral shows the dot product indicating that the computed energy depends on the surface that is perpendicular to the Poynting vector.

We can show that the volume energy densities of the magnetic and electric fields (in a sine wave electromagnetic field) are equal to each others. By dividing

$$\frac{w_E}{w_H} = \frac{\epsilon\,E^2}{\mu\,H^2}. \tag{7.34}$$

Recall that $E^2/H^2 = \mu/\epsilon$, so (7.35) reads

$$\frac{w_E}{w_H} = 1. \tag{7.35}$$

The quotient is indicating that the electric and magnetic energy densities are equal.

We can also calculate the energy flowing with the electromagnetic waves according to Poynting's equation. The waves travel at c, 300 million meters per second. So the energy passing a square meter per second is equal to c times the volume energy density. That is

$$c \left(\frac{\epsilon E^2 + \mu H^2}{2} \right). \tag{7.36}$$

Recall that

$$E = H \sqrt{\frac{\mu}{\epsilon}}, \quad H = E \sqrt{\frac{\epsilon}{\mu}}, \quad c = \frac{1}{\sqrt{\epsilon \mu}}.$$

Then

$$c \frac{\epsilon E^2}{2} = \frac{EH}{2}, \quad c \frac{\epsilon H^2}{2} = \frac{EH}{2}.$$

Resulting in

$$c \frac{\epsilon E^2}{2} + c \frac{\epsilon H^2}{2} = EH. \tag{7.37}$$

The total energy in the electromagnetic wave passing through one square meter each second is

$$HE = \sqrt{\frac{\epsilon}{\mu}} E^2 = 2,65 \; 10^{-3} E^2 \; \frac{\text{watts}}{\text{meter}^2 \; \text{sec}}.$$

7.5. Energy Flow in Electric Circuit

Thus far in this study of Maxwell's and Poynting's equations only the radiation of electromagnetic energy was considered. The equations are also applicable to the design of transformers, motors and generators. We will now study an example of the place where radiation theory and circuit theory meet. Two methods for calculating the distribution of energy will be used: First, the ordinary engineering method and then second, the Pointing vector method.

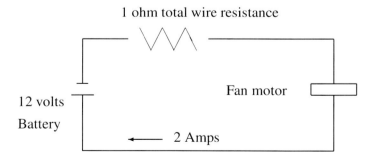

Figure 7.3. Energy flow in electric circuit.

In Figure 7.3 assume that the motor and battery do not have significant resistance and the total resistance, R, of the wiring is one ohm. The voltage drop across the wire resistance is 2 volts. The problem is to calculate the power dissipated in the wire and the power put into the motor. The power dissipated in the wire is

$$i^2 R = \frac{U^2}{R} = i\,U = 4 \text{ watts}. \tag{7.38}$$

The power put into the motor is

$$U\,i = (12 - 2)\,2 = 20 \text{ watts}. \tag{7.39}$$

Now, as indicated by the Poynting equation, calculate the power distribution in the circuit. The Poynting equation shows that

$$\oint_{Sutface} \mathbf{E} \times \mathbf{H} \bullet d\mathbf{a}$$

is the energy leaving the **E** and **H** fields. But, as applied to the circuit, what E and H fields are involved and how are they to be found and what area is involved for the power leaving the fields? The engineering mathematics found that the power lost in the wiring is 4 watts. So we will apply the Poynting energy equation to the wiring to see what S is obtained. Consider Figure 7.4. The one ohm wire resistance of Figure 9.3 is assumed to be lumped in one length, L. The wire ends are cut off perpendicular to the wire length.

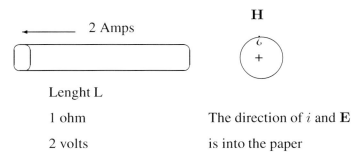

Figure 7.4. **H** flux in electric circuit.

Notice that the direction of **E** is in the same direction as the voltage,V, that is applied to the wire to cause the current, i, to flow. Therefore **E**, V and i all point in the same direction. This direction is represented by the arrow tail going into the paper in Figure 7.4.

Now we have to determine the direction of the **H** vector. The **H** field that surrounds the current flow is determined by the right-hand rule. Grasp the wire in the right hand with the thumb pointed in the direction of current flow. The curl of the fingers points in the direction of **H**. Place a small arrow on the circumference of an **H** flux line in the direction of **H**. Now we can determine the direction of the Poynting vector, **S**, by again using a right-hand rule. Place the right forefinger in the direction of **E**. Place the mid-finger in the direction of H, then the thumb points in the direction of **S**. If we fold or 'cross' the **E** finger into the **H** the result we obtain is the **S** vector pointing in the same direction as the thumb. When we apply the right hand rule next to the wire conductor we see that the **S** vector always points radially into the round surface of the wire. The vector, **S**, points in toward the center of the wire throughout the length of the wire.

We now must find the energy leaving the volume of the fields and entering the surface of the wire as determined by Poynting's area integral

$$\oint_{Surface} (\mathbf{E} \times \mathbf{H}) \bullet d\mathbf{a} . \tag{7.40}$$

Recall that the cross product of two vectors is the product of their absolute values times the sine of the angle between them. Here the angle between **E** and **H** is ninety degrees and the sine of that angle is one. Therefore,

$$| \mathbf{E} \times \mathbf{H} | = | \mathbf{E} | | \mathbf{H} | \sin 90^0 = EH . \tag{7.41}$$

The unit of E is in volts per meter. This is 2 volts over the length of the wire in meters:

$$E = \frac{U}{L} \frac{\text{volts}}{\text{meters}} . \tag{7.42}$$

Consider a circle of **H** almost touching the wire and essentially at the radius distance, r, in meters, from the center of the wire. Then

$$H = \frac{i}{2\pi r} \frac{\text{ampers}}{\text{meters}} . \tag{7.43}$$

As pointed out above the area of the curved surface of the wire is perpendicular to vector **S**, and of course, perpendicular to $(\mathbf{E} \times \mathbf{H})$ since **S** points radially into the wire. So the normal (perpendicular) to the area (curved surface of the wire) and **S** are parallel. The angle between them is zero degrees, with a cosine of 1. So to perform the dot product and the integration we can simply multiply $| \mathbf{E} \times \mathbf{H} |$ (which is their absolute values as shown above) by the curved area of the wires to find the total power leaving the volume of the fields and entering the surface area of the wire. That is

$$\oint_{Surface} (\mathbf{E} \times \mathbf{H}) \bullet d\mathbf{a} = \frac{U}{L} \frac{i}{2\pi r} (2\pi r \, L) = U \, i , \tag{7.44}$$

or

$$U = 2 \text{ volts}, \quad i = 2 \text{ ampers}, \quad U \, i = 4 \text{ vatts} .$$

So by using the Poynting equation we have obtained the answer of 4 watts being dissipated in the wire. We have shown that the energy entered the wire from its round surface. The engineering mathematics, by assuming that the current of 2 amperes immediately entered the end of the wire from the battery without having any fields or the Poynting vector involved, also gave the correct answer of 4 watts dissipated in the wire. But the Poynting equation method shows that the energy dissipated by the wire originated outside of the wire. The Poynting method arrives at the correct formula for calculating the energy but depends on the engineering method to measure the voltage and current so that we may solve the problem.

Both the engineering mathematics method and the Poynting equation method are correct, accept for a discrepancy in the present Poynting equation which we will now correct. Let us look at the Poynting equation that we have evolved thus far

$$\oint_{Surface} (\mathbf{E} \times \mathbf{H}) \bullet d\mathbf{a} = -\frac{d}{dt} \int_{Volume} \left(\frac{\epsilon E^2}{2} + \frac{\mu H^2}{2} \right) dv. \tag{7.45}$$

But the mathematics showed that all of the energy leaving the surface integral is being dissipated in the wire. We can see from the equation that all of the energy in the volume integral is leaving the volume through the area of the surface integral. This leaves no energy left over in the volume integral to be conducted into some electrical device or to be propagated further. The reason for this discrepancy is that we did not incorporate the current density term, $\mathbf{J}$, from Maxwell's equation III, into Poynting's equation.

7.6. The Complete Poynting Equation

Recall the equation (7.5) for the divergence of the Poynting vector:

$$\nabla \bullet (\mathbf{E} \times \mathbf{H}) = [\mathbf{H} \bullet (\nabla \times \mathbf{E}) - \mathbf{E} \bullet (\nabla \times \mathbf{H})], \tag{7.46}$$

and Maxwell's equations III and IV:

$$\nabla \times \mathbf{H} = \mathbf{J} + \epsilon \frac{\partial \mathbf{E}}{\partial t}. \tag{7.47}$$

$$\nabla \times \mathbf{E} = -\mu \frac{\partial \mathbf{H}}{\partial t}. \tag{7.48}$$

In the following derivation to account for charged particles within the volume of the fields (in the circuit example charged particles are free electrons in the copper wire) we must use both terms on the right side of the equal sign of eq. (7.47). Then

$$\nabla \bullet (\mathbf{E} \times \mathbf{H}) = -\mathbf{E} \bullet \mathbf{J} - \frac{\partial}{\partial t} \left(\mu \frac{\mathbf{H} \bullet \mathbf{H}}{2} + \epsilon \frac{\mathbf{E} \bullet \mathbf{E}}{2} \right). \tag{7.49}$$

Integrate both sides of the equation over the same volume:

$$\int_{Volume} \nabla \bullet (\mathbf{E} \times \mathbf{H}) dv$$
$$= -\int_{Volume} \mathbf{E} \bullet \mathbf{J} dv - \frac{d}{dt} \int_{Volume} \left(\mu \frac{\mathbf{H} \bullet \mathbf{H}}{2} + \epsilon \frac{\mathbf{E} \bullet \mathbf{E}}{2} \right) dv. \tag{7.50}$$

Use the divergence theorem to change

$$\int_{Volume} \nabla \bullet (\mathbf{E} \times \mathbf{H}) \, dv \quad \text{to} \quad \oint_{Surface} (\mathbf{E} \times \mathbf{H}) \, d\mathbf{a}.$$

Then

$$\oint_{Surface} (\mathbf{E} \times \mathbf{H}) \, d\mathbf{a}$$
$$= -\int_{Volume} \mathbf{E} \bullet \mathbf{J} \, dv - \frac{d}{dt} \int_{Volume} \left(\mu \frac{\mathbf{H} \bullet \mathbf{H}}{2} + \epsilon \frac{\mathbf{E} \bullet \mathbf{E}}{2} \right) dv. \tag{7.51}$$

This is Poynting's energy equation. It is based on taking the divergence of the Poynting vector and incorporating the identities found in Maxwell's equations. The Poynting equation is incomplete for determining all the energies associated with the volume that contains the electromagnetic energy. In addition to the Poynting equation, thermal energy, energy due to non electrical forces and energy put into the volume by moving charges outside of the volume will be considered in advanced studies using a more complete electromagnetic energy equation.

The Poynting equation above shows that the energy leaving the volume of the fields is composed of two parts. The first term of the volume integral accounts for the energy absorbed by electric charges within the volume. In the circuit energy allocation example this term will be the energy dissipated in the wire resistance. The charges that receive the energy are the free electrons in the copper wire. The second term is energy propagating further along into circuit loads or radiation.

What energy is contained in

$$\int_{Volume} \mathbf{E} \bullet \mathbf{J} \, dv \ ? \tag{7.52}$$

Vector $\mathbf{J}$ is current density and flows in the same direction as $\mathbf{E}$. So the dot product is again just the product of the absolute values E and J. J is current density in amperes per square meter. Therefore J times the wire cross-sectional area is the current i in amperes flowing in the wire. E is the voltage across the wire divided by the length of the wire:

$$\int_{Volume} \mathbf{E} \bullet \mathbf{J} \, dv = \int_{Volume} \frac{U}{L} \frac{i}{\pi r^2} \, dv \ . \tag{7.53}$$

This integral summed over the volume of the wire is

$$\int_{Volume} \mathbf{E} \bullet \mathbf{J} \, dv = \frac{U}{L} \frac{i}{\pi r^2} (L \, \pi r^2) = U \, i \ . \tag{7.54}$$

in watts dissipated in the wire.

This shows that the theories supporting engineering mathematics and the Poynting vector method are equally valid. But, of course, engineering mathematics is applicable to circuit problems and Poynting equations are applicable to radiation problems.

7.7. Poynting Equation Conclusions

The complete radiating electromagnetic energy equation used in advanced course work is based on Poynting's equation. That equation does include the terms of Poynting's equation, thermal energy, non-electrical energy and energy supplied from outside Poynting's equation boundaries.

The starting point for that complete energy equation is Poynting's equation

$$\oint_{Surface} (\mathbf{E} \times \mathbf{H}) \, d\mathbf{a}$$
$$= -\int_{Volume} \mathbf{E} \bullet \mathbf{J} \, dv - \frac{d}{dt} \int_{Volume} \left(\mu \frac{\mathbf{H} \bullet \mathbf{H}}{2} + \epsilon \frac{\mathbf{E} \bullet \mathbf{E}}{2} \right) dv \ . \tag{7.55}$$

This equation shows that all the energy leaving the right-hand integral is passing through the surface indicated by the left-hand integral. It is to be noted that this new information concerning the energy in electromagnetic radiation is obtained by simply taking the divergence of the Poynting vector, $\mathbf{E} \times \mathbf{H}$, and incorporating Maxwell's equations.

The new information was found to be:
1. The electric field volume energy density is

$$\frac{1}{2}\epsilon E^2 \ .$$

Volume energy density is in joules per cubic meter.
2. The magnetic field volume energy density is

$$\frac{1}{2}\mu H^2 \ .$$

3. The volume energy density is the same for both E and H fields.
4. The volume energy densities of the E and H fields are independent of their boundaries.
5. The total energy in the electromagnetic wave passing through one square meter each second is

$$H\ E = (E\sqrt{\frac{\epsilon}{\mu}})E = \sqrt{\frac{\epsilon}{\mu}}E^2 = 2.65 \times 10^{-3}\ E^2\ \frac{\text{watts}}{\text{meter}^2\ \text{sec}}\ ;\ .$$

6. In space the energy density of the electromagnetic field is

$$\frac{E^2}{\text{impedance of free space}}\ .$$

This correlates with engineering mathematics dimensions of electrical power measurement.
7. In the examples of applying Poynting's equation to circuit wiring power dissipation, the same result was obtained with the engineering mathematics and Poynting's equation. In the first case the term

$$\oint_{Surface} \mathbf{E} \times \mathbf{H} \bullet d\mathbf{a}$$

was used. In the second case the term

$$\int_{Volume} (\mathbf{E} \bullet \mathbf{J})\ dv$$

was used. The Poynting equation showed that the power input to the wire is due to the electric and magnetic fields surrounding the wire. This result does not correlate with the engineering mathematics method that assumes that the current directly enters the wire from the wire end. None of the Poynting vector energy enters the wire from its ends. The theory supporting the engineering mathematics and the theory supporting the Poynting method are equally valid. Each theory is more applicable to certain problems than the other. There is a common example where applying the engineering mathematics shows that energy enters the wire from its sides as shown when using the Poynting vector. That case occurs when a wire is moved through a magnetic field so that a voltage is induced in the wire and then it appears at the wire ends. The engineering mathematics solves the problem by calculating lines of flux cut per second without stating how the energy gets into the wire. The Poynting vector theory shows that the energy does enter the wire through the surface of the wire. Another example wherein the Poynting equations correct our intuitive reasoning is the need to consider the overhead high voltage power lines. We intuitively believe that the fields about the wire are somehow being dragged along the wire by the current in the wire. But the Poynting equations show that the wire (that is the free electrons in the wire) just serve as a guide for the location of fields. The fields, through the Poynting vector concept, transport the energy long distances and also supply the losses in the wire.

In: An Essential Guide to Electrodynamics
Editor: Norma Brewer

ISBN: 978-1-53615-705-5
© 2019 Nova Science Publishers, Inc.

Chapter 3

MAXWELL ELECTROMAGNETIC EQUATIONS IN THE UNIFORM MEDIUM

E. M. Ovsiyuk[1],* V. Balan[2],† O. V. Veko[3],‡
Ya. A. Voynova[3]§ and V. M. Red'kov[3]¶
[1]Mozyr State Pedagogical University, Mozyr, Belarus
[2]University Politehnica of Bucharest, Romania
[3]Institute of Physics, National Academy of Sciences of Belarus,
Minsk, Belarus

Abstract

Two known, alternative to each other, forms of presenting the Maxwell electromagnetic equations in a moving uniform medium are discussed. The commonly used Minkowski approach is based on two tensors, and the constitutive equations relating these tensors change their form under Lorentz transformations, so that Minkowski equations depend upon the 4-velocity of the moving inertial reference frame. In this approach, the wave equation for the electromagnetic 4-potential has a form which explicitly involves this 4-velocity vector of a moving frame. Hence, the Minkowski electrodynamics in a sense implies the absolute nature of mechanical motion in medium. An alternative formalism may be constructed, when the Maxwell equations are written with the use of only one tensor. This form of Maxwell equations exhibits symmetry under modified Lorentz transformations, in which instead of the vacuum speed of light c one uses the speed of light $c' < c$ in the medium. Due to this symmetry, the formulation of Maxwell theory in this medium becomes invariant under the mechanical motion of the reference frame, while the transition velocity must follow modified Lorentz formulas. The transition to 4-potential leads to a simple wave equation which does not contain any additional 4-velocity parameter; also this equation describes waves which propagate in space with light velocity kc, which is invariant under the modified Lorentz transformations. In connection with these two alternative schemes, an essential issue must be stressed: it seems more reasonable to perform the Poincaré-Einstein clock

---

*Author's E-mail: e.ovsiyuk@mail.ru.
†Author's E-mail: vladimir.balan@upb.ro.
‡Author's E-mail: vekoolga@mail.ru.
§Author's E-mail: voinyuschka@mail.ru.
¶Author's E-mail: redkov@ifanbel.bas-net.by.

synchronization in the uniform medium with the use of real light signals influenced by the medium, which leads us to modified Lorentz symmetry.

Keywords: electromagnetic theory, uniform medium, Minkowski approach, modified Lorentz symmetry

MSC2010: 35Q61, 35Q60, 83C50, 78A25

PACS numbers: 1130, 0230, 0365

1. Maxwell Equations in the Medium, New Variables

Maxwell's equations in the uniform medium with two characteristics, $\epsilon > 1$ and $\mu > 1$, have the form [11]

$$\text{div } \mathbf{E} = \frac{1}{\epsilon\epsilon_0}\rho, \quad \text{div } \mathbf{B} = 0,$$
$$\frac{1}{\mu\mu_0}\text{rot } \mathbf{B} = \mathbf{J} + \epsilon\epsilon_0 \frac{\partial \mathbf{E}}{\partial t}, \quad \text{rot } \mathbf{E} = -\frac{\partial \mathbf{B}}{\partial t}. \qquad (1)$$

In the variables

$$\mathbf{B}/\mu\mu_0 = \mathbf{H}, \quad \epsilon\epsilon_0 \mathbf{E} = \mathbf{D},$$

the equations (1) read

$$\text{div } \mathbf{D} = \rho, \quad \text{div } \mathbf{H} = 0,$$
$$\text{rot } \mathbf{H} = \mathbf{J} + \frac{\partial}{\partial t}\mathbf{D}, \quad \frac{1}{\epsilon\epsilon_0}\text{rot } \mathbf{D} = -\mu\mu_0 \frac{\partial}{\partial t}\mathbf{H}. \qquad (2)$$

The four parameters $\epsilon_0, \mu_0, \epsilon, \mu$ enter the Maxwell equations in the form of two products, $\epsilon_0\epsilon$ and $\mu_0\mu$. This means that besides the charge-current density $(\rho, \mathbf{J})$ and the fields $(\mathbf{D}, \mathbf{H})$, these equations include only two independent additional parameters. This may be manifestly revealed by introducing two quantities, the light velocity in vacuum c and the (inverse) refraction coefficient k of the medium:

$$c = \frac{1}{\sqrt{\epsilon_0\mu_0}}, \quad k = \frac{1}{\sqrt{\epsilon\mu}} < 1. \qquad (3)$$

Therefore, from (2) we get

$$\text{div } \mathbf{D} = \rho, \quad \text{div } \mathbf{H} = 0,$$
$$\text{rot } \mathbf{H} = \mathbf{J} + \frac{\partial}{\partial t}\mathbf{D}, \quad \text{rot } \mathbf{D} = -\frac{1}{k^2 c^2}\frac{\partial}{\partial t}\mathbf{H}. \qquad (4)$$

The light velocity in a medium c_{med} is less than the vacuum velovity c and the coefficient k describes this decrease: $c_{med} = k\, c$.

An essential point is that Eqs. (4) may be re-written as

$$\text{div } \mathbf{D} = \rho, \quad \text{div } \frac{\mathbf{H}}{kc} = 0,$$
$$\text{rot } \frac{\mathbf{H}}{kc} = \frac{\mathbf{J}}{kc} + \frac{\partial}{\partial(kct)}\mathbf{D}, \quad \text{rot } \mathbf{D} = -\frac{\partial}{\partial(kct)}\frac{\mathbf{H}}{kc}. \qquad (5)$$

Therefore, instead of the variables $(t, x^i), (\rho, \mathbf{J}), (\mathbf{D}, \mathbf{H})$, one may define new ones $(x^0, x^i), (\rho, \mathbf{j}), (\mathbf{d}, \mathbf{h})$ by the formulas:

$$x^0 = kc\,t, \quad j^0 = \rho\,, \mathbf{j} = \frac{\mathbf{J}}{kc}\,, \quad \mathbf{d} = \mathbf{D}\,, \quad \mathbf{h} = \frac{\mathbf{H}}{kc}\,. \tag{6}$$

Hence, Maxwell's equations (5) will take the form

$$\operatorname{div} \mathbf{d} = j^0\,, \quad \operatorname{div} \mathbf{h} = 0\,,$$
$$\operatorname{rot} \mathbf{h} = \mathbf{j} + \frac{\partial}{\partial x^0}\mathbf{d}\,, \quad \operatorname{rot} \mathbf{d} = -\frac{\partial}{\partial x^0}\mathbf{h}\,. \tag{7}$$

The correctness of the relations (7) can be additionally checked through dimensional considerations:

$$[\,x^0\,] = [\,x^{\,i}\,] = \text{meter}\,, \quad [\,\rho\,] = \frac{\text{coulomb}}{\text{meter}^3}\,,$$
$$[\,j^i\,] = \frac{\text{coulomb}}{\text{meter}^3}\,, \quad [\,d^{\,i}\,] = [\,h^{\,i}\,] = \frac{\text{coulomb}}{\text{meter}^2}\,.$$

2. Lorentz Transformations in Vacuum and Medium

By using the notations

$$\partial_0 = \frac{\partial}{\partial x^0}\,, \quad \partial_i = \frac{\partial}{\partial x^i}\,, \quad i = 1, 2, 3\,,$$

the Maxwell equations (7) can be written down in explicit form as follows

$$(I) \quad \partial_1 d^1 + \partial_2 d^2 + \partial_3 d^3 = j^{\,0}\,,$$

$$(II) \quad \partial_1 h^1 + \partial_2 h^2 + \partial_3 h^3 = 0\,,$$

$$(III) \quad \partial_2 h^3 - \partial_3 h^2 = j^1 + \partial_0 d^1\,,$$
$$\partial_3 h^1 - \partial_1 h^3 = j^2 + \partial_0 d^2\,,$$
$$\partial_1 h^2 - \partial_2 h^1 = j^1 + \partial_0 d^1\,,$$

$$(IV) \quad \partial_2 d^3 - \partial_3 d^2 = -\,\partial_0 h^1\,,$$
$$\partial_3 d^1 - \partial_1 d^3 = -\,\partial_0 h^2\,,$$
$$\partial_1 d^2 - \partial_2 d^1 = -\,\partial_0 h^1\,.$$

$$\tag{8}$$

Now we introduce certain linear transformations over the quantities which enter the Maxwell equations (the *modified Lorentz transformations*):

$$\begin{aligned}
x^{'0} &= \cosh \sigma x^0 - \sinh \sigma x^1,\, x^{'1} = -\sinh \sigma x^0 + \cosh \sigma x^1,\, x^{'2} = x^2, x^3 = x^3,\\
j^{'0} &= \cosh \sigma j^0 - \sinh \sigma j^1,\, j^{'1} = -\sinh \sigma j^0 + \cosh \sigma j^1, j^{'2} = j^2, j^{'3} = j^3,\\
d^{'1} &= +d^1, d^{'2} = \cosh \sigma d^2 - \sinh \sigma h^3, h^{'3} = -\sinh \sigma d^2 + \cosh \sigma h^3,\\
h^{'1} &= +h^1, d^{'3} = \cosh \sigma d^3 + \sinh \sigma h^2, h^{'2} = +\sinh \sigma d^3 + \cosh \sigma h^2.
\end{aligned} \tag{9}$$

Now the task is to show that if one transforms equations (8) to the new (primed) variables

$$(x^0, x^i), (j^0, j^i), (d^i, h^i) \implies (x'^0, x'^i), (j'^0, j'^i), (d'^i, h'^i),$$

then as a result one will obtain again equations of the form (8), but with a single difference: all quantities become primed ones.

In essence, this is the main statement of Lorentz and Poincaré on the symmetry properties of Maxwell theory [4, 9]. Before proceeding further, we derive transformation formulas for the derivatives with respect to the coordinates:

$$\partial_0 = \cosh\sigma\, \partial'_0 - \sinh\sigma\, \partial'_1,\ \partial_1 = -\sinh\sigma\, \partial'_0 + \cosh\sigma\, \partial'_1,\ \partial_2 = \partial'_2,\ \partial_3 = \partial'_3, \qquad (10)$$

After simple calculations we infer

(I) $\cosh\sigma[(\partial'_1 d'^1 + \partial'_2 d'^2 + \partial'_3 d'^3) - j'^0] - \sinh\sigma[-(\partial'_2 h'^3 - \partial'_3 h'^2) + j'^1 + \partial'_0 d'^1] = 0$,

(II) $\cosh\sigma(\partial'_1 h'^1 + \partial'_2 h'^2 + \partial'_3 h'^3) - \sinh\sigma[\partial'_0 h'^1 + (\partial'_2 d'^3 - \partial'_3 d'^2)] = 0$,

$(III)_1$ $\cosh\sigma[(\partial'_2 h'^3 - \partial'_3 h'^2) - j'^1 - \partial'_0 d'^1] - \sinh\sigma[-(\partial'_1 d'^1 + \partial'_2 d'^2 + \partial'_3 d'^3) + j'^0] = 0$,

$(III)_2$ $\partial'_3 h'^1 - \partial'_1 h'^3 = j'^2 + \partial'_0 d'^2$,

$(III)_3$ $\partial'_1 h'^2 - \partial'_2 h'^1 = j'^3 + \partial'_0 d'^3$,

$(IV)_1$ $\cosh\sigma[(\partial'_2 d'^3 - \partial'_3 d'^2) + \partial'_0 h'^1] - \sinh\sigma(\partial'_1 h'^1 + \partial'_2 h'^2 + \partial'_3 h'^3) = 0$,

$(IV)_2$ $\partial'_3 d'^1 - \partial'_1 d'^3 = -\partial'_0 h'^2$,

$(IV)_3$ $\partial'_1 d'^2 - \partial'_2 d'^1 = -\partial'_0 h'^3$.

Let us combine (II) and $(IV)_1$:

$$\cosh\sigma\, (II) + \sinh\sigma\, (IV)_1 = (II)' \implies \partial'_1 h'^1 + \partial'_2 h'^2 + \partial'_3 h'^3 = 0,$$

$$\sinh\sigma\, (II) + \cosh\sigma\, (IV)_1 = (IV)'_1 \implies \partial'_2 d'^3 - \partial'_3 d'^2 = -\partial'_0 h'^1.$$

Analogously, by combining (I) and $(III)_1$, we get

$$\cosh\sigma\, (I) + \sinh\sigma\, (III)_1 = (I)' \implies \partial'_1 d'^1 + \partial'_2 d'^2 + \partial'_3 d'^3 = j'^0,$$

$$\sinh\sigma\, (I) + \cosh\sigma\, (III)_1 = (IV)'_1 \implies \partial'_2 d'^3 - \partial'_3 d'^2 = -\partial'_0 h'^1.$$

Conclusion. From the previous, there follow the Maxwell equations in primed variables:

$$(I)\qquad \partial'_1 d'^1 + \partial'_2 d'^2 + \partial'_3 d'^3 = j'^0,$$

$$(II)\qquad \partial'_1 h'^1 + \partial'_2 h'^2 + \partial'_3 h'^3 = 0,$$

$$(III)\qquad \partial'_2 h'^3 - \partial'_3 h'^2 = j'^1 + \partial'_0 d'^1,$$

$$\partial'_3 h'^1 - \partial'_1 h'^3 = j'^2 + \partial'_0 d'^2,$$

$$\partial'_1 h'^2 - \partial'_2 h'^1 = j'^1 + \partial'_0 d'^1,$$

$$(IV)\qquad \partial'_2 d'^3 - \partial'_3 d'^2 = -\partial'_0 h'^1,$$

$$\partial'_3 d'^1 - \partial'_1 d'^3 = -\partial'_0 h'^2,$$

$$\partial'_1 d'^2 - \partial'_2 d'^1 = -\partial'_0 h'^1.$$

In vector form they read

$$\text{div}' \mathbf{d}' = j'^0, \quad \text{div}' \mathbf{h}' = 0,$$
$$\text{rot}' \mathbf{h}' = \mathbf{j}' + \frac{\partial}{\partial x'^0} \mathbf{d}', \quad \text{rot}' \mathbf{d}' = -\frac{\partial}{\partial x'^0} \mathbf{h}'. \tag{11}$$

3. Interpretation of Lorentz Transformations in Vacuum and Medium

The question is how do the Maxwell equations behave when the reference frame changes from K to a moving one, K'.

For the situation when velocity is small enough, we must obtain a simple solution in the form of Galileo formulas for coordinate transforms:

$$t' = t, \quad x' = x - Vt, \quad y' = y, \quad z' = z.$$

Let us consider the Lorentz transformation

$$x'^0 = \cosh \sigma x^0 - \sinh \sigma x^1, x'^1 = -\sinh \sigma x^0 + \cosh \sigma x^1, x'^2 = x^2, x'^3 = x^3$$

at a very small σ. We use the Taylor expansions

$$e^{+\sigma} = 1 + \frac{\sigma}{1!} + \frac{\sigma^2}{2!} + ..., \quad e^{-\sigma} = 1 - \frac{\sigma}{1!} + \frac{\sigma^2}{2!} - ...,$$

and at $\sigma << 1$ we get the approximating formulas

$$\cosh \sigma = 1 + \frac{\sigma^2}{2!} + ... \approx 1, \quad \sinh \sigma = \frac{\sigma}{1!} + \frac{\sigma^3}{3!} + ... \approx \sigma.$$

Thus, the Lorentz transformation at small σ will take the form

$$kc\, t' \approx kct - \sigma x \approx kct,$$

whence

$$t' = t, \quad x' = -\sigma kc\, t + x = x - Vt, \quad \text{if} \quad \sigma = \frac{V}{kc}.$$

Hence the physical sense of the parameter σ (at its small values $\sigma << 1$) is expressed by relations

$$\sigma << 1 \quad \Longrightarrow \quad \sigma = \frac{V}{kc} = \frac{V}{c_{med}}. \tag{12}$$

One might further need to generalize (12) for arbitrary values of V. This is achieved as follows

$$0 < |V| < kc \quad \Longleftrightarrow \quad 0 < |\sigma| < 1,$$
$$\cosh \sigma = \frac{1}{\sqrt{1 - (V/kc)^2}}, \quad \sinh \sigma = \frac{(V/kc)}{\sqrt{1 - (V/kc)^2}}. \tag{13}$$

At small V, relations (13) coincide with (12). It is useful to translate tthe formulas for Lorentz transformations to ordinary units:

$$t' = \frac{t - Vx/k^2c^2}{\sqrt{1 - (V/kc)^2}}, \quad x' = \frac{x - Vt}{\sqrt{1 - (V/kc)^2}}. \tag{14}$$

4. Behavior of the Light Velocity under Modified Lorentz Transformations

While finding the modified Lorentz transformations, a simple kinematical problem may be immediately solved, which provides us with a postulate on the constancy of light velocity (kc), the crucial logical element in Einstein's construction of Special Relativity.

Let us observe a material point in a fixed rest reference frame K, which starts its history at
$$t_1 = 0, \quad x_1 = 0, \quad y_1 = 0, \quad z_1 = 0,$$
and moves in the plane (x, y) along direction $(\cos\phi, \sin\phi)$ with velocity
$$W_x = W\cos\phi, \quad W_y = W\sin\phi, \quad W_z = 0, \quad W = \sqrt{W_1^2 + W_2^2}.$$
At the moment $t_2 > 0$, its coordinates become
$$t_2, \quad x_2 = W\cos\phi \, t_2, \quad y_2 = W\sin\phi \, t_2, \quad z_2 = 0.$$
The same can be re-written in the form
$$W_x = \frac{x_2 - x_1}{t_2 - t_1}, \quad W_y = \frac{y_2 - y_1}{t_2 - t_1}, \quad W_z = 0.$$

Let us find how the velocity behaves under the modified Lorentz transformation. To this end, we should calculate
$$W'_x = \frac{x'_2 - x'_1}{t'_2 - t'_1}, \quad W'_y = \frac{y'_2 - y'_1}{t'_2 - t'_1}.$$
With the use of
$$t'_1 = 0, \quad x'_1 = 0, \quad t'_2 = \frac{t_2 - Vx_2/k^2c^2}{\sqrt{1 - V^2/k^2c^2}}, \quad x'_2 = \frac{x_2 - Vt_2}{\sqrt{1 - V^2/k^2c^2}},$$
for W'_x we get
$$W'_x = \frac{x'_2 - x'_1}{t'_2 - t'_1} = \frac{(x_2 - Vt_2)}{(t_2 - Vx_2/k^2c^2)}$$
$$= \frac{(x_2 - x_1) - V(t_2 - t_1)}{(t_2 - t_1) - V(x_2 - x_1)/k^2c^2} = \frac{W_x - V}{1 - VW_x/k^2c^2}.$$
Now we turn to W'_y:
$$W'_y = \frac{y'_2 - y'_1}{t'_2 - t'_1} = \frac{y_2}{t'_2} = \sqrt{1 - V^2/k^2c^2} \, \frac{W_y t_2}{t_2 - V x_2/k^2c^2}$$
$$= \frac{\sqrt{1 - V^2/k^2c^2} \, W_y (t_2 - t_1)}{(t_2 - t_1) - V(x_2 - x_1)/k^2c^2},$$
whence it follows
$$W'_y = \frac{\sqrt{1 - V^2/k^2c^2}}{1 - VW_x/k^2c^2} W_y.$$

For the W_z component, we have a trivial result

$$W'_z = \frac{z'_2 - z'_1}{t'_2 - t'_1} = \frac{z_2 - z_1}{t'_2 - t'_1} = 0 = W_z \,.$$

Thus, the velocity vector $\mathbf{W} = (W_x, W_y, 0)$ transforms as follows

$$W'_x = \frac{W_x - V}{1 - VW_x/k^2 c^2}, \; W'_y = \frac{\sqrt{1 - V^2/k^2 c^2}}{1 - VW_x/k^2 c^2} \, W_y \,, \;\; W'_z = W_z = 0 \,. \qquad (15)$$

This is a modified version of the rule for velocity summing by Lorentz-Poincaré-Einstein.

For simplifying the appearance of many formulas to come, we convene on the following notation[1]

$$\frac{V}{kc} \iff V \,, \quad \frac{W_x}{kc} \iff W_x \,, \quad \text{etc}\,.$$

Then formulas (15) become

$$W'_x = \frac{W_x - V}{1 - VW_x} \,, \quad W'_y = \frac{\sqrt{1 - V^2}}{1 - VW_x} \, W_y \,, \quad W'_z = W_z \,. \qquad (16)$$

We shall further point out a series of *consequences* which follow from the developed framework.

By applying (16) to the light ray propagating along the x axis in fixed (unmoving) reference frame K, we get the result:

Consequence 1. The light ray propagating along the x axis in the moving reference frame K', has the velocity

$$\begin{array}{l} W_x = 1 \\ W_y = 0 \\ W_z = 0 \end{array} \implies \begin{array}{l} W'_x = \frac{1}{1-VW_x} (W_x - V) = \frac{1-V}{1-V} = 1 \,, \\ W'_y = \frac{\sqrt{1-V^2}}{1-VW_x} W_y = 0, \\ W'_z = 0 \,, \end{array}$$

that is

$$W'_x = 1 \,, \qquad W'_y = 0 \,, \qquad W'_z = 0 \,.$$

The Lorentz transformation along the x axis does not change the value of light propagation along the direction of x.

By applying the (16) to a light ray propagating along arbitrary ϕ-direction in a rest reference frame K, and using Eqs. (16), we infer:

Consequence 2. The velocity vector in the moving reference frame K' is given by

$$\begin{array}{l} W_x^2 + W_y^2 = 1, \\ W_z = 0 \end{array} \implies \begin{array}{l} W'_x = \frac{1}{1-VW_x} (W_x - V), \\ W'_y = \frac{\sqrt{1-V^2}}{1-VW_x} W_y \,, \\ W'_z = 0 \,, \end{array}$$

---

[1] Sometimes this will be formulated as the use of a special system with unit light velocity.

and for the squared modulus of the vector velocity, we get

$$W'^2_x + W'^2_y = \frac{W_x^2 - 2VW_x + V^2 + (1-V^2)(1-W_x^2)}{(1-VW_x)^2} = +1.$$

The Lorentz transformation along the axis x does not change the modulus of the light velocity vector.

Consequence 3. We note that the same result holds true even for $W_z \neq 0$, since

$$(W_x^2 + W_y^2) + W_z^2 = 1 \;\Rightarrow\; W'^2_x + W'^2_y + W'^2_z$$

$$= \frac{W_x^2 - 2VW_x + V^2 + (1-V^2)(1-W_x^2)}{(1-VW_x)^2} = 1 \,.$$

Consequence 4. The invariance of the modulus of the velocity vector under Lorentz transformations concerns only the light velocity-which has unit length in the fixed (non-moving) frame.

We have, indeed,

$$W_x^2 + W_y^2 = W^2, \;\; W_z = 0 \,,$$

whence we get

$$W'^2 = \frac{W_x^2 - 2VW_x + V^2 + (1-V^2)(W^2-W_x^2)}{(1-VW_x)^2} = 1 + (W^2-1)\frac{1-V^2}{(1-VW_x)^2} \,.$$

Consequence 5. The above formulas of the velocity vector transforms can be expressed in terms of angular variables. In the case of the light,

$$(W_x, W_y, 0) \,, \qquad W_x^2 + W_y^2 = 1 \,, \qquad \cos\phi = W_x \qquad \sin\phi = W_y \,,$$

so that from (16) there follow the formulas for the light aberration:

$$\cos\phi' = \frac{\cos\phi - V}{1 - V\cos\phi} \,, \qquad \sin\phi' = \sin\phi \frac{\sqrt{1-V^2}}{1-V\cos\phi} \,.$$

5. The Lorentz-Poincaré-Einstein

It was Lorentz [4] who first established a remarkable property of Maxwell equations, namely its (approximate) symmetry under special mathematical transformations when plugging into involved quantities, time and space coordinates, charge-current density, and electromagnetic fields.

Poincaré introduced exactness and clarity [9] into the Lorentz initial formulas, and revealed its mathematical (so-called) group structure. Undoubtedly, the first deciding steps on the road to Special Relativity theory were made by Lorentz and this was stressed by Poincaré more than once. At the same time, Lorentz never ascribes to himself the merit all along this road, and willingly appreciated the role of Poincaré's contribution.

Unfortunately, afterwards, in connection with Special Relativity, there arose both controversy and misunderstanding on the question – who is the creator of this theory: Lorentz,

Poincaré, or Einstein [2]. To the present day this dispute is still present. We shall not join any side of the debaters, and assert that in our opinion all three, Lorentz, Poincaré, and Einstein, are the creators of the theory[2].

The first was Lorentz, then Poincaré sided with him, and next Einstein started his work on creating Special Relativity [2], mainly on its physical interpretation, comprehension, and logical reconstruction. The question – who is the *main* creator – is false. Lorentz formulated his view on this matter concisely and definitely [12]: the same what we had deduced from Maxwell's equations, Einstein has postulated.

So, the development of Special Relativity arose along two lines, which are absurd to be considered as completely independent. One line goes upward to Special Relativity from the symmetry property of the Maxwell Theory. This is an *inductive* way and it is historically the first.

The second line, though logically independent in appearance, is a *deductive* construction of the theory by going down from a special postulate [2]. But the postulate itself can be regarded as a logical mathematical result of the Lorentz-Poincaré analysis of Maxwell's theory.

The logical treatment suggested by Einstein seems for many people simpler and clearer, so that it may be explained even to a person without any special education. This circumstance assists in the promotion of Einstein's treatment of Special Relativity and its notability for the general public.

However, the creating by deductive way to construct Special Relativity does not provide grounds to assign to A. Einstein the main or single creator role for the theory: there were Lorentz and Poincaré who provided us as well with the inductive way to this theory - and this way was historically the first. Both approaches to Special Relativity are legitimate and mutually complementary.

6. Deriving the Lorentz Formulas from Einstein Postulate of Light Modified Velocity Constancy

Consider the known Einstein postulate:

> The (modified) light velocity is the same in both reference frames, K (the rest frame) and the other one K' (which moves with velocity V).

Problem.

> What kind of linear transformation between (t, x) and (t', x') agrees with the postulate of (modified) light velocity constancy.

The invariance condition for the light velocity is

$$kc = \frac{x_2 - x_1}{t_2 - t_1}, \quad kc = \frac{x'_2 - x'_1}{t'_2 - t'_1},$$

$$0 = k^2 c^2 (t_2 - t_1)^2 - (x_2 - x_1)^2 = k^2 c^2 (t'_2 - t'_1)^2 - (x'_2 - x'_1)^2.$$

---
[2]We have to give credit to a great number of physicists as well and their enormous and laborious work, still emphasizing that these three great names are the historic epistemic promoters

We enforce the following special convention: let the light ray start propagating at $t_1 = 0, x_1 = 0$ in the positive direction of the axis x in the K-frame. Due to the linearity of the transformation law, we have in the reference frame K' that $t'_1 = 0, x'_1 = 0$, as well. Therefore, the previous relation will take the form[3]:

$$k^2 c^2 t^2 - x^2 = k^2 c^2 (t')^2 - (x')^2 , \qquad (x^0)^2 - x^2 = (x'^0)^2 - (x')^2 .$$

So, the task is to find a linear transformation which leaves invariant such an unusual length of the vector (x^0, x). However, the solution to this problem is already known:

$$x'^0 = \cosh \sigma \, x^0 - \sinh \sigma \, x , \qquad x' = -\sinh \sigma \, x^0 + \cosh \sigma \, x ,$$

or, using (13)

$$t' = \frac{t - Vx/k^2 c^2}{\sqrt{1 - V^2/k^2 c^2}} , \qquad x' = \frac{x - Vt}{\sqrt{1 - V^2/k^2 c^2}} .$$

In other words, there exists a logical relationship between the Einstein postulate of (modified) light velocity constancy and the (modified) Lorentz formulas for coordinate transformations.

Here it is appropriate to remember the words of Lorentz [12]: *the same what we had deduced from Maxwell's equations Einstein has postulated.* Evidently, this is exactly the case.

7. On Maxwell Theory in the Uniform Medium, Minkowski Approach

Although A. Einstein's main work on Special Relativity [2] dated 1905 is titled "*On the Electrodynamics of Moving Bodies*", in this paper Maxwell equations had been considered only in vacuum[4] in fact, and the symmetry properties of these equations have been used. In accordance with this, throughout the whole theoretical construction, from the very beginning, only a certain universal light velocity in vacuum was used and only for that very velocity was advanced a postulate regarding its constancy irrespective of the motion of the reference frame.

Later in 1908, H. Minkowski gave [6] a more detailed and accurate treatment of the Maxwell Theory in the uniform medium ($\epsilon \neq 1, \mu \neq 1$) with respect to the requirements of Special Relativity. Two issues of his study should be emphasized:

Minkowski elaborated a very convenient and still actively exploited mathematical technique, the so called 4-dimensional tensor formalism [5].

Minkowski found the way to describe symmetry properties of the Maxwell equations in a uniform medium with the use of Lorentz formulas on the base of the light velocity c in vacuum[6]. In this work Minkowski achieved in fact a certain unification between Einstein's earlier analysis and electrodynamics in medium.

---

[3]Here, the index 2 is omitted in x'_2 and in x_2.

[4]A medium with trivial values $\epsilon = 1, \mu = 1$.

[5]To be exact, H.Poincaré had proposed and developed in some aspects the same technique before Minkowski [4].

[6]This issue is most significant in the context established above: the presence of symmetry of the Maxwell theory in a medium under modified Lorentz transformations involving the light velocity kc in the medium.

Here it might be specially mentioned that the logical construction of Special Relativity by Einstein does not formally depend on the numerical value of light velocity – this might be 300000 km/sec as well as 3 sm/sec. The essential thing here is only the existence of a (light) signal which goes through the space with the same velocity, for all inertial observers. Moreover, in this context we should admit that operating with a light signal of velocity c in a medium is a fiction; in fact any real light can move through a uniform medium with velocity $\tilde{c} = kc$. So it might seem as well-taken the requirement to perform clock synchronization in uniform media with the help of real light signals influenced by the medium. However this was not done by Minkowski. On the contrary, Minkowski found the way to speak about relativistic symmetry of Maxwell Theory in a medium and to use only the Lorentz formulas with the vacuum light velocity c.

Below we shall introduce Minkowski's approach [6] without following it in detail.

8. Standard Lorentz Symmetry of the Maxwell Equations in the Medium

We start from the Maxwell equations in the form

$$\operatorname{div} \mathbf{D} = J^0, \quad \operatorname{rot} \frac{\mathbf{H}}{c} = \frac{\mathbf{J}}{c} + \frac{\partial \mathbf{D}}{\partial ct}, \qquad (17)$$

$$\operatorname{div} c\mathbf{B} = 0, \quad \operatorname{rot} \mathbf{E} = -\frac{\partial c\mathbf{B}}{\partial ct},$$

with $(x^0, x^i) = (ct, x^i)$, $(J^0 = \rho, \frac{\mathbf{J}}{c})$, $\mathbf{E}$ and $\mathbf{D}$ represent electric field, $\mathbf{B}$ and $\mathbf{H}$ represent magnetic field, $\mathbf{J}$ is the total current density, ρ is the total charge density.

Here the equations are divided into two groups, regarding the vectors $(\mathbf{D}, \mathbf{H}/c)$ and the vectors $(\mathbf{E}, c\mathbf{B})$. Note that the source fields $(J^0 = \rho, \mathbf{J}/c)$ enter only the first group. Also, one issue to emphasize is that equations (17) do not include the parameters of permittivity and of permeability of the free space. However as a peculiar compensation for this, we have to use two sets of electromagnetic vectors: $(\mathbf{D}, \mathbf{H}/c)$ and $(\mathbf{E}, c\mathbf{B})$.

It is readily established that if Eqs. (17) are subjected to the (ordinary) Lorentz transformation (with the light velocity c in the vacuum and correspondingly with the variable $x^0 = ct$)

$$x'^0 = \cosh\beta\, x^0 - \sinh\beta\, x^1, x'^1 = -\sinh\beta\, x^0 + \cosh\beta\, x^1, x'^2 = x^2, x'^3 = x^3,$$

$$J'^0 = \cosh\beta\, J^0 - \sinh\beta\, c^{-1} J^1,$$

$$c^{-1} J'^1 = -\sinh\beta\, J^0 + \cosh\beta\, c^{-1} J^1, J'^2 = J^2, J'^3 = J^3,$$

$$D'^1 = +D^1, D'^2 = \cosh\beta\, D^2 - \sinh\beta\, c^{-1} H^3, D'^3 = \cosh\beta\, D^3 + \sinh\beta\, c^{-1} H^2,$$

$$H'^1 = +H^1, c^{-1} H'^2 = +\sinh\beta\, D^3 + \cosh\beta\, c^{-1} H^2,$$

$$c^{-1} H'^3 = -\sinh\beta\, D^2 + \cosh\beta\, c^{-1} H^3,$$

$$E'^1 = +E^1, E'^2 = \cosh\beta\, E^2 - \sinh\beta\, cB^3, E'^3 = \cosh\beta\, E^3 + \sinh\beta\, cB^2,$$

$$B'^1 = +B^1, cB'^2 = +\sinh\beta\, E^3 + \cosh\beta\, cB^2\, cB'^3 = -\sinh\beta\, E^2 + \cosh\beta\, cB^3,$$

where

$$\cosh \beta = \frac{1}{\sqrt{1-(V/c)^2}}, \quad \sinh \beta = \frac{(V/c)}{\sqrt{1-(V/c)^2}},$$

we shall again obtain equations in Maxwell's form:

$$\operatorname{div} {}'\mathbf{D}' = J'^0, \quad \operatorname{rot} {}'\frac{\mathbf{H}'}{c} = \frac{\mathbf{J}'}{c} + \frac{\partial \mathbf{D}'}{\partial ct'},$$
$$\operatorname{div} {}'c\mathbf{B}' = 0, \quad \operatorname{rot} {}'\mathbf{E}' = -\frac{\partial c\mathbf{B}'}{\partial ct'}. \qquad(18)$$

The issue of first importance is that the modified Lorentz transformations used in (9) generate significantly different formulas. For convenience, we shall write them down[7]:

The modified relations (the light velocity in a medium kc and $x^0 = kct$).

$$x'^0 = \cosh\sigma\, x^0 - \sinh\sigma\, x^1, \quad x'^1 = -\sinh\sigma\, x^0 + \cosh\sigma\, x^1, \quad x'^2 = x^2, \quad x'^3 = x^3,$$

$$J'^0 = \cosh\sigma J^0 - \sinh\sigma(kc)^{-1}J^1,$$
$$(kc)^{-1}J'^1 = -\sinh\sigma J^0 + \cosh\sigma(kc)^{-1}J^1, \quad J'^2 = J^2, \quad J'^3 = J^3,$$
$$D'^1 = D^1, \quad D'^2 = \cosh\sigma D^2 - \sinh\sigma(kc)^{-1}H^3, \quad D'^3 = \cosh\sigma D^3 + \sinh\sigma(kc)^{-1}H^2,$$
$$H'^1 = H^1, \quad (kc)^{-1}H'^2 = \sinh\sigma D^3 + \cosh\sigma(kc)^{-1}H^2,$$
$$(kc)^{-1}H'^3 = -\sinh\sigma D^2 + \cosh\sigma(kc)^{-1}H^3,$$
$$E'^1 = E^1, \quad E'^2 = \cosh\sigma E^2 - \sinh\sigma kcB^3, \quad E'^3 = \cosh\sigma E^3 + \sinh\sigma kcB^2,$$
$$B'^1 = B^1, \quad kcB'^2 = \sinh\sigma E^3 + \cosh\sigma kcB^2, \quad kcB'^3 = -\sinh\sigma E^2 + \cosh\sigma kcB^3,$$
$$\cosh\sigma = \frac{1}{\sqrt{1-(V/kc)^2}}, \quad \sinh\beta = \frac{(V/kc)}{\sqrt{1-(V/kc)^2}}.$$

So, regarding the Maxwell equations, we face the rather peculiar situation, in which two different symmetries are revealed at the same time: a symmetry with respect to the ordinary Lorentz transformations L; another symmetry, with respect to the modified Lorentz transformations L^{mod}, in which there appears a medium dependent light velocity kc. The explicit transforms both for space-time coordinates and for electromagnetic quantities (t, x^i), (J^0, J^i), (E^i, B^i), (D^i, H^i) differ for these two cases.

Then there occur the following open questions: which symmetry of these two is more adequate to consider? Which is the in-depth meaning of the simultaneous existence of two symmetries for Maxwell equations in a medium? Which of them closer corresponds to the physical reality? Do there exist any criteria to pick out only one of two logical possibilities?

From purely theoretical view point, considering the need to synchronize clocks with the help of real light signals in a medium, as being imperative one must use the modified version of the Lorentz transformations in the medium.

---

[7]In order to to distinguish them, instead of β we shall use the symbol σ; from formal viewpoint, the whole difference reduces to the emerging modified quantity $\tilde{c} = kc$.

9. Constitutive Relations and Minkowski Equations

We further examine the following problem: which will be the form of the additional constraints

$$D^i = \epsilon_0 \epsilon \, E^i, \quad H^i = \frac{1}{\mu_0 \mu} B^i, \qquad (19)$$

after the Lorentz transformation to a moving reference frame? This problem was firstly considered by H. Minkowski in 1908 [6]. For simplicity we will consider the simplest Lorentz formulas that correspond to a moving reference frame along the x axis.

We first consider the ordinary Lorentz transforms:

$$D^i = \epsilon_0 \epsilon \, E^i \implies$$

$$\begin{aligned}
D'^1 &= \epsilon_0 \epsilon \, E'^1, \\
\cosh \beta \, D'^2 + \sinh \beta \, \frac{H'^3}{c} &= \epsilon_0 \epsilon \, (\cosh \beta \, E'^2 + \sinh \beta \, cB'^3), \\
\cosh \beta \, D'^3 - \sinh \beta \, \frac{H'^2}{c} &= \epsilon_0 \epsilon \, (\cosh \beta \, E'^3 - \sinh \beta \, cB'^2);
\end{aligned} \qquad (20)$$

$$H^i = \frac{1}{\mu_0 \mu} B^i \implies$$

$$\begin{aligned}
H'^1 &= \frac{1}{\mu_0 \mu} B^1, \\
\sinh \beta \, D'^3 - \cosh \beta \, \frac{H'^2}{c} &= \frac{1}{\mu_0 \mu} \frac{1}{c^2} (\sinh \beta \, E'^3 - \cosh \beta \, cB'^2), \\
\sinh \beta \, D'^2 + \cosh \beta \, \frac{H'^3}{c} &= \frac{1}{\mu_0 \mu} \frac{1}{c^2} (\sinh \beta \, E'^2 + \cosh \beta \, cB'^3);
\end{aligned} \qquad (21)$$

we notice that

$$\frac{1}{\mu_0 \mu c^2} = \frac{\epsilon_0 \mu_0}{\mu_0 \mu} = \frac{\epsilon_0 \epsilon}{\epsilon \mu} = \epsilon_0 \epsilon \, k^2.$$

The relations (20) and (21) are just what we call *Minkowski equations* [6] written down in a particular simple case. Let us change them to another form. To this end, they should be rewritten as three pairs of linear systems in the variables (D'^1, H'^1), $(D'^2, H'^3/c)$, $(D'^3, H'^2/c)$:

$$D'^1 = \epsilon_0 \epsilon \, E'^1, \quad H'^1 = \frac{1}{\mu_0 \mu} B^1;$$

$$\begin{aligned}
\cosh \beta \, D'^2 + \sinh \beta \, H'^3/c &= \epsilon_0 \epsilon (\cosh \beta \, E'^2 + \sinh \beta \, cB'^3), \\
\sinh \beta \, D'^2 + \cosh \beta \, H'^3/c &= \epsilon_0 \epsilon k^2 (\sinh \beta \, E'^2 + \cosh \beta \, cB'^3);
\end{aligned} \qquad (22)$$

$$\begin{aligned}
\cosh \beta \, D'^3 - \sinh \beta \, \frac{H'^2}{c} &= \epsilon_0 \epsilon (\cosh \beta \, E'^3 - \sinh \beta \, cB'^2), \\
\sinh \beta \, D'^3 - \cosh \beta H'^2/c &= \epsilon_0 \epsilon k^2 (\sinh \beta \, E'^3 - \cosh \beta \, cB'^2).
\end{aligned}$$

Solutions of (22)[8] are

$$D'^1 = \epsilon_0 \epsilon \, E'^1 \, ,$$
$$D'^2 = \epsilon_0 \epsilon \, [\, (\cosh^2 \beta - k^2 \sinh^2 \beta) \, E'^2 + \sinh \beta \, \cosh \beta \, (1 - k^2) \, cB'^3 \,] \, , \qquad (23)$$
$$D'^3 = \epsilon_0 \epsilon \, [\, (\cosh^2 \beta - k^2 \sinh^2 \beta) \, E'^3 - \sinh \beta \, \cosh \beta \, (1 - k^2) \, cB'^2 \,] \, ;$$

$$H'^1/c = \epsilon_0 \epsilon k^2 \, cB^1 \, ,$$
$$H'^2/c = \epsilon_0 \epsilon \, [\, (k^2 \cosh^2 \beta - \sinh^2 \beta) \, cB'^2 - \sinh \beta \, \cosh \beta \, (k^2 - 1) \, E'^3 \,] \, , \qquad (24)$$
$$H'^3/c = \epsilon_0 \epsilon \, [\, (k^2 \cosh^2 \beta - \sinh^2 \beta) \, cB'^3 + \sinh \beta \, \cosh \beta \, (k^2 - 1) \, E'^2 \,] \, .$$

The relations (23)–(24) say that the simple connections (19) which exist between electromagnetic vectors in initial (rest) reference frame after translating to a moving reference frame become rather complex ones: they involve now the velocity as an external parameter. In other words, this means that the field relations (19) are reference frame dependent.

However, we can see that in the vacuum case when $k = 1$, the formulas (23)–(24) will take the same form from which we initially started:

$$k = 1 \, , \quad D'^i = \epsilon_0 \, E'^i \, , \quad H'^i = \frac{1}{\mu_0} \, B^i \, . \qquad (25)$$

Now, we consider the field relations, while using modified Lorentz transformations. We shall easily see that the properties of these relations under the modified Lorentz theory are different and much more attractive: they turn out to be Lorentz invariant. Indeed, let us start with[9]:

$$D^i = \epsilon_0 \epsilon \, E^i \quad \Longrightarrow \quad D'^1 = \epsilon_0 \epsilon \, E'^1 \, ,$$
$$\cosh \sigma \, D'^2 + \sinh \sigma \, \frac{H'^3}{kc} = \epsilon_0 \epsilon \, (\cosh \sigma \, E'^2 + \sinh \sigma \, kcB'^3) \, ,$$
$$\cosh \sigma \, D'^3 - \sinh \sigma \, \frac{H'^2}{kc} = \epsilon_0 \epsilon \, (\cosh \sigma \, E'^3 - \sinh \sigma \, kcB'^2) \, ;$$
$$H^i = \frac{1}{\mu_0 \mu} B^i \quad \Longrightarrow \quad H'^1 = \frac{1}{\mu_0 \mu} B'^1 \, , \quad -\sinh \sigma \, D'^3 + \cosh \sigma \, \frac{H'^2}{kc}$$
$$= \frac{1}{\mu_0 \mu} \frac{1}{k^2 c^2} \, (-\sinh \sigma \, E'^3 + \cosh \sigma \, kcB'^2) \, ,$$
$$\sinh \sigma \, D'^2 + \cosh \sigma \, \frac{H'^3}{kc} = \frac{1}{\mu_0 \mu} \frac{1}{k^2 c^2} \, (\sinh \sigma \, E'^2 + \cosh \sigma \, kcB'^3) \, .$$

They may be rewritten as

$$D'^1 = \epsilon_0 \epsilon \, E'^1 \, , \quad H'^1 = \frac{1}{\mu_0 \mu} B'^1 \, ;$$
$$\cosh \sigma \, D'^2 + \sinh \sigma \, \tfrac{H'^3}{kc} = \epsilon_0 \epsilon \, (\cosh \sigma \, E'^2 + \sinh \sigma \, kcB'^3) \, ,$$
$$\sinh \sigma \, D'^2 + \cosh \sigma \, \tfrac{H'^3}{kc} = \tfrac{1}{\mu_0 \mu} \tfrac{1}{k^2 c^2} \, (\sinh \sigma \, E'^2 + \cosh \sigma \, kcB'^3) \, ;$$
$$\cosh \sigma \, D'^3 - \sinh \sigma \, \tfrac{H'^2}{kc} = \epsilon_0 \epsilon \, (\cosh \sigma \, E'^3 - \sinh \sigma \, kcB'^2) \, ,$$
$$-\sinh \sigma \, D'^3 + \cosh \sigma \, \tfrac{H'^2}{kc} = \tfrac{1}{\mu_0 \mu} \tfrac{1}{c^2} \, (-\sinh \sigma \, E'^3 + \cosh \sigma \, kcB'^2) \, .$$

---
[8]These are the same Minkowski relations only translated to another form.
[9]Everywhere instead of c, there appears kc.

and further with the help of

$$\frac{1}{\mu_0 \mu k^2 c^2} = \frac{\epsilon_0 \epsilon \mu_0 \mu}{\mu_0 \mu} = \epsilon_0 \epsilon \,,$$

they become

$$\cosh \sigma \, D'^2 + \sinh \sigma \, \tfrac{H'^3}{kc} = \epsilon_0 \epsilon \left(\cosh \sigma \, E'^2 + \sinh \sigma \, kcB'^3 \right),$$
$$\sinh \sigma \, D'^2 + \cosh \sigma \, \tfrac{H'^3}{kc} = \epsilon_0 \epsilon \left(\sinh \sigma \, E'^2 + \cosh \sigma \, kcB'^3 \right);$$

$$\cosh \sigma \, D'^3 - \sinh \sigma \, \tfrac{H'^2}{kc} = \epsilon_0 \epsilon \left(\cosh \sigma \, E'^3 - \sinh \sigma \, kcB'^2 \right),$$
$$-\sinh \sigma \, D'^3 + \cosh \sigma \, \tfrac{H'^2}{kc} = \epsilon_0 \epsilon \left(-\sinh \sigma \, E'^3 + \cosh \sigma \, kcB'^2 \right).$$

So we get

$$D'^2 = \epsilon_0 \epsilon E'^2,\; D'^3 = \epsilon_0 \epsilon E'^3,\; H'^3 = \frac{1}{\mu_0 \mu} B'^3,\; H'^2 = \frac{1}{\mu_0 \mu} B'^2. \tag{26}$$

We have arrived at an unexpected and most attractive result: the field equations (19) turn out to be invariant under the *modified Lorentz transformations*. This is a significant theoretical argument in favor of the Lorentz symmetry involving the light velocity in a medium with light velocity kc instead of the light velocity in the vacuum c.

10. Tensor Formalism

The two Maxwell equations with sources

$$\operatorname{div} \mathbf{D} = J^0, \quad \operatorname{rot} \frac{\mathbf{H}}{c} = \frac{\mathbf{J}}{c} + \frac{\partial \mathbf{D}}{\partial ct} \tag{27}$$

can be presented in a very compact and simple form, if one introduces special notations with the use of indices taking over four values:

$$x^a = (x^0 = ct;\; x^i), \quad \partial_a = \frac{\partial}{\partial x^a}, \quad j^a = (J^0, J^i/c),$$

$$(H^{ab}) = \begin{vmatrix} 0 & -D^1 & -D^2 & -D^3 \\ +D^1 & 0 & -H^3/c & +H^2/c \\ +D^1 & +H^3/c & 0 & -H^1/c \\ +D^3 & -H^2/c & +H^1/c & 0 \end{vmatrix}. \tag{28}$$

In the following, we will apply the rule of changing the location (at bottom or at top) of any index-symbol:

$$A^0 = +A_0,\; A^i = -A_i,\, (i = 1, 2, 3),\; \mathbf{A} = (A^1, A^2, A^3) = (-A_1, -A_2, -A_3).$$

We further accept Einstein's convention on assumed summation over any two repeated indexes: the special sign of summing $\sum$ will not be written if two identical indexes are encountered in a formula, e.g., $C^a B_a = C^0 B_0 - \mathbf{C}\,\mathbf{B}$.

Now the main assertion is that Eqs. (27) are equivalent to the tensor equation

$$\partial_b H^{ba} = j^a . \qquad (29)$$

Indeed, we have

$$a = 0 \implies \begin{array}{l} \partial_b H^{b0} = j^0 , \quad \partial_1 H^{10} + \partial_2 H^{20} + \partial_3 H^{30} = j^0 , \\ \partial_1 D^1 + \partial_2 D^2 + \partial_3 D^3 = \rho , \quad \text{div } \mathbf{D} = \rho . \end{array}$$

In the same way, when $a = 1, 2, 3$, we get

$$a = 1 \implies \begin{array}{l} \partial_b H^{b1} = j^1 , \quad \partial_0 H^{01} + \partial_2 H^{21} + \partial_3 H^{31} = j^1 , \\ \partial_2 H^3 - \partial_3 H^2 = \frac{\partial}{\partial t} D^1 + J^1 ; \end{array}$$

and so on.

Now let us consider two remaining Maxwell equations

$$\text{div } c\mathbf{B} = 0 , \quad \text{rot } \mathbf{E} = -\frac{\partial c\mathbf{B}}{\partial ct} . \qquad (30)$$

In order to deal with these equations, Minkowski introduced another tensor F^{ab}:

$$(F^{ab}) = \begin{vmatrix} 0 & -E^1 & -E^2 & -E^3 \\ +E^1 & 0 & -cB^3 & +cB^2 \\ +E^1 & +cB^3 & 0 & -cB^1 \\ +E^3 & -cB^2 & +cB^1 & 0 \end{vmatrix} . \qquad (31)$$

The main assertion here is that Eqs. (30) are equivalent to the tensor equation

$$\partial_c F_{ab} + \partial_a F_{bc} + \partial_b F_{ca} = 0 . \qquad (32)$$

In the left side of (32) we have a 3-index quantity which is skew-symmetric with respect to any pair of indexes, so that in (32) we have only four different equations corresponding to the 4 combinations:

$$(cab) \quad = \quad (123), (012), (023), (013) .$$

From (32) we can derive the following equations:

$$(123), \partial_1 F_{23} + \partial_2 F_{31} + \partial_3 F_{12} = 0 , -\partial_1 B^1 - \partial_2 B^2 - \partial_3 B^3 = 0, \text{ div } \mathbf{B} = 0 ;$$
$$(012), \partial_0 F_{12} + \partial_1 F_{20} + \partial_2 F_{01} = 0 , \partial_1 E^2 - \partial_2 E^1 = -\partial_0 B^3 , (\text{ rot } \mathbf{E})^3 = -(\partial_0 \mathbf{B})^3$$

and so on.

Thus, we have arrived at the compact tensor form of the Maxwell equations:

$$\partial_b H^{ba} = j^a , \quad \partial_c F_{ab} + \partial_a F_{bc} + \partial_b F_{ca} = 0 , \qquad (33)$$

where two electromagnetic tensors are used.

The Maxwell equations in other variables (in which they exhibit symmetry under modified Lorentz transformations), $x^0 = kct$, $j^0 = \rho$, $\mathbf{j} = \mathbf{J}/kc$, $\mathbf{d} = \mathbf{D}$, $\mathbf{h} = \mathbf{H}/kc$, , are

$$\text{div } \mathbf{d} = j^0, \quad \text{rot } \mathbf{h} = \mathbf{j} + \frac{\partial \mathbf{d}}{\partial x^0}, \quad \text{div } \mathbf{h} = 0, \quad \text{rot } \mathbf{d} = -\frac{\partial \mathbf{h}}{\partial x^0},$$

and they may be rewritten with the use of only one electromagnetic tensor[10]:

$$(f^{AB}) = \begin{vmatrix} 0 & -d^1 & -d^2 & -d^3 \\ +d^1 & 0 & -h^3 & +h^2 \\ +d^1 & +h^3 & 0 & -h^1 \\ +d^3 & -h^2 & +h^1 & 0 \end{vmatrix}, \tag{34}$$

in the form of two equations for one tensor

$$\partial_B f^{BA} = j^A, \quad \partial_C f_{AB} + \partial_A f_{BC} + \partial_B f_{CA} = 0. \tag{35}$$

It should be noted that the Maxwell equations, invariant under modified Lorentz transformations, may be rewritten with the help of two tensors as well. Indeed, these equations are presented as

$$\text{div } kc\mathbf{B} = 0, \text{ rot } \mathbf{E} = -\frac{\partial}{\partial kct} kc\mathbf{B},$$

$$\text{div } \mathbf{D} = J^0, \quad \text{rot } \frac{\mathbf{H}}{kc} = \frac{\mathbf{J}}{kc} + \frac{\partial}{\partial kct} \mathbf{D}. \tag{36}$$

From here, by introducing the (modified) electromagnetic tensors

$$(H^{AB}) = (\mathbf{D}, \mathbf{H}/kc), \quad (F^{AB}) = (\mathbf{E}, kc\mathbf{B}) \tag{37}$$

we note that (36) can be readily written as

$$\partial_B H^{BA} = j^A, \quad \partial_C F_{AB} + \partial_A F_{BC} + \partial_B F_{CA} = 0. \tag{38}$$

11. Ordinary and Modified 4-Vectors and 4-Tensors

The main significant property of the notation introduced by Poincaré–Minkowski is that it allows us to determine in detail the correct transformation formulas for any physical quantity. The recipe is easy: the transformation rules for any quantity are determined in a straightforward manner by its tensorial nature.

Consider several examples. Any physical entity with one upper index, first-order tensor, behaves like the coordinate 4-vector x^a:

$$\begin{aligned} x'^0 &= \cosh \beta \, x^0 - \sinh \beta \, x^1, \\ x'^1 &= -\sinh \beta \, x^0 + \cosh \beta \, x^1, \\ x'^2 &= x^2, \quad x'^3 = x^3, \end{aligned}$$

---
[10]Here and in the following, the capital letters as tensor indexes means that such quantities transform in accordance with the modified Lorentz symmetry.

i.e., 1-rank tensor A^a transforms as follows

$$A'^0 = \cosh\beta\ A^0 - \sinh\beta\ A^1,$$
$$A'^1 = -\sinh\beta\ A^0 + \cosh\beta\ A^1,$$
$$A'^2 = A^2, \qquad A'^3 = A^3.$$

The same situation, but in matrix form, looks as $A'^a = L'^a{}_b(\beta)\ A^b$. In accordance with the above assertion, a 2-rank tensor will be translated by the formulas

$$K'^{ab} = L^a{}_m(\beta) L^b{}_n(\beta) K^{mn}.$$

When considering modified Lorentz symmetry, the whole the tensor technique remains the same, except of small alterations, like: $x'^B = L'^B{}_C(\sigma)\ x^C$, etc.

12. Minkowski's Constitutive Relations in Covariant Form

We consider Minkowski's constraints generated by transformation of the reference frame along the axis x:

$$D^i = \epsilon_0 \epsilon\ E^i \implies$$

$$D'^1 = \epsilon_0 \epsilon\ E'^1,$$
$$\cosh\beta\ D'^2 + \sinh\beta\ \frac{H'^3}{c} = \epsilon_0 \epsilon\ (\cosh\beta\ E'^2 + \sinh\beta\ cB'^3),$$
$$\cosh\beta\ D'^3 - \sinh\beta\ \frac{H'^2}{c} = \epsilon_0 \epsilon\ (\cosh\beta\ E'^3 - \sinh\beta\ cB'^2);$$

$$H^i = \frac{1}{\mu_0 \mu}\ B^i \implies$$

$$H'^1 = \frac{1}{\mu_0 \mu}\ B^1,$$
$$-\sinh\beta\ D'^3 + \cosh\beta\ \frac{H'^2}{c} = \frac{1}{\mu_0 \mu}\frac{1}{c^2}(-\sinh\beta\ E'^3 + \cosh\beta\ cB'^2),$$
$$\sinh\beta\ D'^2 + \cosh\beta\ \frac{H'^3}{c} = \frac{1}{\mu_0 \mu}\frac{1}{c^2}(\sinh\beta\ E'^2 + \cosh\beta\ cB'^3),$$

The trick which we shall use below is simple but useful and often applicable. It is based on the following property of the tensor formalism: if we think (know) that a certain physical equation must be Lorentz invariant and an explicit form of the equation is given only in some particular reference frame then its invariant form may be found with the help of Lorentz transformations. The same may be achieved if we can derive from the particular equation its general tensor form.

One special notion, 4-vector of velocity, is needed for the following. It may be introduced using the following simple formal considerations. Let a material particle move steadily in some inertial reference frame:

$$x^0 = ct, \qquad x^i = v^i\ t.$$

Since x^a is a 4-vector, its differential dx^a will be a 4-vector too; moreover, we have a scalar quantity (which is called an *interval*) with respect to the Lorentz group transformations:

$$s^2 = (x^0)^2 - (x^1)^2 - (x^2)^2 - (x^3)^2,$$

$$ds^2 = (dx^0)^2 - dl^2 \,, \quad ds = c\,dt\,\sqrt{1 - \frac{v^2}{c^2}}\,.$$

Dividing dx^a by ds, we get the velocity 4-vector:

$$u^a = \frac{dx^a}{ds} = \frac{(cdt, dx^1, dx^2, dx^3)}{cdt\,\sqrt{1 - v^2/c^2}} = \left(\frac{1}{\sqrt{1 - v^2/c^2}}, \frac{v^i/c}{\sqrt{1 - v^2/c^2}}\right)\,.$$

Now will show that we are ready to obtain the tensor form of Minkowski equations. To this end, we take a particular 4-velocity vector:

$$u^a = \left(\frac{1}{\sqrt{1 - v^2/c^2}}, \frac{-v/c}{\sqrt{1 - v^2/c^2}}, 0, 0\right) = (\cosh\beta, -\sinh\beta, 0, 0)\,. \tag{39}$$

We will show that the first tensor relation we need is [1]

$$H^{ab} u_b = \epsilon_0 \epsilon\, F^{ab} u_b\,. \tag{40}$$

When u^a is given by (39), relations (40) will take the form

$a = 0,\ H^{01} u_1 = \epsilon_0 \epsilon\, F^{01} u_1,\ D^1 \sinh\beta = \epsilon_0 \epsilon \sinh\beta E^1,\ D^1 = \epsilon_0 \epsilon E^1;$

$a = 1,\ H^{10} u_0 = \epsilon_0 \epsilon F^{10} u_0,\ -D^1 \cosh\beta = -\epsilon_0 \epsilon \cosh\beta E^1,\ D^1 = \epsilon_0 \epsilon E^1;$

$a = 2,\ H^{20} u_0 + H^{21} u_1 = \epsilon_0 \epsilon (F^{20} u_0 + F^{21} u_1),$
$\quad D^2 \cosh\beta + \frac{H^3}{c} \sinh\beta = \epsilon_0 \epsilon (E^2 \cosh\beta + cB^3 \sinh\beta);$

$a = 3,\ H^{30} u_0 + H^{31} u_1 = \epsilon_0 \epsilon (F^{30} u_0 + F^{31} u_1),$
$\quad D^3 \cosh\beta - \frac{H^2}{c} \sinh\beta = \epsilon_0 \epsilon (E^3 \cosh\beta - cB^2 \sinh\beta).$

Therefore, the tensor relation (40) is equivalent to (20). The second tensor relation we need is [1]

$$(H^{ab} u^c + H^{bc} u^a + H^{ca} u^b) = \frac{1}{c^2 \mu \mu_0} (F^{ab} u^c + F^{bc} u^a + F^{ca} u^b)\,. \tag{41}$$

With u^a as in (39), relation (41) gives

$$(123)\,,\quad (H^{12} u^3 + H^{23} u^1 + H^{31} u^2)$$

$$= \frac{1}{c^2 \mu \mu_0} (F^{12} u^3 + F^{23} u^1 + F^{31} u^2),\quad H^1 = \frac{1}{\mu \mu_0} B^1\,;$$

$$(023)\,,\quad (H^{02} u^3 + H^{23} u^0 + H^{30} u^2)$$

$$= \frac{1}{c^2 \mu \mu_0} (F^{02} u^3 + F^{23} u^0 + F^{30} u^2)\,,\quad H^1 = \frac{1}{\mu \mu_0} B^1\,;$$

$$(012)\,,\quad (H^{01} u^2 + H^{12} u^0 + H^{20} u^1)$$

$$= \frac{1}{c^2 \mu \mu_0} (F^{01} u^2 + F^{12} u^0 + F^{20} u^1)\,,\,(\frac{H^3}{c} \cosh\beta + D^2 \sinh\beta)$$

$$= \frac{1}{c^2\mu\mu_0} (cB^3 \cosh\beta + E^2 \sinh\beta) ;$$

$$\underline{(013)}, \quad (H^{01}u^3 + H^{13}u^0 + H^{30}u^1) = \frac{1}{c^2\mu\mu_0} (F^{01}u^3 + F^{13}u^0 + F^{30}u^1),$$

$$(\frac{H^2}{c} \cosh\beta - D^3 \sinh\beta) = \frac{1}{c^2\mu\mu_0} (cB^2 \cosh\beta - E^3 \sinh\beta),$$

and these coincide with the above formulas (21). Thus, all six Minkowski equations are equivalent to the tensor ones:

$$H^{ab}u_b = \epsilon_0\epsilon F^{ab}u_b , \tag{42}$$

$$H^{ab}u^c + H^{bc}u^a + H^{ca}u^b = \frac{1}{c^2\mu\mu_0} (F^{ab}u^c + F^{bc}u^a + F^{ca}u^b) . \tag{43}$$

Although we have verified them only with the use of a special Lorentz transformation, but the tensor form itself guarantees that they will be correct for any arbitrary Lorentz one.

In the vacuum case, when $\epsilon = 1, \mu = 1$, the equations (42) and (43) may be rewritten differently

$$H^{ab}u_b = \epsilon_0 F^{ab}u_b , \tag{44}$$

$$H^{ab}u^c + H^{bc}u^a + H^{ca}u^b = \epsilon_0(F^{ab}u^c + F^{bc}u^a + F^{ca}u^b) . \tag{45}$$

These equations admit a simple solution. Indeed, let us multiply (45) by u_c (considering $u^c u_c = +1$). Then

$$H^{ab} = \epsilon_0(F^{ab} + F^{bc}u_c u^a + F^{ca}u_c u^b) - H^{bc}u_c u^a - H^{ca}u_c u^b ;$$

whence, having in view (44), we infer

$$H^{ab} = \epsilon_0 F^{ab} . \tag{46}$$

which in component form is expressed by six relations (25):

$$D^i = \epsilon_0 E^i , \qquad H^i = \frac{1}{\mu_0} B^i . \tag{47}$$

However, in the case of the medium, similar calculation leads to a very different result. Indeed, let us multiply (43) by u_c:

$$H^{ab} + H^{bc}u_c u^a + H^{ca}u_c u^b = \frac{1}{c^2\mu\mu_0} (F^{ab} + F^{bc}u_c u^a + F^{ca}u_c u^b) .$$

From this, taking (42), we get

$$H^{ab} = \epsilon_0\epsilon k^2 F^{ab} + \epsilon_0\epsilon (k^2 - 1) \left(F^{bc}u_c u^a - F^{ac}u_c u^b \right) . \tag{48}$$

Take notice that while using the Maxwell theory with only one (modified) tensor f^{AB}, no additional condition between electromagnetic tensors is needed at all.

Some addition clarifying analysis can be easily done. While using the modified Minkowski relations (c is changed by kc):

$$D'^1 = \epsilon_0\epsilon E'^1,$$
$$\cosh\sigma D'^2 + \sinh\sigma \frac{H'^3}{kc} = \epsilon_0\epsilon(\cosh\sigma E'^2 + \sinh\sigma kcB'^3), \quad (49)$$
$$\cosh\sigma D'^3 - \sinh\sigma \frac{H'^2}{kc} = \epsilon_0\epsilon(\cosh\sigma E'^3 - \sinh\sigma kcB'^2),$$

$$H'^1 = \frac{1}{\mu_0\mu} B^1,$$
$$-\sinh\sigma D'^3 + \cosh\sigma \frac{H'^2}{kc} = \frac{1}{\sigma_0\mu}\frac{1}{k^2c^2}(-\sinh\sigma E'^3 + \cosh\sigma kcB'^2), \quad (50)$$
$$\sinh\sigma D'^2 + \cosh\sigma \frac{H'^3}{kc} = \frac{1}{\mu_0\mu}\frac{1}{k^2c^2}(\sinh\sigma E'^2 + \cosh\sigma kcB'^3).$$

Now a modified 4-velocity U^A is needed:

$$U^A = \frac{dx^a}{ds} = \frac{(kcdt, dx^1, dx^2, dx^3)}{kcdt\sqrt{1 - v^2/k^2c^2}} = \left(\frac{1}{\sqrt{1 - v^2/k^2c^2}}, \frac{v^i/kc}{\sqrt{1 - v^2/k^2c^2}}\right),$$

and its particular form

$$U^A = \left(\frac{1}{\sqrt{1 - v^2/k^2c^2}}, \frac{-v/kc}{\sqrt{1 - v^2/k^2c^2}}, 0, 0\right) = (\cosh\sigma, -\sinh\sigma, 0, 0).$$

According to (37), the tensor representation of (49) and (50) is

$$H^{AB}U_B = \epsilon_0\epsilon F^{AB}U_B,$$
$$H^{AB}U^C + H^{BC}U^A + H^{CA}U^B \quad (51)$$
$$= \frac{1}{k^2c^2\mu_0}(F^{AB}U^C + F^{BC}U^A + F^{CA}U^B).$$

The latter equation may be rewritten as

$$H^{AB}U^C + H^{BC}U^A + H^{CA}U^B = \epsilon_0\epsilon\left(F^{AB}U^C + F^{BC}U^A + F^{CA}U^B\right).$$

Multiplying it by U_C:

$$H^{AB} + H^{BC}U_CU^A + H^{CA}U_CU^B = \epsilon_0\epsilon\left(F^{AB} + F^{BC}U_CU^A + F^{CA}U_CU^B\right),$$

whence with (51) it follows

$$H^{AB} = \epsilon_0\epsilon\, F^{AB}. \quad (52)$$

which, in components, looks as claimed

$$D^i = \epsilon_0\epsilon\, E^i, \quad H^i = \frac{1}{\mu_0\mu} B^i.$$

In our opinion, there is no ground for feeling enthusiastic about the existence of the relativistically invariant formulas (42) and (43):

$$H^{ab}u_b = \epsilon_0 \epsilon F^{ab} u_b \,,$$
$$H^{ab}u^c + H^{bc}u^a + H^{ca}u^b = \frac{1}{c^2\mu\mu_0}(F^{ab}u^c + F^{bc}u^a + F^{ca}u^b)\,.$$

These formulas involve the 4-velocity vector, which characterizes the motion of the medium under the inertial reference frame. In other words, these formulas show an explicit dependence of the basic electrodynamic equations on the absolute velocity of a moving body. This contradicts the initial principles and the main claims of Special Relativity Theory.

13. Potentials in a Medium

In remains to see which peculiarities arise from the presence of a uniform medium, when we try do describe electromagnetic fields in terms of the scalar and vector potentials φ, $\mathbf{A}$:

$$\{\, \mathbf{E},\, \mathbf{B},\, \mathbf{D},\, \mathbf{H}\, \} \implies \{\, \varphi,\, \mathbf{A}\, \}\,.$$

We might anticipate some difficulties because the Maxwell theory in the medium exhibits symmetry under ordinary (the vacuum light velocity based) Lorentz transformations, only if the two electromagnetic tensors, H^{ab} and F^{ab}, are used.

Let us write down the Maxwell equations again

$$\nabla \bullet \mathbf{B} = 0\,, \qquad \nabla \times \mathbf{E} = -\frac{\partial \mathbf{B}}{\partial t}\,, \tag{53}$$

$$\nabla \bullet \mathbf{E} = \frac{1}{\epsilon\epsilon_0}\rho\,, \qquad \frac{1}{\mu\mu_0}\nabla \times \mathbf{B} = \mathbf{J} + \epsilon\epsilon_0 \frac{\partial \mathbf{E}}{\partial t}\,. \tag{54}$$

The most general substitution for potentials φ, $\mathbf{A}$ which transforms the first two equations (53) into identities, has the form

$$\mathbf{B} = d\,\nabla \times \mathbf{A}\,, \qquad \mathbf{E} = -n\,\nabla\varphi - d\,\frac{\partial \mathbf{A}}{\partial t}\,, \tag{55}$$

where d, n are some yet unknown parameters. The first equation in (54) yields

$$(-n\nabla^2\varphi + dm\frac{\partial^2\varphi}{\partial t^2}) = \frac{1}{\epsilon\epsilon_0}\rho + d\frac{\partial}{\partial t}(\nabla \bullet \mathbf{A} + m\frac{\partial\varphi}{\partial t})\,,$$

or

$$(-\nabla^2\varphi + \frac{dm}{n}\frac{\partial^2\varphi}{\partial t^2}) = \frac{1}{n\epsilon\epsilon_0}\rho + \frac{d}{n}\frac{\partial}{\partial t}(\nabla \bullet \mathbf{A} + m\frac{\partial\varphi}{\partial t})\,, \tag{56}$$

The second equations in (54) leads to

$$\frac{d}{\mu\mu_0}\nabla \times (\nabla \times \mathbf{A}) = \mathbf{J} + \epsilon\epsilon_0 \frac{\partial}{\partial t}(-n\nabla\varphi - d\frac{\partial \mathbf{A}}{\partial t})\,,$$

or
$$\nabla \times (\nabla \times \mathbf{A}) = \frac{\mu \mu_0}{d} \mathbf{J} + \frac{\epsilon \epsilon_0 \mu_0 \mu}{d} \frac{\partial}{\partial t} (-n \nabla \varphi - d \frac{\partial \mathbf{A}}{\partial t}).$$

From this, with the help of the identity
$$\nabla \times (\nabla \times \mathbf{A}) = -\nabla^2 \mathbf{A} + \nabla (\nabla \bullet \mathbf{A}),$$

we get
$$-\nabla^2 \mathbf{A} + \nabla (\nabla \bullet \mathbf{A}) = \frac{\mu \mu_0}{d} \mathbf{J} + \frac{\epsilon_0 \epsilon \mu_0 \mu}{d} \frac{\partial}{\partial t} (-n \nabla \varphi - d \frac{\partial \mathbf{A}}{\partial t}),$$

or
$$(-\nabla^2 \mathbf{A} + \epsilon_0 \epsilon \mu_0 \mu \frac{\partial^2}{\partial t^2} \mathbf{A}) = \frac{\mu \mu_0}{d} \mathbf{J} - \nabla (\nabla \bullet \mathbf{A} + \frac{\epsilon_0 \epsilon \mu_0 \mu\, n}{d} \frac{\partial \varphi}{\partial t}). \qquad (57)$$

Comparing (56) and (57), we see that it suffices to impose the constraints
$$\frac{md}{n} = \epsilon_0 \epsilon \mu_0 \mu \quad \Longrightarrow \quad m = \frac{\epsilon_0 \epsilon \mu_0 \mu\, n}{d},$$

so that (56) and (57) will have quite symmetrical form with the same wave operator on the left:
$$(-\nabla^2 \varphi + \epsilon_0 \epsilon \mu_0 \mu \frac{\partial^2 \varphi}{\partial t^2}) = \frac{1}{n \epsilon \epsilon_0} \rho + \frac{d}{n} \frac{\partial}{\partial t}(\nabla \bullet \mathbf{A} + m \frac{\partial \varphi}{\partial t}),$$
$$(-\nabla^2 \mathbf{A} + \epsilon_0 \epsilon \mu_0 \mu \frac{\partial^2 \mathbf{A}}{\partial t^2}) = \frac{\mu \mu_0}{d} \mathbf{J} - \nabla(\nabla \bullet \mathbf{A} + m \frac{\partial \varphi}{\partial t}).$$

With the use of the constants c and k, defined by:
$$c^2 = \frac{1}{\epsilon_0 \mu_0}, \quad k^2 = \frac{1}{\epsilon \mu},$$

the previous equations become
$$\frac{1}{k^2 c^2} = \frac{dm}{n}, \qquad (58)$$
$$(-\nabla^2 \varphi + \frac{1}{k^2 c^2} \frac{\partial^2 \varphi}{\partial t^2}) = \frac{1}{n \epsilon \epsilon_0} \rho + \frac{d}{n} \frac{\partial}{\partial t}(\nabla \bullet \mathbf{A} + m \frac{\partial \varphi}{\partial t}), \qquad (59)$$
$$(-\nabla^2 \mathbf{A} + \frac{1}{k^2 c^2} \frac{\partial^2 \mathbf{A}}{\partial t^2}) = \frac{\mu \mu_0}{d} \mathbf{J} - \nabla(\nabla \bullet \mathbf{A} + m \frac{\partial \varphi}{\partial t}). \qquad (60)$$

The relation (58) admits several distinct solutions. The most symmetrical and simplifying all formulas substitution seems to be
$$m = \frac{1}{kc}, \ n = \frac{1}{\epsilon_0 \epsilon}, \ \text{whence } d = \frac{n}{kc} = \frac{1}{kc\, \epsilon_0 \epsilon}, \ \frac{\mu \mu_0}{d} = \mu \mu_0 \epsilon \epsilon_0\, kc = \frac{1}{kc}.$$

Then (59) and (60) lead to

$$\begin{aligned}(-\nabla^2\varphi + \frac{1}{k^2c^2}\frac{\partial^2\varphi}{\partial t^2}) &= \rho + \frac{1}{kc}\frac{\partial}{\partial t}(\nabla\bullet\mathbf{A} + \frac{1}{kc}\frac{\partial\varphi}{\partial t}),\\ (-\nabla^2\mathbf{A} + \frac{1}{k^2c^2}\frac{\partial^2\mathbf{A}}{\partial t^2}) &= \frac{\mathbf{J}}{kc} - \nabla(\nabla\bullet\mathbf{A} + \frac{1}{kc}\frac{\partial\varphi}{\partial t}).\end{aligned} \quad (61)$$

Moreover, (61) may be rewritten as a (modified) tensor equation:

$$\partial^B\partial_B A^C = j^C + \partial^C(\partial_B A^B), \quad (62)$$

$$x^C = (kct, x^i),\ A^C = (\varphi, A^i),\ j^C = (\rho, \frac{J^i}{kc}). \quad (63)$$

Then (62) evidently shows its invariance under modified Lorentz transformations constructed on the base of the light velocity kc in the medium. The initial relations (55) (they introduced the electromagnetic potentials) lead to the following relations

$$\mathbf{B} = d\,\nabla\times\mathbf{A} = \frac{1}{\epsilon\epsilon_0 kc}\nabla\times\mathbf{A} \implies \nabla\times\mathbf{A} = \frac{\mathbf{B}}{\mu_0\mu\ kc} = \mathbf{h},$$

$$\mathbf{E} = \frac{1}{\epsilon_0\epsilon}\nabla\varphi - \frac{1}{\epsilon\epsilon_0 kc}\frac{\partial\mathbf{A}}{\partial t} \implies -\nabla\varphi - \frac{1}{kc}\frac{\partial\mathbf{A}}{\partial t} = \epsilon_0\epsilon\,\mathbf{E} = \mathbf{d},$$

they may be readily translated to tensor form

$$f_{BC} = \partial_B A_C - \partial_C A_B, \quad (64)$$

where f_{ab} is the electromagnetic tensor for Maxwell equations in modified variables:

$$x^0 = kct,\ j^0 = \rho,\ \mathbf{j} = \frac{\mathbf{J}}{kc},\ \mathbf{d} = \mathbf{D},\ \mathbf{h} = \frac{\mathbf{H}}{kc},\ f^{AB} = (\mathbf{d}, \mathbf{h}).$$

The tensorial relation (64) easily reveals a gauge freedom in determining of electromagnetic potentials:

$$A'_B = A_B + \partial_B\Lambda \implies f'_{BC} = f_{BC}.$$

Often the Lorentz gauge condition $\partial_B A^B = 0$ is taken to be the most convenient.

Thus, as a result of the change of variables:

$$(\mathbf{E}, \mathbf{D}, \mathbf{B}, \mathbf{H}) \implies (\varphi, \mathbf{A}) = A^C,$$

simplicity has been achieved: all the Maxwell electrodynamics is formally equivalent to one single equation for 1-order tensor A^B.

In the Lorentz gauge, the Maxwell electrodynamics looks most simple and beautiful: namely it reduces to the wave equation:

$$\partial^B\partial_B\ A^C = j^C,\quad \partial_B A^B = 0. \quad (65)$$

One should note that to solutions of Eq. (65), there correspond wave processes propagating with the velocity of the light in the medium, not in the vacuum.

14. Potentials in a Medium, Ordinary Treatment

Now we consider an alternative way of introducing potentials into electrodynamics, in presence of a medium which has its origin in earlier investigations by Minkowski[11].

The most noticeable feature of this method consists in the following: in this approach we are able to formulate the Maxwell electrodynamics in a medium in terms of potentials with the use of the ordinary Lorentz symmetry based on the vacuum light velocity c only. Concurrently, the mathematical equations achieved look more complicated, and also these equations explicitly involve the velocity of the medium under the reference frame. The latter might be considered as return to prehistory of Special Relativity with all the search of some absolute velocities. Though, we are not going to be submerged in such metaphysical subtleties.

Nevertheless, one issue should be emphasized: there exist two alternative ways to develop potential approach for electrodynamics in medium – one developed in the previous Section, and another exposed below. The two ways are completely equivalent in mathematical sense. The first is much more simpler technically but it presumes invariance under modified Lorentz symmetry based on the light velocity kc. There arises the question, if the more complicated technique is more likely to be used only because of its concomitant treatment of ordinary Lorentz symmetry.

The Minkowski's constitutive relations in covariant tensor form are

$$H^{ab}u_b = \epsilon_0 \epsilon F^{ab} u_b \;,$$

$$H^{ab}u^c + H^{bc}u^a + H^{ca}u^b = \frac{F^{ab}u^c + F^{bc}u^a + F^{ca}u^b}{c^2 \mu \mu_0} \;.$$

It was shown in (48) that these are equivalent to

$$H^{ab} = \epsilon_0 \epsilon k^2 \, F^{ab} + \epsilon_0 \epsilon \, (k^2 - 1) \, (\, F^{bc} u_c u^a - F^{ac} u_c u^b \,) \;,$$

or

$$H_{ab} = [\, \epsilon_0 \epsilon k^2 \, g_{am} g_{bn} + \epsilon_0 \epsilon \, (k^2 - 1) \, u_n (g_{bm} u_a - g_{am} u_b) \,] \, F^{mn} \;, \tag{66}$$

which can be rewritten as:

$$H_{ab} = \Delta_{abmn} F^{mn} \;, \tag{67}$$

$$\Delta_{abmn} = \epsilon_0 \epsilon k^2 g_{am} g_{bn} + \epsilon_0 \epsilon (k^2 - 1) u_n (g_{bm} u_a - g_{am} u_b) \;. \tag{68}$$

Firstly, the 4-order tensor connecting H^{ab} and F^{ab} was introduced (for a more general case of anisotropic medium) by Tamm and Mandel'stam [5, 13]. For a uniform medium, according to Watson-Yauch-Riazanov [3, 10] the tensor Δ_{abmn} (68) may be taken in another form, leading to $H_{ab} = \Delta_{abmn} F^{mn}$, where

$$\Delta_{abmn} = A \, (g_{am} + B u_a u_m)(g_{bn} + B u_b u_n) \;. \tag{69}$$

---

[11] Just this variant is being used mainly; see for instance the thorough review [1].

Although this Δ_{abmn} contains a term of fourth order in velocity, only terms of second order give non-zero contributions into the formula (69). Let us demonstrate that (68) and (69) are the same in different forms for certain given A and B. From (69) it follows

$$H_{ab} = A\, g_{am}\, g_{bn}\, F^{mn} + AB\, g_{am} B u_b u_n\, F^{mn} + AB\, u_a u_m g_{bn}\, F^{mn}$$
$$= A\, g_{am}\, g_{bn}\, F^{mn} + AB\, u_n\, (g_{am} u_b - g_{bm} u_a)\, F^{mn}. \qquad (70)$$

Comparing this with (68), we get

$$A = \epsilon_0 \epsilon k^2, \quad -AB = \epsilon_0 \epsilon (k^2 - 1),$$

whence it follows

$$A = \epsilon_0 \epsilon k^2, \quad B = \frac{1 - k^2}{k^2} = \epsilon\mu - 1.$$

Therefore, equations (69) become $H_{ab} = \Delta_{abmn}\, F^{mn}$, where

$$\Delta_{abmn} = \epsilon_0 \epsilon k^2 [\, g_{am} + (\epsilon\mu - 1)\, u_a u_m\,][\, g_{bn} + (\epsilon\mu - 1)\, u_b u_n\,]. \qquad (71)$$

This very representation for 4-rank tensor relating H^{ab} to F^{ab} in a uniform medium is given in the review [1].

Now we are ready to introduce 4-potentials. The Maxwell equations are

$$\partial_a\, H^{ab} = j^b \implies \partial_a\, (\Delta^{abmn} F_{mn}) = j^b, \qquad (72)$$
$$\partial_c\, F_{ab} + \partial_a\, F_{bc} + \partial_b\, F_{ca} = 0. \qquad (73)$$

The potentials A_b are defined in such a way that equations (73) turn into identities:

$$F_{ab} = \partial_a A_b - \partial_b A_a.$$

With the use of (66), the tensor H^{ab} may be rewritten as

$$H^{ab} = \epsilon_0 \epsilon k^2\, (\partial^a A^b - \partial^b A^a) + \epsilon_0 \epsilon\, (k^2 - 1)(\, u^a \partial^b - u^b \partial^a\,)\, (u^n A_n).$$

Therefore, (72) leads us to

$$\epsilon_0 \epsilon k^2\, \partial_a(\partial^a A^b - \partial^b A^a) + \epsilon_0 \epsilon (k^2 - 1) \partial_a (u^a \partial^b - u^b \partial^a)(u^n A_n) = j^b. \qquad (74)$$

This is the main equation for electromagnetic 4-potentials in a medium, invariant under the ordinary Lorentz transformations. For purely vacuum case, the factor $(k^2 - 1)$ equals to zero and (74) takes the more familiar form

$$\partial_a \partial^a A^b - \partial^b (\partial_a A^a) = \mu_0\, j^b. \qquad (75)$$

So when we take into account the presence of a medium, the equation (75) extends to (74).

15. Dirac Equation in the Uniform Medium

It is well-known the possibility to simulate some media in electrodynamics with the help of Riemannian geometry (see the recent consideration and bibliography in [7]). In this context, let us consider one special form of the metrical tensor:

$$g_{\alpha\beta} = \begin{vmatrix} a^2 & 0 & 0 & 0 \\ 0 & -b^2 & 0 & 0 \\ 0 & 0 & -b^2 & 0 \\ 0 & 0 & 0 & -b^2 \end{vmatrix}, \quad g^{\alpha\beta} = \begin{vmatrix} a^{-2} & 0 & 0 & 0 \\ 0 & -b^{-2} & 0 & 0 \\ 0 & 0 & -b^{-2} & 0 \\ 0 & 0 & 0 & -b^{-2} \end{vmatrix}, \quad (76)$$

where a^2 and b^2 are arbitrary (positive) numerical parameters. The constitutive equations generated by this geometry are

$$D^i = \epsilon^{ik} E_k,$$

$$(\epsilon^{ik}) = \epsilon_0 \sqrt{-g}\, g^{00} \begin{vmatrix} g^{11} & 0 & 0 \\ 0 & g^{22} & 0 \\ 0 & 0 & g^{33} \end{vmatrix} = \epsilon_0 \frac{b}{a} \begin{vmatrix} -1 & 0 & 0 \\ 0 & -1 & 0 \\ 0 & 0 & -1 \end{vmatrix},$$

$$H^i = \mu^{ik} B_k,$$

$$(\mu^{ik}) = \frac{1}{\mu_0} \sqrt{-g} \begin{vmatrix} g^{22}g^{33} & 0 & 0 \\ 0 & g^{33}g^{11} & 0 \\ 0 & 0 & g^{11}g^{22} \end{vmatrix} = \frac{1}{\mu_0} \frac{a}{b} \begin{vmatrix} 1 & 0 & 0 \\ 0 & 1 & 0 \\ 0 & 0 & 1 \end{vmatrix},$$

or differently

$$D^i = -\epsilon_0 \frac{b}{a} E_i, \quad H^i = \frac{1}{\mu_0} \frac{a}{b} B_i. \quad (77)$$

The Maxwell equations

$$\text{div } \mathbf{B} = 0, \quad \text{rot } \mathbf{E} = -\frac{\partial \mathbf{B}}{\partial t}, \quad \text{div } \mathbf{D} = \rho, \quad \text{rot } \mathbf{D} = \mathbf{J} + \frac{\partial \mathbf{D}}{\partial t},$$

in such a geometry induced medium should be studied with additional constraints

$$\mathbf{D} = \epsilon_0 \frac{b}{a} \mathbf{E}, \quad \mathbf{H} = \frac{1}{\mu_0} \frac{a}{b} \mathbf{B},$$

from which it follows $\epsilon = b/a$, $\mu = b/a$. The corresponding metric tensor (76) is

$$g_{\alpha\beta}(x) = b^2 \begin{vmatrix} k^2 & 0 & 0 & 0 \\ 0 & -1 & 0 & 0 \\ 0 & 0 & -1 & 0 \\ 0 & 0 & 0 & -1 \end{vmatrix}, \quad k = \frac{1}{\sqrt{\epsilon\mu}} = \frac{a}{b}. \quad (78)$$

For simplicity, we further assume that

$$b = 1, \quad \epsilon = \mu = \frac{1}{a} = k, \quad g^{\alpha\beta} = \begin{vmatrix} k^2 & 0 & 0 & 0 \\ 0 & -1 & 0 & 0 \\ 0 & 0 & -1 & 0 \\ 0 & 0 & 0 & -1 \end{vmatrix}. \quad (79)$$

Let us specify a generally covariant Dirac equation in space-time determined by the metrics (78)–(79), assuming that in such a way we obtain the description of the Dirac particle in the uniform medium. We start with the general covariant form of the Dirac equation [8]:

$$\left[i\gamma^a \left(e_{(a)}^\alpha \frac{\partial}{\partial x^\alpha} + \frac{1}{2}(\frac{1}{\sqrt{-g}}\frac{\partial}{\partial x^\alpha}\sqrt{-g}e_{(a)}^\alpha)\right) - \frac{mc}{\hbar}\right]\Psi(x) = 0. \qquad (80)$$

with the tetrad

$$e_{(a)}^\beta = \begin{vmatrix} k^{-1} & 0 & 0 & 0 \\ 0 & 1 & 0 & 0 \\ 0 & 0 & 1 & 0 \\ 0 & 0 & 0 & 1 \end{vmatrix};$$

and then (80) takes the form

$$\left(i\gamma^0 \frac{1}{kc}\frac{\partial}{\partial t} + i\gamma^j \frac{\partial}{\partial x^j} - \frac{mc}{\hbar}\right)\Psi(x) = 0.$$

The Klein–Fock–Gordon equation related with the metrics (78)–(79) has the form

$$\left(\frac{1}{k^2 c^2}\frac{\partial^2}{\partial t^2} - \nabla^2 + \frac{m^2 c^2}{\hbar^2}\right)\Phi = 0.$$

This approach to scalar and spinor fields, equally as the electromagnetic field, is consistent with assumption on the Poincaré–Einstein clock synchronization in uniform media with the help of real light signals influenced by the medium, which leads us to modified Lorentz symmetry.

References

[1] B. M. Bolotowskij, C. N. Stoliarov, *Contamporain state of electrodynamics of moving medias (unlimited media)*, Einstein collection, 1974; Nauka, Moscow, 1976. 179–275.

[2] A. Einstein, Zur Elektrodynamik bewegter Körper [Electrodynamics of moving bodies], *Ann. Phys.* **17** (1905), 892-921; Relativitätsprinzip und die aus demselben gezogenen [Relativity principle and drawn from the same], *Folgerungen Jahrbuch der Radioaktivität* **4** (1907), 411-462.

[3] J. M. Jauch, K. M. Watson, Phenomenological Quantum Electrodynamics. Part II. Interaction of the Field with Charges, *Phys. Rev.* **74** (1948), 950-957, 1485-1494.

[4] H. A. Lorentz, Elektromagnetische verschijnselen in een stelsel dat zich met willekeurige snelheid, kleiner dan die van het licht, beweegt [Electromagnetic phenomena in a system that occurs with random speed, less than that of the light, moves] (in tijdschrift: *Verhandelingen der Koninklijke* Akademie van wetenschappen, Afdeling Natuurkunde, 2-de sectie, vol. 12, Amsterdam 1904.

[5] L. I. Mandel'stam, I. E. Tamm, Elektrodynamik der anisotropen Medien und der speziallen Relativitatstheorie [Electrodynamics of anisotropic media and special relativity theory], *Math. Annalen* **95** (1925), 154–160.

[6] H. Minkowski, Die Grundgleichungen fr die elektromagnetischen Vorgnge in bewegten Krpern, [The basic equations for the electromagnetic processes in moving bodies], *Nachrichten von der Königlichen Gesellschaft der Wissenschaften zu Göttingen, mathematisch-physikalische Klasse II Math. Phys., Kl.* **2** (1908), 53-111; see also reprint in *Math. Ann.* **68** (1910), 472-525; *Raum und Zeit*, Jahresbericht der Deutschen Mathematiker-Vereinigung, **18** (1909), 75–88; *Raum und Zeit, Phys. Zeit* **10** (1909), 104–111; Das Relativitätsprinzip [The relativity principle], *Annalen der Physik* **47** (1915), 927–938.

[7] E. M. Ovsiyuk, V. V. Kisel, V. M. Red'kov, *Maxwell Electrodynamics and boson fields in spaces of constant curvature*, Nova Science Publishers Inc., New York, 2014.

[8] O. V. Veko, E. M. Ovsiyuk, A. Oana, M. Neagu, V. Balan, V. M. Red'kov, *Spinor Structures in Geometry and Physics*, Nova Science Publishers Inc., New York, 2014–2015, in press.

[9] H. Poincaré, Sur la dynamique de l'éelectron, *Comptes Rendus Acad. Sci.* **140** (1905), 1504-1508; H. Poincaré, Sur la dynamique de l'éelectron, *Rendiconti del Circolo Matematico di Palermo* **21** (1906), 129-176.

[10] M. I. Ryazanov, *J.E.T.P.* **32** (1957), 1244-1246.

[11] J. A. Stratton, *Electromagnetic Theory*, New York, McGraw-Hill, 1941.

[12] H. A. Lorentz, De l'influence du mouvement de la terre sur les phénom ènes lumineux [From the influence of the movement of the earth on luminous phenomena], *Archives Néerlandaises des Sciences Exactes et Naturelles* **21** (1886), 103–176; La théorie électromagnétique de Maxwell et son application aux corps mouvants [The electromagnetic theory of Maxwell and its application to moving bodies], *Archives néerlandaises des sciences exactes et naturelles* **25** (1892), 363–552; Versuch einen Theorie der elektrischen und optischen Erscheinungen in bewegten Körpern Leiden, Brill, 1895 [Simplified theory of electrical and optical phenomena in moving systems], *Proceedings of the Section of Sciences, Koninklijke Akademie van Wetenschappen te Amsterdam*. **1** (1889), 427–442; Electromagnetic phenomena in a system moving with any velocity less than that of light, *Proceedings of the Section of Sciences, Koninklijke Akademie van Wetenschappen te Amsterdam*, **6** (1904), 809–831.

[13] I. E. Tamm, Electrodynamics of an anisotropic medium and the Special Theory of Relativity (in Russian), *Zh. Russ. Fiz.-Khim. O-va Chast. Fiz.* **56**, 2-3 (1924), 248–262; Crystal optics in the theory of relativity and its relationship to the geometry of a biquadratic form, (in Russian), *Zh. Russ. Fiz.-Khim. O-va Chast. Fiz.* **57**, 3-4 (1925), 209–240.

In: An Essential Guide to Electrodynamics
Editor: Norma Brewer
ISBN: 978-1-53615-705-5
© 2019 Nova Science Publishers, Inc.

Chapter 4

HIDDEN ASPECTS OF THE ELECTROMAGNETIC FIELD

Ivanhoe B. Pestov[*]
Bogoliubov Laboratory of Theoretical Physics,
Joint Institute for Nuclear Research, Dubna, Moscow Region, Russia

Abstract

In this chapter our goal is to represent the electromagnetic field from different and unknown points of view. The duality of natural Time is considered. It is established that there are two different Times in nature and the dual Time is tightly connected with natural rotation. On this ground, equations of dual electrodynamics are established and a new theoretical approach to the world of leptons and dual particles (quarks) is formulated, which gives a simple and evident explanation of lepton-quark symmetry, quark confinement and baryon number conservation.

It is shown that the electromagnetic field emerges as a ground state of the simplest form of energy called the generalized electromagnetic field, and the formulated theory of this field predicts the existence of the shadow world that consists of particles (heavy photons) which do not interact at all with any detectors and the influence of which becomes apparent only in the form of gravity and a new form of an energy flow.

We show that the ground state of the electromagnetic field itself has a hidden structure which takes a form of a complex scalar field associated with a one-dimensionally extended object called a charged string. The theory which predicts the existence of these objects is formulated.

On the basis of the principles of general relativity and the natural concept of Time the motion of a charged massive point particle in the external gravitational and electromagnetic fields is considered. The general covariant and reparametrization invariant Newton equations are derived from the equations of geodesic motion. The general covariant and reparametrization invariant expressions for the physical momentum, velocity and energy of a massive and charged point particle are presented. It is marked that the change of the orientation of a path and the operation of the charge conjugation are connected. The existence of the gravitational force that is defined by the momentum of the gravitational field and is not trivial only for the non-static gravitational fields is predicted. An adequate solution to the well known problem of zero Hamiltonian is manifested as well.

[*]Author's E-mail: pestov@theor.jinr.ru.

Keywords: space and time, duality of time, lepton-quark symmetry, generalized electromagnetic field, Newton equations

Introduction

The Standard Model provides an excellent description of what goes on in the physics of elementary particles. However, we begin to get into trouble when we ask the question of why the Standard Model has the features that it does. Hence, we need to explore a deeper level of nature to find the answer and have a clear guidance to the best place to look for physics beyond the Standard Model. Here we should like to exhibit some physical evidences of the existence of this deeper level and to this end we start from representation of the electromagnetic field from different and unknown points of view. It is well known that the Maxwell theory has a hidden intrinsic property that gives rise to special relativity. With the establishment of general relativity, it became unclear how to harmonize the fundamental physical concepts of the electric and magnetic fields with the first principles of this theory. The concept of natural Time (which can be considered as a hidden intrinsic property of the Einstein-Maxwell theory), originated in the work [1], provides an opportunity to clear up these difficulties (and many others) and opens new perspectives in this field. In Sect. II, we exhibit the concept of dual electrodynamics tightly connected with the existence of dual Time in nature. After that, we go to a picture on which the electromagnetic field emerges as a ground state of the generalized electromagnetic field or heavy light. This situation is explained in detail in Sect. III. We especially pay attention to the geometrical and physical sense of heavy light and the problem of gauge fixing in the context of the idea that the gauge fixing is an intrinsic property of local (gauge) symmetry itself. The hidden structure of the ground state of the electromagnetic field and possible extension of the electrodynamics in the context of the problem of gauge fixing is also a subject of conversation in this section. This extension can be very important for understanding, for example, superconductivity at the fundamental (field-theoretical) level. At last, in Sect. IV, we investigate the action of the gravitational and electromagnetic field on a massive charged particle and derive Newton's equations from the equations of the geodesic motion in the very general frameworks of the general covariance and reparametrization invariance. This gives a solution of the so-called zero-Hamiltonian problem and discovers an unusual property of massive charged particles: if the motion from one point of the path to another point of the path is possible, then the inverse motion is possible only under the condition that a particle changes the sign of charge. Asymmetry between matter and antimatter is clearly visible in this case and, evidently, may have far reaching consequences.

Dual Time and Dual Electrodynamics

In this section, we exhibit a very general and a deep property of natural Time introduced in the work [1] and give the derivation of equations of dual electrodynamics defined by the duality of Time. We start from the definition of natural geometry and natural Time to demonstrate the duality of Time.

Since the field of real numbers R is continuous and unconditional to all forms of physical matter, we can define on this ground continuous and unconditional natural geometry R^n, in which a point is defined as a n-tuple of real numbers

$$x = (x^1, x^2, \cdots x^n),$$

and the distance function is introduced as usual

$$d(x,y) = \sqrt{(x^1 - y^1)^2 + (x^2 - y^2)^2 \cdots (x^n - y^n)^2}.$$

It is clear that R^1, R^2, R^3 can be considered as models of the Euclidian straight line, plane and space, respectively. However, R^1, R^2, R^3 admit simple and clear generalization and hence R^n is a very important geometry which can be considered as the underlying structure of any investigation. On the basis of natural geometry more complicated geometries may be constructed in which a point is defined as a point of some n-dimensional surface in the space R^N, $n < N$. These generalized geometries can be put in correspondence to the states of a full system of fields (a system which include the gravitational field) [1].

It is no doubt that the variables (Cartesian coordinates) x^1, x^2, $\cdots x^n$ in the definition of point R^n should be considered on an absolutely equal footing. Hence, it is unclear how to introduce the so-called space coordinates and time coordinate (a space-time structure) in the framework of natural geometry alone. To do this, we need to give a definition of natural Time as entity which is tightly connected with all natural dynamical processes and is as simple as possible from a geometrical point of view. To make it easier to perceive the definition given in [1] and below, let us appeal to physical intuition. We know very well the physical phenomena connected with the temperature and pressure difference. We speak about the gradient of temperature and pressure and presuppose that values of these physical quantities are known for any point of some region of the Euclidian space. From a geometrical point of view we deal with a scalar field that is invariant with respect to all admissible transformations of coordinates. Now it is natural to suppose that there are a field of moments of Time and an area of phenomena defined by the gradient of Time.

Definition: a moment of natural Time is a number that we put in correspondence to any point of the reference space R^n. Hence, a moment of Time is defined by the equation

$$t = f(x^1, x^2, \cdots x^n) = f(x).$$

For understanding the nature of Time, it is very important to clarify that a moment of Time is invariant with respect to general coordinate transformations

$$\bar{x}^i = \bar{x}^i(x^1, x^2, \cdots, x^n), \quad x^i = x^i(\bar{x}^1, \bar{x}^2, \cdots, \bar{x}^n), \quad i = 1, 2, \cdots n,$$

where x^1, x^2, $\cdots$, x^n are the Cartesian coordinates and $\bar{x}^1$, $\bar{x}^2$, $\cdots$, $\bar{x}^n$ are the new ones.

All points of the reference space that correspond to the same moment of Time constitute physical space $S(t)$. A point of $S(t)$ is defined by the equation

$$f(x^1, x^2, \cdots x^n) = f(x) = t = constant.$$

This one-parameter family of physical spaces defines causality or determinism of physical reality.

The gradient of Time is the vector field **t** with the components $t^i = (\nabla f)^i = g^{ij}\partial_j f = g^{ij}t_j$, where g^{ij} are the contravariant components of the Riemann positive definite metric g_{ij}, which we have put in correspondence to the gravitational field [1].

The gradient of Time defines fundamental discrete internal symmetry- bilateral symmetry. A pair of vector fields **v** and **v̄** has bilateral symmetry if the sum of these fields is collinear to the gradient of Time and their difference is orthogonal to it, $\overline{\mathbf{v}} + \mathbf{v} = \lambda\mathbf{t}$, $(\overline{\mathbf{v}}|\mathbf{t}) = (\mathbf{v}|\mathbf{t})$, where $(\mathbf{v}|\mathbf{w}) = g_{ij}v^i w^j = v^i w_i$ is a scalar product. The bilateral symmetry may be represented as a linear transformation (reflection) $\overline{v}^i = R^i_j v^j$, where $R^i_j = 2t^i t_j - \delta^i_j$. We have a natural result: $R^i_j t^j = t^i$ and $R^i_k R^j_l g_{ij} = g_{kl}$.

The bilateral symmetry defines the causal structure of the physical world since the auxiliary metric

$$\overline{g}_{ij} = 2t_i t_j - g_{ij} = g_{ik}R^k_j, \quad \overline{g}^{ij} = 2t^i t^j - g^{ij}$$

provides a straightforward method of consideration of the dynamical processes through the introduction of natural Time into the Lagrangians (and the equations) of the fundamental physical fields [1]. We put forward an idea that bilateral symmetry is strict symmetry of nature and hence in all physical processes one cannot distinguish the right-hand sided physical quantity from the left-hand sided one.

From the consideration of the bilateral symmetry it follows that in the geometrical (coordinate independent) form the time reversal invariance means that a theory is invariant with respect to the transformations

$$T: \quad t^i \to -t^i.$$

It is clear that the transformation T has meaning if and only if domains of values of the potentials $f(x)$ and $-f(x)$ coincide. It is clear that the theory will be time reversal invariant if the gradient of the temporal field appears in all formulae only as an even number of times, like $t^i t^j$.

Since the scalar temporal field enters into the Lagrangians of the physical fields in the form of the gradient of the scalar field

$$t_i = \partial_i f(u),$$

the laws of the unified physics are invariant with respect to transformations of the form

$$f(x) \Rightarrow f(x) + a,$$

where a is a constant. This symmetry defines the law of energy conservation as a fundamental physical law of the universe which is true in all cases.

The potential $f(x)$ of natural Time is a solution to the equation

$$D_{\mathbf{t}}f = (\nabla f)^2 = g^{ij}\frac{\partial f}{\partial x^j}\frac{\partial f}{\partial x^j} = 1, \qquad (1)$$

which can be considered as the definition of uniformity of natural Time. Other mathematical arguments in favour of this equation are also impressible. Let $dx^i = t^i dt$, $f(x + dx) = t + dt$, $f(x) = t$, then $df(x) = t_i t^i dt = dt$ and hence $t_i t^i = 1$. This is equation (1). Further, we consider the differential operator $D_{\mathbf{t}} = t^i \partial_i$ defined by the gradient of natural Time t^i and its exponent $\exp(aD_{\mathbf{t}}) = 1 + aD_{\mathbf{t}} + \frac{a^2}{2}(D_{\mathbf{t}})^2 + \cdots$. We put forward the natural demand that transformation $f(x) \Rightarrow f(x) + a$ is generated by the exponent of the gradient of natural Time t^i, and from equation $\exp(aD_t)f(x) = f(x) + a$ we again derive equation (1). Taking into account the possibility of changing the scale, we also subordinate the potential $f(x)$ of natural Time to the equation

$$f(\lambda x^1, \lambda x^2, \cdots \lambda x^n) = \lambda f(x^1, x^2, \cdots x^n). \qquad (2)$$

The fundamental (from a physical point of view) observation reads that equation (1) has not only general solution but also a special solution known as the function of geodesic distance. This means that there are two different Times in nature and hence two different kinds of dynamical processes. Below we consider a strict realization of this statement.

In what follows we consider the four-dimensional reference space R^4 with the metric $ds^2 = g_{ij}dx^i dx^j = (dx^1)^2 + (dx^2)^2 + (dx^3)^2 + (dx^4)^2$, $g_{ij} = \delta_{ij}$. We consider equations (1) and (2) under these conditions and hence look for solutions to the equations

$$\left(\frac{\partial f}{\partial x^1}\right)^2 + \left(\frac{\partial f}{\partial x^2}\right)^2 + \left(\frac{\partial f}{\partial x^3}\right)^2 + \left(\frac{\partial f}{\partial x^4}\right)^2 = 1,$$

$$f(\lambda x^1, \lambda x^2, \lambda x^3, \lambda x^4) = \lambda f(x^1, x^2, x^3, x^4).$$

In accordance with our general statement these equations have two solutions

$$f(x) = a_i x^i = a_1 x^1 + a_2 x^2 + a_3 x^3 + a_4 x^4,$$

where $\mathbf{a} = (a^1, a^2, a^3, a^4)$ is a unit constant vector $(\mathbf{a}|\mathbf{a}) = 1$, and

$$f(x) = \sqrt{(x^1)^2 + (x^2)^2 + (x^3)^2 + (x^4)^2}.$$

From the equations

$$f(x) = a_i x^i = a_1 x^1 + a_2 x^2 + a_3 x^3 + a_4 x^4 = t = constant,$$

and

$$f(x) = \sqrt{(x^1)^2 + (x^2)^2 + (x^3)^2 + (x^4)^2} = \tau = constant$$

we see that in one case physical space is the familiar three-dimensional Euclidian space E^3 and in the other case new physical space is three-dimensional Riemannian space of constant positive curvature, i.e. the 3d-sphere S^3. The physical (mass) points are to be identified with the points belonging to the three-dimensional Euclidian space E^3, but the points belonging to the 3d-sphere S^3 should be put in correspondence to the Spherical Tops. Indeed, the symmetries of the Euclidian space can be composed of translations and rotations and the symmetries of the 3d-sphere S^3 coincide with those of the Spherical Top which will be considered below. In other words, geometrical points in the Euclidian and Riemannian spaces have a different physical meaning. Thus, from the duality of Time it follows that any known particle can be put in correspondence to a dual particle moving in the dual Time. We see that there are two space-time structures (two different times) on the same reference space R^4. It will be shown that the first space-time structure uncovers the essence of special theory of relativity, and the dual space-time structure describes natural rotation as a motion in the dual Time. Hence, it is natural to put forward the idea of dual approach to the world of elementary particles which can explain the existence of leptons and quarks, lepton-quark symmetry and confinement (if we identify dual particles with quarks).

Now we exhibit some interesting properties of the reference space R^4. The concept of quaternion is heuristic in this case and we adopt our notation to this situation. We demonstrate that the angular momenta of the Spherical Top are tightly connected with a natural geometrical structure in the reference space R^4. The points of R^4 have the vector

$$\mathbf{q} = (q^1, q^2, q^3, q^4)$$

and the quaternion representations

$$q = q^1 i + q^2 j + q^3 k + q^4 1,$$

with the usual linear structure. The quaternion algebra is defined as ordinary

$$i^2 = j^2 = k^2 = -1, \quad ij = -ji = k, \quad jk = -kj = i, \quad ki = -ik = j.$$

The scalar product

$$(\mathbf{p}|\mathbf{q}) = p^1 q^1 + p^2 q^2 + p^3 q^3 + p^4 q^4$$

can be written in the quaternion form in two ways

$$(\mathbf{p}|\mathbf{q}) = \frac{1}{2}(p\bar{q} + q\bar{p}) = \frac{1}{2}(\bar{p}q + \bar{q}p),$$

where $\bar{q} = -q^1 i - q^2 j - q^3 k + q^4 1$. The scalar product is invariant with respect to the right and left turn dilations

$$q \Rightarrow \tilde{q} = s\,q, \quad q \Rightarrow \tilde{q} = q\,\bar{t},$$

since

$$(\tilde{\mathbf{p}}|\tilde{\mathbf{q}}) = s\bar{s}\,(\mathbf{p}|\mathbf{q}), \quad (\tilde{\mathbf{p}}|\tilde{\mathbf{q}}) = t\bar{t}\,(\mathbf{p}|\mathbf{q}).$$

We suppose that q and λq, where λ is a positive number, are equivalent. For a given $q \neq 0$, the equations $q = sq$, $q = q\bar{t}$ have only trivial solutions $s = \bar{t} = 1$ and the absence of fixed points under turn dilations exhibits a fundamental property of the reference space R^4: the existence of two simply transitive groups of transformations under the condition $q \neq 0$. This duality inherent in R^4 is geometrically exhibited as follows.

First of all, we introduce two natural frames intrinsically connected with R^4. The standard frame

$$\mathbf{s}_1 = (1, 0, 0, 0) \quad \mathbf{s}_2 = (0, 1, 0, 0) \quad \mathbf{s}_3 = (0, 0, 1, 0) \quad \mathbf{s}_4 = (0, 0, 0, 1),$$

$$s_1 = i, \quad s_2 = j, \quad s_3 = k, \quad s_4 = 1$$

gives rise to a pair of right-handled dual mobile frames

$$m_1 = iq, \quad m_2 = jq, \quad m_3 = kq, \quad m_4 = 1q,$$

$$n_1 = qi, \quad n_2 = qj, \quad n_3 = qk, \quad n_4 = q1,$$

Hidden Aspects of the Electromagnetic Field

$$\begin{aligned}
\mathbf{m}_1 &= (\ q^4,\ -q^3,\ q^2,\ -q^1\ ) \\
\mathbf{m}_2 &= (\ q^3,\ q^4,\ -q^1,\ -q^2\ ) \\
\mathbf{m}_3 &= (\ -q^2,\ q^1,\ q^4,\ -q^3\ ) \\
\mathbf{m}_4 &= (\ q^1,\ q^2,\ q^3,\ q^4\ )
\end{aligned}$$

$$\begin{aligned}
\mathbf{n}_1 &= (\ q^4,\ q^3,\ -q^2,\ -q^1\ ) \\
\mathbf{n}_2 &= (\ -q^3,\ q^4,\ q^1,\ -q^2\ ) \\
\mathbf{n}_3 &= (\ q^2,\ -q^1,\ q^4,\ -q^3\ ) \\
\mathbf{n}_4 &= (\ q^1,\ q^2,\ q^3,\ q^4\ ).
\end{aligned}$$

It is easy to see that

$$(\mathbf{m}_a|\mathbf{m}_b) = q\bar{q}\,\delta_{ab}, \quad (\mathbf{n}_a|\mathbf{n}_b) = q\bar{q}\,\delta_{ab}, \quad (a,b = 1,2,3,4).$$

After that, let us consider the running point $T(q^1, q^2, q^3, q^4)$, and the twelve coherent points

$$A(q^4, -q^3, q^2, -q^1), \quad B(q^3, q^4, -q^1, -q^2), \quad C(-q^2, q^1, q^4, -q^3),$$
$$K(q^4, q^3, -q^2, -q^1), \quad L(-q^3, q^4, q^1, -q^2), \quad M(q^2, -q^1, q^4, -q^3),$$
$$\bar{A}(-q^4, q^3, -q^2, q^1), \quad \bar{B}(-q^3, -q^4, q^1, q^2), \quad \bar{C}(q^2, -q^1, -q^4, q^3),$$
$$\bar{K}(-q^4, -q^3, q^2, q^1), \quad \bar{L}(q^3, -q^4, -q^1, q^2), \quad \bar{M}(-q^2, q^1, -q^4, q^3).$$

The distance function is defined as usual

$$d^2_{PQ} = (p^1 - q^1)^2 + (p^2 - q^2)^2 + (p^3 - q^3)^2 + (p^4 - q^4)^2,$$

where d_{PQ} is the distance between the points P and Q. With this it is easy to see that

$$d^2_{AB} = d^2_{AC} = d^2_{BC} = d^2_{TA} = d^2_{TB} = d^2_{TC} = 2q\bar{q},$$
$$d^2_{\bar{A}\bar{B}} = d^2_{\bar{A}\bar{C}} = d^2_{\bar{B}\bar{C}} = d^2_{T\bar{A}} = d^2_{T\bar{B}} = d^2_{T\bar{C}} = 2q\bar{q}$$

and

$$d^2_{KL} = d^2_{KM} = d^2_{LM} = d^2_{TK} = d^2_{TL} = d^2_{TM} = 2q\bar{q},$$
$$d^2_{\bar{K}\bar{L}} = d^2_{\bar{K}\bar{M}} = d^2_{\bar{L}\bar{M}} = d^2_{T\bar{K}} = d^2_{T\bar{L}} = d^2_{T\bar{M}} = 2q\bar{q}.$$

We see a pair of regular tetrahedrons and a mirror one with a common vertex T: $TABC$ and $TKLM$, $T\bar{A}\bar{B}\bar{C}$ and $T\bar{K}\bar{L}\bar{M}$. These tetrahedrons give a visual representation of the dual mobile frames $(\mathbf{m}_a,\ \mathbf{n}_a)$ and $(-\mathbf{m}_a,\ -\mathbf{n}_a)$ and discover the origin of natural rotation.

Really, let $\mathbf{q} = \mathbf{q}(t)$ be a trajectory in R^4. When point T moves along this trajectory, the tetrahedrons $TABC$ and $TKLM$ (defined identically by T) pulse and rotate with respect to each other. And the same for the mirror tetrahedrons $T\bar{A}\bar{B}\bar{C}$ and $T\bar{K}\bar{L}\bar{M}$. The matrix of scalar products

$$\mathrm{P}_{\mu\nu} = (\mathbf{m}_\mu|\mathbf{n}_\nu), \quad (\mu, \nu = 1, 2, 3)$$

describes this relative natural rotation.

Further, the scalar products of the tangent vector $\dot{\mathbf{q}} = d\mathbf{q}/dt$ with the vectors of dual frames $\mathbf{m}_a$ and $\mathbf{n}_a$, $(a = 1, 2, 3, 4)$

$$(\mathbf{m}_1|\dot{\mathbf{q}}) = q^4 \frac{dq^1}{dt} - q^3 \frac{dq^2}{dt} + q^2 \frac{dq^3}{dt} - q^1 \frac{dq^4}{dt},$$

$$(\mathbf{m}_2|\dot{\mathbf{q}}) = q^3 \frac{dq^1}{dt} + q^4 \frac{dq^2}{dt} - q^1 \frac{dq^3}{dt} - q^2 \frac{dq^4}{dt},$$

$$(\mathbf{m}_3|\dot{\mathbf{q}}) = -q^2 \frac{dq^1}{dt} + q^1 \frac{dq^2}{dt} + q^4 \frac{dq^3}{dt} - q^3 \frac{dq^4}{dt},$$

$$(\mathbf{n}_1|\dot{\mathbf{q}}) = q^4 \frac{dq^1}{dt} + q^3 \frac{dq^2}{dt} - q^2 \frac{dq^3}{dt} - q^1 \frac{dq^4}{dt},$$

$$(\mathbf{n}_2|\dot{\mathbf{q}}) = -q^3 \frac{dq^1}{dt} + q^4 \frac{dq^2}{dt} + q^1 \frac{dq^3}{dt} - q^2 \frac{dq^4}{dt},$$

$$(\mathbf{n}_3|\dot{\mathbf{q}}) = q^2 \frac{dq^1}{dt} - q^1 \frac{dq^2}{dt} + q^4 \frac{dq^3}{dt} - q^3 \frac{dq^4}{dt},$$

$$(\mathbf{m}_4|\dot{\mathbf{q}}) = (\mathbf{n}_4|\dot{\mathbf{q}}) = q^1 \frac{dq^1}{dt} + q^2 \frac{dq^2}{dt} + q^3 \frac{dq^3}{dt} + q^4 \frac{dq^4}{dt}$$

are invariant with respect to the left and right turn dilatations (q and λq, where λ is a positive number, are equivalent). The invariants

$$\Omega_\mu = \frac{1}{2}(\mathbf{m}_\mu|\dot{\mathbf{q}}), \quad \tilde{\Omega}_\mu = \frac{1}{2}(\mathbf{n}_\mu|\dot{\mathbf{q}}), \quad (\mu = 1, 2, 3)$$

can be considered as components of angular velocity of rotation of tetrahedron $TABC$ with respect to tetrahedron $TKLM$ and vice versa. Thus, the kinematics of natural rotation has an adequate representation in the four dimensions as an amazingly regular and natural association of thirteen points.

To quantize natural rotation, let us introduce the $4d$ operator ∇

$$\nabla_4 = \left(\frac{\partial}{\partial q^1}, \frac{\partial}{\partial q^2}, \frac{\partial}{\partial q^3}, \frac{\partial}{\partial q^4} \right)$$

and setting

$$M_\nu = \frac{1}{2}(\mathbf{m}_\nu|\nabla_4), \quad N_\nu = \frac{1}{2}(\mathbf{n}_\nu|\nabla_4), \quad (\nu = 1, 2, 3)$$

we have dual anti hermitian operators of angular momenta of the Spherical Top. Factor $\frac{1}{2}$ is essential since natural commutation relations hold valid

$$M_1 M_2 - M_2 M_1 = M_3, \quad N_1 N_2 - N_2 N_1 = -N_3$$

and so on. The operator of dilatations

$$D = (\mathbf{m}_4|\nabla_4) = (\mathbf{n}_4|\nabla_4) = q^1 \frac{\partial}{\partial q^1} + q^2 \frac{\partial}{\partial q^2} + q^3 \frac{\partial}{\partial q^3} + q^4 \frac{\partial}{\partial q^4}$$

has an important meaning as well since it commutes with the operators of angular momenta of the Spherical Top

$$DM_\nu - M_\nu D = 0, \quad DN_\nu - N_\nu D = 0, \quad (\nu = 1, 2, 3)$$

and will play an important role in the description of natural rotation at the quantum level. We see that the symmetry of the Spherical Top is the structure which is tightly associated with a moving point particle.

Now we will consider the equations of electrodynamics associated with two different Times. At first, we consider familiar electrodynamics from the new point of view. Let

$$\mathbf{a} = (a^1, a^2, a^3, a^4)$$

be a gradient of natural time $t = a_i q^i$, then a global frame in R^4 is defined as follows:

$$\mathbf{E}_0 = (a^1, a^2, a^3, a^4), \quad \mathbf{E}_1 = (-a^4, -a^3, a^2, a^1),$$
$$\mathbf{E}_2 = (a^3, -a^4, -a^1, a^2), \quad \mathbf{E}_3 = (-a^2, a^1, -a^4, a^3).$$

We put

$$D_0 = (\mathbf{E}_0|\nabla_4), \quad D_1 = (\mathbf{E}_1|\nabla_4), \quad D_2 = (\mathbf{E}_2|\nabla_4), \quad D_3 = (\mathbf{E}_3|\nabla_4),$$

and in accordance with the general consideration [2], the Dirac equation with respect to the first space-time structure reads

$$i\gamma^\mu D_\mu \psi = \frac{mc}{\hbar}\psi. \tag{3}$$

Since

$$\gamma^\mu D_\mu = \gamma^0 D_0 + \gamma^1 D_1 + \gamma^2 D_2 + \gamma^3 D_3,$$

then to get a regular transition from the Dirac equation (3) to the original Dirac equation, we need to introduce the system of coordinates x^1, x^2, x^3, t which is defined by the gradient of Time in question.

To this end, we need to solve the system of equations

$$\frac{dq^i}{dt} = a^i$$

and find the lines of Time [1]. The general solution is a straight line that goes through the fixed point $\mathbf{q}_0 = (q_0^1, q_0^2, q_0^3, q_0^4)$:

$$\mathbf{q}(t) = \mathbf{a}(t - t_0) + \mathbf{q}_0. \tag{4}$$

Further, following [1] we consider the 3d surface

$$t_0 = f(q_0^1, q_0^2, q_0^3, q_0^4) = (\mathbf{a}|\mathbf{q}_0) \tag{5}$$

in the space of initial data. The general solution to equation (5) has the form

$$\mathbf{q}_0 = t_0 \mathbf{E}_0 + x\mathbf{E}_1 + y\mathbf{E}_2 + z\mathbf{E}_3.$$

Substituting this representation into formula (4) we have

$$\mathbf{q} = t\mathbf{E}_0 + x\mathbf{E}_1 + y\mathbf{E}_2 + z\mathbf{E}_3.$$

The Dirac equation in the coordinates x, y, z, t has the ordinary form

$$i(\gamma^0 \frac{\partial}{\partial t} + \gamma^1 \frac{\partial}{\partial x} + \gamma^2 \frac{\partial}{\partial y} + \gamma^3 \frac{\partial}{\partial z})\psi = \frac{mc}{\hbar}\psi.$$

One can work in either the coordinates q^1, q^2, q^3, q^4 (that are considered on equal footing) or the coordinates x, y, z, t but the first approach looks like more fundamental because the direction of the vector $\mathbf{a}$ can be arbitrary, and this distinctive degeneration is not visible in the second approach. In view of this we formulate the Maxwell equation in this more general approach.

In accordance with the general theory formulated in [1], this problem is solved as follows. Let A_i be the vector potential of the electromagnetic field. The gauge invariant tensor of the electromagnetic field is defined as usual $F_{ij} = \partial_i A_j - \partial_j A_i$. The strength of the electric field is a general covariant and gauge invariant quantity that is defined by the equation $E_i = a^k F_{ik}$.

The rotor of the vector field $\mathbf{A} = (A_1, A_2, A_3, A_4)$ is defined as a vector product of ∇_4 and $\mathbf{A}$

$$\operatorname{rot} \mathbf{A} = \nabla_4 \times \mathbf{A}, \quad (\operatorname{rot} \mathbf{A})^i = e^{ijkl} a_j \partial_k A_l = \frac{1}{2} e^{ijkl} a_j (\partial_k A_l - \partial_l A_k),$$

where e^{ijkl} are the contravariant components of the Levi-Civita tensor normalized as $e_{1234} = 1$. The general covariant and gauge invariant definition of the magnetic field strength is given by the formula $\mathbf{H} = \operatorname{rot} \mathbf{A}$, $H^i = (\operatorname{rot} \mathbf{A})^i$. Thus, $H_i = a^k \overset{*}{F}_{ik}$, where $\overset{*}{F}_{ij} = g_{ik} g_{jl} \overset{*}{F}^{kl} = \frac{1}{2} g_{ik} g_{jl} e^{klmn} F_{mn}$. It is evident that vectors $\mathbf{E}$ and $\mathbf{H}$ are orthogonal to $\mathbf{a}$

$$(\mathbf{a}|\mathbf{E}) = 0, \quad (\mathbf{a}|\mathbf{H}) = 0.$$

The Maxwell equations read:

$$D\mathbf{H} = -\operatorname{rot} \mathbf{E}, \tag{6}$$

$$D\mathbf{E} = \operatorname{rot} \mathbf{H} + e\mathbf{J}, \tag{7}$$

$$(\nabla_4|\mathbf{E}) = e\bar{\psi}\gamma^0\psi, \quad (\nabla_4|\mathbf{H}) = 0, \tag{8}$$

where the current $\mathbf{J}$ is given by the expression

$$\mathbf{J} = \mathbf{E}_1 \bar{\psi}\gamma^1\psi + \mathbf{E}_2 \bar{\psi}\gamma^2\psi + \mathbf{E}_3 \bar{\psi}\gamma^3\psi, \quad D = a^1 \frac{\partial}{\partial q^1} + a^2 \frac{\partial}{\partial q^2} + a^3 \frac{\partial}{\partial q^3} + a^4 \frac{\partial}{\partial q^4}.$$

Now it is important to show the definition of the interval in the space in question. The interval in R^4 is defined as follows. Let

$$\mathbf{q}_s = 2(\mathbf{a}|\mathbf{q})\,\mathbf{a} - \mathbf{q}$$

be the vector symmetrical to the vector **q** with respect to the vector **a**. Then in the coordinates q^1, q^2, q^3, q^4 the interval can be written as follows:

$$s^2 = (\mathbf{q}|\mathbf{q}_s) = 2(\mathbf{a}|\mathbf{q})^2 - (\mathbf{q}|\mathbf{q}) = (\mathbf{q}|\mathbf{q})\cos 2\theta,$$

where θ is an angle between **a** and **q**. It is easy to see that in the coordinates x, y, z, t

$$s^2 = t^2 - x^2 - y^2 - z^2.$$

We see that the first space-time structure and the bilateral symmetry presuppose the existence of the known space-time structure. The special theory of relativity is discovered here as a spontaneous breaking of isotropy of the four-dimensional Euclidian space. The bilateral symmetry defines the auxiliary metric as usual

$$\bar{g}_{ij} = \eta_{\mu\nu} E_i^\mu E_j^\nu = 2a_i a_j - \delta_{ij}.$$

Thus, when we put in correspondence to any point of R^4 a moment of time t by the equation

$$t = a_i q^i,$$

we uncover the underlying structure of the special theory of relativity. We can take another gradient of Time b_i which is different from a_i and consider the Dirac equation associated with b_i. One can show that these equations are equivalent since there is one-to-one and smooth transformation of R^4 onto itself which translates b_i into a_i.

Below we formulate the equations of dual electrodynamics associated with the dual Time. Let

$$\mathbf{q} = (q^1, q^2, q^3, q^4)$$

be a radius-vector, then a natural local frame (mobile frame) in the reference space R^4 can be represented as a quadruplet of orthogonal unit vector fields

$$\mathbf{E}_0 = \left(\frac{q^1}{\tau}, \frac{q^2}{\tau}, \frac{q^3}{\tau}, \frac{q^4}{\tau}\right), \quad \mathbf{E}_1 = \left(\frac{-q^4}{\tau}, \frac{-q^3}{\tau}, \frac{q^2}{\tau}, \frac{q^1}{\tau}\right),$$

$$\mathbf{E}_2 = \left(\frac{q^3}{\tau}, \frac{-q^4}{q}, \frac{-q^1}{\tau}, \frac{q^2}{\tau}\right), \quad \mathbf{E}_3 = \left(\frac{-q^2}{\tau}, \frac{q^1}{\tau}, \frac{-q^4}{\tau}, \frac{q^3}{\tau}\right),$$

where

$$\tau = \sqrt{(\mathbf{q}|\mathbf{q})} = \sqrt{(q^1)^2 + (q^2)^2 + (q^3)^2 + (q^4)^2}$$

is the potential of the dual Time.

We again put

$$D_0 = (\mathbf{E}_0|\tilde{\nabla}_4), \quad D_1 = (\mathbf{E}_1|\tilde{\nabla}_4), \quad D_2 = (\mathbf{E}_2|\tilde{\nabla}_4), \quad D_3 = (\mathbf{E}_3|\tilde{\nabla}_4)$$

but here the operator $\tilde{\nabla}_4$ is defined as follows:

$$\tilde{\nabla}_4 = \nabla_4 - \frac{3}{2q^2}\mathbf{q}$$

since the vector part U^i_{ik} of the torsion tensor $U^i_{jk} = E^i_\mu(\partial_j E^\mu_k - \partial_k E^\mu_j)$ is not equal to zero in this case [2]. The Dirac equation describing the behavior of charged particles with spin one half in the dual Time takes the following form:

$$i\gamma^\mu D_\mu \psi = \frac{mc}{\hbar}\psi. \tag{9}$$

Let us consider how to equip the reference space with a space-time structure in this case. The general solution of the system of equations

$$\frac{dq^i}{d\tau} = \frac{q^i}{\sqrt{(q^1)^2 + (q^2)^2 + (q^3)^2 + (q^4)^2}}$$

can be written as follows:

$$q^i(\tau) = q^i_0 \frac{\tau}{\tau_0}, \quad \tau \in (0, \infty),$$

where the initial data belong in this case to the 3d sphere

$$(\mathbf{q}_0|\mathbf{q}_0) = \tau_0^2.$$

The physical space section in question can be parameterized by the Euler angles α, β, γ. In the coordinates α, β, γ, τ we have

$$D_0 = \frac{\partial}{\partial \tau} - \frac{3}{2\tau}, \quad D_1 = \frac{1}{\tau}\left(-\cot\beta\cos\alpha\frac{\partial}{\partial\alpha} - \sin\alpha\frac{\partial}{\partial\beta} + \frac{\cos\alpha}{\sin\beta}\frac{\partial}{\partial\gamma}\right),$$

$$D_2 = \frac{1}{\tau}\left(-\cot\beta\sin\alpha\frac{\partial}{\partial\alpha} + \cos\alpha\frac{\partial}{\partial\beta} + \frac{\sin\alpha}{\sin\beta}\frac{\partial}{\partial\gamma}\right), \quad D_3 = \frac{1}{\tau}\frac{\partial}{\partial\alpha}.$$

The auxiliary metric defined by the dual Time has a simple representation

$$\bar{g}_{ij} = \eta_{\mu\nu}E^\mu_i E^\nu_j = 2t_i t_j - \delta_{ij}, \quad t_i = q_i/\tau$$

The action for the point particle associated with the dual Time can be written in the following form:

$$S = -mc \int_p^q \sqrt{1 - \tau^2\omega^2}d\tau,$$

where $\omega = dl/d\tau$ and dl is the element of the arc on the unit 3d sphere. Really, $(\mathbf{dq}|\mathbf{dq}) = d\tau^2 + \tau^2 dl^2$, and $(\mathbf{q}|\mathbf{dq}) = \tau d\tau$. On this ground one can develop the classical mechanics in the new frameworks, where the concept of Spherical Top is reduced to the concept of point particle. In this connection it should be noted that the Newton equations describing the dynamics of a particle on the three dimensional sphere S^3 are equivalent to the Euler equations for the Spherical Top.

To complete this section, we formulate the Maxwell equations which are defined by the dual space-time structure on R^4. In accordance with the general theory formulated in [1], this problem is solved as follows. Let A_i be the vector potential of the electromagnetic field. The gauge invariant tensor of the electromagnetic field is defined as usual $F_{ij} = \partial_i A_j - \partial_j A_i$. The strength of the electric field is a general covariant and gauge invariant quantity that is defined by the equation $E_i = t^k F_{ik}$, where in our case $t^k = t_k = q^k/\tau$.

The rotor of the vector field $\mathbf{A} = (A_1, A_2, A_3, A_4)$ is defined as a vector product of ∇_4 and $\mathbf{A}$

$$\operatorname{rot} \mathbf{A} = \nabla_4 \times \mathbf{A}, \quad (\operatorname{rot} \mathbf{A})^i = e^{ijkl} t_j \partial_k A_l = \frac{1}{2} e^{ijkl} t_j (\partial_k A_l - \partial_l A_k),$$

where e^{ijkl} are the contravariant components of the Levi-Civita tensor normalized as $e_{1234} = 1$. The general covariant and gauge invariant definition of the magnetic field strength is given by the formula $\mathbf{H} = \operatorname{rot} \mathbf{A}$, $H^i = (\operatorname{rot} \mathbf{A})^i$. Thus, $H_i = t^k \overset{*}{F}_{ik}$, where $\overset{*}{F}_{ij} = g_{ik} g_{jl} \overset{*}{F}^{kl} = \frac{1}{2} g_{ik} g_{jl} e^{klmn} F_{mn}$. It is evident that vectors $\mathbf{E}$ and $\mathbf{H}$ are orthogonal to $\mathbf{q}$

$$(\mathbf{q}|\mathbf{E}) = 0, \quad (\mathbf{q}|\mathbf{H}) = 0.$$

The Maxwell equations of the dual electrodynamics read:

$$\frac{1}{\tau} D \mathbf{H} + \frac{2}{\tau} \mathbf{H} = -\operatorname{rot} \mathbf{E}, \tag{10}$$

$$\frac{1}{\tau} D \mathbf{E} + \frac{2}{\tau} \mathbf{E} = \operatorname{rot} \mathbf{H} + e\mathbf{J}, \tag{11}$$

$$(\nabla_4 | \mathbf{E}) = e\bar{\psi}\gamma^0\psi, \quad (\nabla_4 | \mathbf{H}) = 0, \tag{12}$$

where the current $\mathbf{J}$ is given by the expression

$$\mathbf{J} = \mathbf{E}_1 \bar{\psi}\gamma^1\psi + \mathbf{E}_2 \bar{\psi}\gamma^2\psi + \mathbf{E}_3 \bar{\psi}\gamma^3\psi, \quad D = q^1 \frac{\partial}{\partial q^1} + q^2 \frac{\partial}{\partial q^2} + q^3 \frac{\partial}{\partial q^3} + q^4 \frac{\partial}{\partial q^4}.$$

Rotations are usually considered on the ground of the artifacts in the three-dimensional Euclidian space. That is why we consider below the projection of the four-dimensional picture of symmetry of the Spherical Top onto the three-dimensional space and formulate with this the theory of orbital angular momentum. To complete the picture of natural rotation and throw light on some other questions of the duality of Time, we consider here the properties of natural mappings of the four-dimensional Euclidian space onto the three-dimensional one and start from the general consideration. Let $\varphi(x, y, z)$ be a smooth function of the cartesian coordinates x, y, z of the three-dimensional Euclidian space, and three differentiable functions

$$x = x(q^1, q^2, q^3, q^4), \quad y = y(q^1, q^2, q^3, q^4), \quad z = z(q^1, q^2, q^3, q^4)$$

define a mapping of R^4 onto E^3. Let us calculate the result of the action of the linear differential operator

$$L = \xi^i(q^1, q^2, q^3, q^4) \frac{\partial}{\partial q^i}$$

on the function $\varphi(x, y, z)$. Using the chain rule we have

$$L\varphi(x, y, z) = (Lx)\frac{\partial \varphi}{\partial x} + (Ly)\frac{\partial \varphi}{\partial y} + (Lz)\frac{\partial \varphi}{\partial z}.$$

If the functions

$$Lx = \xi^i \frac{\partial x}{\partial q^i}, \quad Ly = \xi^i \frac{\partial y}{\partial q^i}, \quad Lz = \xi^i \frac{\partial z}{\partial q^i}$$

of the variables q^1, q^2, q^3, q^4 can be presented as functions of the variables x, y, z, then setting
$$Lx = v_x(x, y, z), \quad Ly = v_y(x, y, z), \quad Lz = v_z(x, y, z),$$
one can calculate the result of the action of the operator L with the help of the new differential operator
$$V = v_x \frac{\partial}{\partial x} + v_y \frac{\partial}{\partial y} + v_z \frac{\partial}{\partial z},$$
which can be considered as a transform of the operator L under the mapping in question.

After these general remarks let us consider natural mapping of R^4 onto E^3. It is well known that rotation with dilatation of the vector $v = v_1 i + v_2 j + v_3 k$ can be represented as follows:
$$v \to sv\bar{s}.$$

Let us consider the quaternions
$$R_1 = qi\bar{q}, \quad R_2 = qj\bar{q}, \quad R_3 = qk\bar{q},$$
$$T_1 = \bar{q}iq, \quad T_2 = \bar{q}jq, \quad T_3 = \bar{q}kq.$$

Under the left turn dilatations $q \to sq$, the quaternions R_1, R_2, R_3, transform as follows: $R_\mu \to sR_\mu\bar{s}$, $(\mu = 1, 2, 3)$. Under the right turn dilatations $q \to q\bar{t}$, the quaternions T_1, T_2, T_3 transform similarly to R_μ, $T_\mu \to tT_\mu\bar{t}$, $(\mu = 1, 2, 3)$. We see that the coordinates of the quaternions in question can be considered as cartesian coordinates of E^3. We denote these coordinates as x_μ, y_μ, z_μ, $(\mu = 1, 2, 3)$ and, respectively, $\xi_\mu, \eta_\mu, \zeta_\mu$, $(\mu = 1, 2, 3)$,
$$\mathbf{R}_\mu = (x_\mu, y_\mu, z_\mu), \quad \mathbf{T}_\mu = (\xi_\mu, \eta_\mu, \zeta_\mu).$$

The vectors $\mathbf{R}_\mu$, and $\mathbf{T}_\mu$ have the same length and constitute the right-handled orthogonal bases since
$$\mathbf{R}_1 \times \mathbf{R}_2 = q\bar{q}\mathbf{R}_3, \quad \mathbf{R}_1 \cdot (\mathbf{R}_2 \times \mathbf{R}_3) = (q\bar{q})^3,$$
$$\mathbf{T}_1 \times \mathbf{T}_2 = q\bar{q}\mathbf{T}_3, \quad \mathbf{T}_1 \cdot (\mathbf{T}_2 \times \mathbf{T}_3) = (q\bar{q})^3.$$

Here we are slightly detained to give a simple and important geometrical interpretation of the Cartan spinors [3], which is tightly connected with the complex-analytic structures on R^4. To this end, let us consider the complex null vectors
$$\mathbf{W}_1 = \mathbf{R}_2 + \sqrt{-1}\mathbf{R}_3, \quad \mathbf{W}_2 = \mathbf{R}_3 + \sqrt{-1}\mathbf{R}_1, \quad \mathbf{W}_3 = \mathbf{R}_1 + \sqrt{-1}\mathbf{R}_2.$$

Calculating components of these vectors, we have
$$\mathbf{W}_1 = (u_1, v_1, w_1,) = (2\xi_1\xi_2, \quad \xi_1^2 - \xi_2^2, \quad -\sqrt{-1}\xi_1^2 - \sqrt{-1}\xi_2^2),$$
where $\xi_1 = q^2 + \sqrt{-1}q^3$, $\xi_2 = q^1 + \sqrt{-1}q^4$,
$$\mathbf{W}_2 = (u_2, v_2, w_2,) = (-\sqrt{-1}\eta_1^2 - \sqrt{-1}\eta_2^2, \quad 2\eta_1\eta_2, \quad \eta_1^2 - \eta_2^2),$$
where $\eta_1 = q^3 + \sqrt{-1}q^1$, $\eta_2 = q^2 + \sqrt{-1}q^4$,

$$\mathbf{W}_3 = (u_3, v_3, w_3,) = (\zeta_1^2 - -\zeta_2^2, \quad -\sqrt{-1}\zeta_1^2 - \sqrt{-1}\zeta_2^2, \quad 2\zeta_1\zeta_2)$$

where $\zeta_1 = q^1 + \sqrt{-1}q^2$, $\zeta_2 = q^3 + \sqrt{-1}q^4$.

Studying the behavior of the pairs (ξ_1, ξ_2), (η_1, η_2), (ζ_1, ζ_2) under the turn dilatations, we conclude that these pairs are the Cartan spinors. To introduce the spinor with the so called dotted indices, one simply needs to consider the vectors

$$\overline{\mathbf{W}}_1 = \mathbf{R}_2 - \sqrt{-1}\mathbf{R}_3, \quad \overline{\mathbf{W}}_2 = \mathbf{R}_3 - \sqrt{-1}\mathbf{R}_1, \quad \overline{\mathbf{W}}_3 = \mathbf{R}_1 - \sqrt{-1}\mathbf{R}_2.$$

It is also evident that the Cartan spinor is simply the system of complex coordinates on R^4. Actually, it is shown that there are three canonical systems of complex coordinates defined by the complex structures i, j, k. The turn dilatations in the complex coordinates coincide with the spinor transformations.

Now it is time to prolong and write out expressions for the coordinates x_μ, y_μ, z_μ, ($\mu = 1, 2, 3$) and ξ_μ, η_μ, ζ_μ, ($\mu = 1, 2, 3$). We have

$$x_1 = (q^1)^2 - (q^2)^2 - (q^3)^2 + (q^4)^2, \quad y_1 = 2q^1q^2 + 2q^3q^4, \quad z_1 = 2q^1q^3 - 2q^2q^4,$$
$$x_2 = 2q^1q^2 - 2q^3q^4, \quad y_2 = -(q^1)^2 + (q^2)^2 - (q^3)^2 + (q^4)^2, \quad z_2 = 2q^1q^4 + 2q^2q^3,$$
$$x_3 = 2q^1q^3 + 2q^2q^4, \quad y_3 = -2q^1q^4 + 2q^2q^3, \quad z_3 = -(q^1)^2 - (q^2)^2 + (q^3)^2 + (q^4)^2.$$

and

$$(\xi_1, \eta_1, \zeta_1) = (x_1, x_2, x_3), \quad (\xi_2, \eta_2, \zeta_2) = (y_1, y_2, y_3), \quad (\xi_3, \eta_3, \zeta_3) = (z_1, z_2, z_3).$$

Thus, all natural mappings of R^4 onto E^3 are presented.

Now it is interesting to find transforms of the dual operators $\mathbf{M}$ and $\mathbf{N}$ of the angular momenta of the Spherical Top. We put $\mathbf{M} = (M_1, M_2, M_3.)$ and $\mathbf{N} = (N_1, N_2, N_3.)$ Below, the results of calculations will be presented only for one case (with comments only with respect to other situations). For obviousness, let us put $x_1 = x$, $y_1 = y$, $z_1 = z$. After some calculations the following results can be presented:

$$M_1 \varphi(x, y, z) = 0,$$

$$-M_2 \varphi(x, y, z) = x_3 \frac{\partial \varphi}{\partial x} + y_3 \frac{\partial \varphi}{\partial y} + z_3 \frac{\partial \varphi}{\partial z}, \quad M_3 \varphi(x, y, z) = x_2 \frac{\partial \varphi}{\partial x} + y_2 \frac{\partial \varphi}{\partial y} + z_2 \frac{\partial \varphi}{\partial z}.$$

It is visible that the operators in question have no transforms. In other case a picture is more interesting since

$$N_1 \varphi(x, y, z) = 0 \frac{\partial \varphi}{\partial x} - z \frac{\partial \varphi}{\partial y} + y \frac{\partial \varphi}{\partial z},$$

$$N_2 \varphi(x, y, z) = z \frac{\partial \varphi}{\partial x} + 0 \frac{\partial \varphi}{\partial y} - x \frac{\partial \varphi}{\partial z}, \quad N_3 \varphi(x, y, z) = -y \frac{\partial \varphi}{\partial x} + x \frac{\partial \varphi}{\partial y} + 0 \frac{\partial \varphi}{\partial z}.$$

The last relations can be written as follows:

$$\mathbf{N}\varphi(x, y, z) = (\mathbf{r} \times \nabla) \, \varphi(x, y, z).$$

These relations are valid for all coordinates x_μ, y_μ, z_μ, ($\mu = 1, 2, 3$). If we consider the coordinates ξ_μ, η_μ, ζ_μ, ($\mu = 1, 2, 3$), then the operators **N** take place of the dual operators **M** and vice versa. The formula

$$\mathbf{M}\varphi(\xi, \eta, \zeta) = -(\mathbf{r} \times \nabla)\,\varphi(\xi, \eta, \zeta)$$

exhibits this exchange. Thus, after the mappings in question we see instead of the dual operators of the angular momentum of the Spherical Top the operators of the orbital angular momentum of a point particle. It is interesting that the relation

$$\frac{1}{2}D\varphi(x,\,y,\,z) = x\frac{\partial \varphi}{\partial x} + y\frac{\partial \varphi}{\partial y} + z\frac{\partial \varphi}{\partial z}$$

holds valid in all instances. Thus, the duality is not conserved under the mapping in question.

We should like to pay attention to the following important things. The Cartesian coordinates of the four-dimensional Euclidian space are not observable, but the picture of natural rotation is very detailed and beautiful in this case and can be represented in the descriptive-geometric form. The Cartesian coordinates of the three-dimensional Euclidian space are quadratic functions of the coordinates of R^4 and are observable, but the harmonic picture of the Spherical Top angular momenta is reduced to the operators of the orbital angular momentum of a point particle. The duality is broken and hidden in this case. Further, there is an interesting problem of half-integer orbital angular momentum which is in the sphere of interests of physicists up to now [3]. From our consideration it follows that eigenfunctions of the dual operators of angular momentum of the natural rotational motion (known as matrices of rotations, see below) can be the eigenfunctions of the operator of the orbital angular momentum only in the case when these functions are even, and this is the hidden reason of the integer eigenvalues.

We pay attention that the results obtained in this section will be summarized in the conclusion of our investigation.

Generalized Electromagnetic Field

In this section, we consider a picture on which the electromagnetic field emerges as a ground state of the simplest form of energy called in [2] the generalized electromagnetic field (or heavy light). Here the theory of the generalized electromagnetic field is put into correspondence with the concept of natural Time [1] and formulated in the final form. We especially pay attention to the geometrical and physical meaning of heavy light and demonstrate that the problem of gauge fixing is an intrinsic property of local (gauge) symmetry itself. The investigation of the ground state of the proper electromagnetic field motivated by the theory of heavy light is also a subject of conversation at this exhibition.

The parallel displacement of vector fields $\overline{\mathbf{v}}$ and $\mathbf{v}$ with bilateral symmetry $\overline{v}^i = R^i_j v^j$ can be produced only by a pair of connections $\overline{P}^i_{jk}$ and P^i_{jk} with bilateral symmetry. From the law of parallel displacement we have

$$\overline{P}^i_{jk} = R^i_m P^m_{jn} R^n_k + R^i_m \partial_j R^m_k.$$

Being generalized the bilateral symmetry takes status of general linear group $GL(n, \mathbf{R})$ with the law of transformation

$$\overline{P}^i_{jk} = S^i_m P^m_{jn} T^n_k + S^i_m \partial_j T^m_k,$$

where T^i_j are the components of the operator S^{-1} inverse to the operator S, $S^i_k T^k_j = \delta^i_j$. For brevity, we use the matrix notation

$$\mathbf{S} = (S^k_l), \quad \mathbf{P}_i = (P^k_{il}), \quad \mathbf{E} = (\delta^k_l), \quad \mathbf{H}_{ij} = (H_{ijl}{}^k), \quad \mathrm{Tr}\,\mathbf{S} = S^k_k.$$

The transformations of internal symmetry in question take the form

$$\overline{\mathbf{P}}_i = \mathbf{S}\mathbf{P}_i\mathbf{S}^{-1} + \mathbf{S}\partial_i\mathbf{S}^{-1} = \mathbf{P}_i + \mathbf{S}\mathbf{D}_i\mathbf{S}^{-1},$$

where $\mathbf{D}_i$ is the natural differential operator associated with this internal symmetry only

$$\mathbf{D}_i\mathbf{S} = \partial_i\mathbf{S} + \mathbf{P}_i\mathbf{S} - \mathbf{S}\mathbf{P}_i = \partial_i\mathbf{S} + [\mathbf{P}_i, \mathbf{S}].$$

The Riemann tensor of P^i_{jk}

$$\mathbf{B}_{ij} = \partial_i\mathbf{P}_j - \partial_j\mathbf{P}_i + [\mathbf{P}_i, \mathbf{P}_j]$$

is reducible with respect to the transformations

$$\overline{\mathbf{B}}_{ij} = \mathbf{S}\mathbf{B}_{ij}\mathbf{S}^{-1},$$

since

$$\mathbf{B}_{ij} = (\mathbf{B}_{ij} - \frac{1}{4}\mathrm{Tr}(\mathbf{B}_{ij})\mathbf{E}) + \frac{1}{4}\mathrm{Tr}(\mathbf{B}_{ij})\mathbf{E}.$$

Hence, the strength tensor of the generalized electromagnetic field is given by the formula

$$\mathbf{H}_{ij} = \mathbf{B}_{ij} - \frac{1}{4}\mathrm{Tr}(\mathbf{B}_{ij})\mathbf{E}, \quad \mathrm{Tr}(\mathbf{H}_{ij}) = 0.$$

The ground states of the generalized electromagnetic field is defined by the equation $\mathbf{H}_{ij} = 0$. This equation has two evident solutions,

$$\mathbf{B}_{ij} = \frac{1}{4}\mathrm{Tr}(\mathbf{B}_{ij})\mathbf{E}$$

and $\mathbf{B}_{ij} = 0$. At first, we give a general solution of the last equation. Let four linear independent vector fields E^i_μ be given and one can construct purely algebraical components of the four covector fields E^μ_i, so that $E^i_\mu E^\mu_j = \delta^i_j$ holds valid. Setting $P^i_{jk} = L^i_{jk}$, where $L^i_{jk} = E^i_\mu \partial_j E^\mu_k$, we get a general solution of the equation in question. Now we can present the solution of the equation which defines the first ground state of the generalized electromagnetic field,

$$P^i_{jk} = L^i_{jk} + A_j\delta^i_k,$$

where A_i is any covector field. Since $\mathrm{Tr}(\mathbf{P}_i) = \partial_i q + nA_i$, where $q = \mathrm{Det}(E^\mu_i)$, then

$$\frac{1}{4}\mathrm{Tr}(\mathbf{B}_{ij}) = \partial_i A_j - \partial_j A_i = F_{ij}.$$

We conclude that the first ground state of the generalized electromagnetic field represents the new form of energy and should be considered independently. The theory of electromagnetic field in the framework of the concept of natural (gradient) Time was considered in [1] and will be discussed below in a different context. The second ground state of the generalized electromagnetic field we use as follows. By the local transformation S^i_j we can reduce the four covector fields E^μ_i to the form of four gradient covector fields $\partial_j \alpha^\mu = S^i_j E^\mu_i$, $\mathrm{Det}(\partial_j \alpha^\mu) \neq 0$. Now we consider the natural system of coordinates defined by the gradient of Time [1], $x^i = \phi^i(x, y, z, t)$. We can take that

$$\alpha^1(x^1, x^2, x^3, x^4) = \alpha^1(\phi^1, \phi^2, \phi^3, \phi^4) = x,$$
$$\alpha^2(x^1, x^2, x^3, x^4) = \alpha^2(\phi^1, \phi^2, \phi^3, \phi^4) = y,$$
$$\alpha^3(x^1, x^2, x^3, x^4) = \alpha^3(\phi^1, \phi^2, \phi^3, \phi^4) = z,$$
$$\alpha^4(x^1, x^2, x^3, x^4) = \alpha^4(\phi^1, \phi^2, \phi^3, \phi^4) = t.$$

Thus, in the system of natural coordinates x, y, z, t the connection in question is trivial, $L^i_{jk} = 0$. We see that the second ground state and gradient of Time define natural gauge fixing identically.

Let us introduce a tensor field $Q^i_{jk} = P^i_{jk} - L^i_{jk}$ and consider the irreducible deviation tensor

$$W^i_{jk} = Q^i_{jk} - \frac{1}{4} Q^l_{jl} \delta^i_k, \quad \mathbf{W}_j = \mathbf{Q}_j - \frac{1}{4} \mathrm{Tr}(\mathbf{Q}_j) \mathbf{E}$$

with the trivial trace $\mathrm{Tr}(\mathbf{W}_j) = 0$.

The Lagrangian of the generalized electromagnetic field takes the form

$$\mathcal{L}_P = -\frac{1}{4} \mathrm{Tr}(\mathbf{H}_{ij} \overline{\mathbf{H}}^{ij}) - \frac{\mu^2}{2} \mathrm{Tr}(\mathbf{W}_i \overline{\mathbf{W}}^i), \qquad (13)$$

where μ is a constant of dimension of cm^{-1},

$$\overline{\mathbf{H}}^{ij} = \overline{g}^{ik} \overline{g}^{jl} \mathbf{H}_{kl}, \quad \overline{\mathbf{W}}^i = \overline{g}^{ik} \mathbf{W}_k, \quad \overline{g}^{ij} = 2 t^i t^j - g^{ij}.$$

By varying the Lagrangian $\mathcal{L}_P$ with respect to $\mathbf{P}_i$, the following equations of the generalized electromagnetic field hold valid

$$\frac{1}{\sqrt{g}} D_i(\sqrt{g} \overline{\mathbf{H}}^{ij}) + \mu^2 \overline{\mathbf{W}}^j = 0, \qquad (14)$$

where $g = \mathrm{Det}(g_{ij})$. From the properties of the operator D_i it is not difficult to see that equations (14) are invariant with respect to the local internal transformations in question. The tensor character of these equations can be seen on the same ground. Equations (14) of the generalized electromagnetic field should be extended by the identity

$$D_i \mathbf{H}_{jk} + D_j \mathbf{H}_{ki} + D_k \mathbf{H}_{ij} = 0. \qquad (15)$$

From the definition of the operator D_i it follows that the left-hand side of relation (15) is a tensor.

We see that in some sense one can treat μ as the effective mass of the heavy photon. Since trace of $\mathbf{H}_{ij}$ equals zero, it is clear why we need to consider an irreducible tensor of deviation. In our case, the trace of $\mathbf{W}^i$ is trivial and the system of equations (14) is compatible. From (14) it follows that $\overline{\mathbf{W}}^i$ has to satisfy the equation

$$\frac{1}{\sqrt{g}} D_i(\sqrt{g}\overline{\mathbf{W}}^i) = 0, \tag{16}$$

because $D_i D_j(\sqrt{g}\overline{\mathbf{H}}^{ij}) = 0$. It is very important that the same equation appears under varying (16) with respect to E^i_μ. But as it is shown above, equations (16) represent sixteen additional invariant constraints on the potential $\mathbf{P}_i$ but not equations for E^i_μ.

Equations (14), (15) and (16) are invariant with respect to the local internal transformations of the group $GL(n, \mathbf{R})$, but in accordance with the explanation given above we have no problems with gauge fixing.

The so-called metric tensor of energy–momentum of the generalized electromagnetic field is quite similar to the energy–momentum tensor of its ground state F_{ij},

$$T_{ij} = -\mathrm{Tr}\,(\mathbf{H}_{ik}\overline{\mathbf{H}}_j{}^k) - \overline{g}_{ij}\mathcal{L}_P + \mu^2 \mathrm{Tr}\,(\mathbf{W}_i\mathbf{W}_j), \tag{17}$$

where $\overline{\mathbf{H}}_j{}^k = \mathbf{H}_{jl}\overline{g}^{kl}$. It is evident that the metric tensor of energy–momentum (17) is invariant with respect to the transformations of the general linear group $GL(n, \mathbf{R})$. One can show that the metric tensor of the energy–momentum satisfies the equation

$$\overline{\nabla}^i T_{ij} = 0, \tag{18}$$

where $\overline{\nabla}_i$ denotes the covariant derivative with respect to the Christoffel connection belonging to the auxiliary metric $\overline{g}_{ij}$ and $\overline{\nabla}^i = \overline{g}^{ij}\overline{\nabla}_j$.

Now we derive from eqs. (14) and (15) the generalized Maxwell equations. To this end, let us introduce the electric and magnetic strength of the generalized electromagnetic field by the direct and inverse mapping

$$\mathbf{J}_i = t^k \mathbf{H}_{ik}, \quad \mathbf{M}_i = \frac{1}{2} e_{ikjl} t^k \mathbf{H}^{jl} = t^k \overset{*}{\mathbf{H}}_{ik},$$

$$\mathbf{H}_{ik} = -t_i \mathbf{J}_k + t_k \mathbf{J}_i - e_{ikjl} t^j \mathbf{M}^l,$$

defined by the gradient of natural Time t^i. Since

$$\overline{\mathbf{H}}^{ij} = \overline{g}^{ik}\overline{g}^{jl}\mathbf{H}_{kl} = 2t^i \mathbf{J}^j - 2t^j \mathbf{J}^i + \mathbf{H}^{ij},$$

equations (14) can be represented in the following form:

$$t^k D_k \mathbf{J}^i - \mathbf{J}^k \partial_k t^i + \varphi \mathbf{J}^i = e^{ijkl} t_j D_k \mathbf{M}_l + \mu^2 \mathbf{S}^i,$$

$$\frac{1}{\sqrt{g}} D_i(\sqrt{g}\mathbf{J}^i) = \mu^2 \mathbf{S},$$

where

$$\mathbf{S}_i = h^j_i \mathbf{W}_j, \quad \mathbf{S} = t^k \mathbf{W}_k, \quad \varphi = \nabla_i t^i.$$

Equations (15) read

$$t^k D_k \mathbf{M}^i - \mathbf{M}^k \partial_k t^i + \varphi \mathbf{M}^i = -e^{ijkl} t_j D_k \mathbf{J}_l,$$

$$\frac{1}{\sqrt{g}} D_i(\sqrt{g} \mathbf{M}^i) = 0.$$

Since

$$\mathrm{Tr}(\mathbf{J}_i) = \mathrm{Tr}(\mathbf{M}_i) = \mathrm{Tr}(\mathbf{W}_i) = 0,$$

the system of equations in question is simultaneous.

This system of equations predicts the existence of a new and simplest forms of matter. We see that in the universe there are two simplest form of energy: the energy of the gravitational field [1] and the energy of the generalized electromagnetic field. These forms of energy exist in full harmony, because it can be verified that the group of diffeomorphisms (symmetry group of the gravitational field) is a group of external automorphisms of the local internal symmetry group $GL(n, \mathbf{R})$ of the generalized electromagnetic field. Thus, these fields are natural candidates for the construction of reasonable cosmological models. The theory of generalized electromagnetic field predicts the existence of the shadow world which consists of particles (heavy photons) that do not interact at all with any detectors and the influence of which becomes apparent only in the form of gravity and a new form of energy flow.

Above we have considered the extension of the theory of electromagnetic field into breadth and now we convert this process into depth. We show that in the reference space with the usual space-time structure

$$\bar{g}_{ij} = \eta_{ij} = 2a_i a_j - \delta_{ij}, \quad a_i = (0, 0, 0, 1)$$

the ground state of the electromagnetic field ($F_{ij} = 0$) has a hidden structure which takes a form of a complex scalar field associated with a one-dimensionally extended object, called a charged string. The string is said to be charged because the complex scalar field describing it interacts with the electromagnetic field. A charged string is characterized by extension of the symmetry group of the charge space to a group of turn dilatations. We propose relativistically invariant and gauge-invariant equations describing the interaction of a charged string with the electromagnetic field, and each their solution corresponds to a charged string. We achieve this by introducing the notion of a charged string index, which, as verified, takes only integer values. We establish equations from which it follows that charged strings fit naturally into the framework of the Maxwell-Dirac electrodynamics. The question of possible extensions of electrodynamics was discussed from various standpoints. The well known concept of a magnetic charge is considered as one of the crucial concepts, although it does not have experimental support. The existence of axions is not also confirmed experimentally, but this idea is currently required in at least three fundamental areas of modern research: quantum chromodynamics, cosmology, and solid state physics. Here, we show that in addition to monopoles and axions, it is worthwhile to consider the concept of a charged string, which is one aspect of the field theory and enters the Maxwell-Dirac electrodynamics as an intrinsic but still hidden element.

In the proposed theory, invariant under the local group of turn dilatations in the charge space, the concept of charged string leads to the observed effects only in the case where the index of a charged string is nonzero. The index of a charged string is equal to the closed path index on the plane of the complex variable, which corresponds to a closed contour in the reference space. Here, we regard a complex scalar field as a map of the reference space to the plane of a complex variable. We show how the concept of a charged string can be included in the Maxwell-Dirac electrodynamics.

The concept of charged strings was introduced there based on the algebra of complex numbers. Our approach is to introduce the concept of a charged string based on a geometric vision and symmetry considerations, with using linear rectangular coordinates on the plane of a complex variable similarly to how they are used in analytic geometry. In our opinion, this introduces the concept of a charged string most naturally.

By convention, we understand that for a complex variable $z = x + iy$, the letter z simultaneously denotes the complex number, the point z representing this complex number on the plane, and the vector z corresponding to this complex number. The modulus of $z = x + iy$ denoted by $|z|$, $|z| = \sqrt{x^2 + y^2}$, is called the length of the vector z. For normalized algebras with unity, the fundamental relation is

$$|zz_1| = |z|\,|z_1|. \tag{19}$$

The two-parameter group G_2 of turn dilatations of the plane of a complex variable is formed by transformations of the form

$$z \to \tilde{z} = \Lambda z, \tag{20}$$

where $\Lambda = \Lambda_1 + i\Lambda_2$ is an arbitrary complex number with a nonzero modulus $|\Lambda|$. Hence, the group space of G_2 is the plane of a complex variable $\Lambda = \Lambda_1 + i\Lambda_2$ with a punctured point $\Lambda = 0$. The action of the group of turn dilatations on the plane of a complex variable $z = x + iy$ is intransitive because the point $z = 0$ remains fixed for all transformations in the group G_2.

A complex number $z/|z|$ of the unit modulus is called the phase of the complex number z. Geometrically, the phase of the complex number z can be found as follows. Complex numbers whose moduli are equal to unity form a unit circle centered at the origin. Drawing a ray from the origin through the point z, we obtain a point representing the complex number $z/|z|$ at the intersection of the ray with the circle. In what follows, it is important that the phase of the complex number is invariant under dilatations of the complex plane because all points on the ray under consideration have identical phases. The phase of the complex number $z = 0$ is undefined because the point representing it coincides with the origin.

The subgroup $U(1)$ of the group G_2 is formed by phase transformations of the form

$$z \to \tilde{z} = \frac{\Lambda}{|\Lambda|} z. \tag{21}$$

It follows from (19) that the product of two phase transformations is again a phase transformation. Considering transformations (20) of the group of turn dilatations, we do not impose the condition $|\tilde{z}|^2 = |z|^2$, which is satisfied only by phase transformations. Because

$$\Lambda = (|\Lambda|)(\Lambda/|\Lambda|),$$

any transformation in the group G_2 can be regarded as a product of a phase transformation and expansion. As can be seen, in the consideration of phase transformations, the notion of the phase angle is not used, which, in our opinion, ensures the most natural and transparent introduction of the concept of a charged string.

The relation
$$z/|z| = \cos\alpha + i\sin\alpha.$$
defines the phase as a periodic function of the phase angle α, and α is therefore any real number. Let this number be positive. Then to find $z/|z|$, we take the thread of length α, fix one end at the point 1 on the unit circle, wind the thread counterclockwise on the circle, and find the number $z/|z|$. If α is a negative number, then to find $z/|z|$, we wind the thread of length $|\alpha|$ clockwise on the circle. This is well known from the theory of elementary trigonometric functions.

Now let the point z moves in the complex plane. This motion defines a path in the complex plane and is described by a function $z(\tau)$, $\tau_0 \leq \tau \leq \tau_1$. If the beginning of the path $z(\tau_0)$ coincides with the end of the path $z(\tau_1)$, then we have $z(\tau_0) = z(\tau_1)$ and say that the path is closed. The process of motion corresponds to a continuous change of the phase angle. This can be achieved as follows. On the complex plane, we draw a ray from the origin O through the point $z(\tau_0)$ intersecting the unit circle centered at the origin at the point P. The length of a thread superimposed on a circle and connecting the points 1 and P is equal to $\alpha(\tau_0)$. When the point z moves along the path $z(\tau)$, the ray OP rotates and on a unit circle winds a thread whose length from the point 1 to the point P is equal to $\alpha(\tau)$. If the closed path encircles the coordinate origin once, then
$$\alpha(\tau_1) - \alpha(\tau_0) = 2\pi,$$
and the considered process of motion is counterclockwise. In general,
$$\alpha(\tau_1) - \alpha(\tau_0) = 2\pi n, \tag{22}$$
where n is an integer.

We now introduce the notion of a charged string, which has the following motivation in the framework of the field theory. From complex numbers, we pass to complex scalar fields
$$\varphi(x) = \varphi_1(x^1, x^2, x^3, x^4) + i\,\varphi_2(x^1, x^2, x^3, x^4),$$
which are defined everywhere and have derivatives of the required order. We regard the complex plane as a charge space and $\varphi_1(x)$ and $\varphi_2(x)$ as rectangular coordinates on the complex plane. A local group of turn dilatations is formed by transformations
$$\varphi(x) \to \tilde{\varphi}(x) = \Lambda(x)\varphi(x), \tag{23}$$
where in this case
$$\Lambda(x) = \Lambda_1(x^1, x^2, x^3, x^4) + i\,\Lambda_2(x^1, x^2, x^3, x^4)$$
is an arbitrary complex scalar field with a nonzero modulus. The modulus of a complex scalar field $\varphi(x)$ is nonzero if the loci M_1 and M_2, where the functions $\Lambda_1(x^1, x^2, x^3, x^4)$ and $\Lambda_2(x^1, x^2, x^3, x^4)$ respectively vanish, are disjoint.

We now consider the region where the phase of the complex scalar field is undefined. As $z \to \varphi(x)$, the point $z = 0$ corresponds to the locus of points defined by the equation $\varphi(x) = 0$ or the equations

$$\varphi_1(x^1, x^2, x^3, x^4) = 0, \quad \varphi_2(x^1, x^2, x^3, x^4) = 0, \tag{24}$$

where we of course assume that the functions are not sign-definite and the gradients of $\varphi_1(x)$ and $\varphi_2(x)$ are noncollinear. In what follows, we assume conditions under which Eqs. (24) define a line in the reference space at every instant.

A line in the reference space on which the phase of a complex scalar field is undefined at a fixed instant is called a charged string. Equations (24) describe the motion of a charged string in space. We must find field equations for the functions $\varphi_1(x)$ and $\varphi_2(x)$ such that each solution of them describes a charged string. For this, we consider the phase of the complex scalar field φ:

$$\chi = \frac{\varphi}{|\varphi|}, \quad \bar{\chi} = \frac{\bar{\varphi}}{|\varphi|}.$$

We note that the introduced notation for the phase of the complex scalar field is convenient for a more compact and transparent form of the formulas. The complex scalar field $\chi = \varphi/|\varphi|$, invariant under local expansions, is called the phase field. As shown above, the considered Abelian gauge symmetry is one of the properties of the theory of complex scalar fields, and the phase field $\chi = \varphi/|\varphi|$ therefore differs from local phase transformations $\Lambda(x)/|\Lambda(x)|$, which do not lead to the observed effects. This circumstance in fact forms the basis of the concept of a charged string.

For the phase field $\chi = \varphi/|\varphi|$, to describe a charged string, we introduce the index θ of a charged string. The charged-string index θ is defined as an integral over a closed contour in the reference space that does not pass through the string:

$$\theta = \frac{1}{2\pi} \oint (i\bar{\chi}\partial_k \chi) dx^k = \frac{1}{2\pi} \oint (i\bar{\chi} d\chi). \tag{25}$$

We can represent the closed contour visually as follows. Through a given string point, we draw the plane perpendicular to the tangent vector at this point. The closed contour is placed in this plane so that it encircles the chosen string point. The utility of the charged string index is that a pure gauge does not give any observable effects. We once again note that a local phase transformation is defined everywhere and the phase of a given complex scalar field is defined everywhere. In contrast, only complex scalar fields with the phase defined in not the entire reference space are suitable for describing charged strings.

We show that the index of a charged string is invariant under local turn dilatations (23). For local turn dilatations (23), the covector field $i\bar{\chi}\partial_k \chi$ is transformed according to the law

$$i\bar{\chi}\partial_k \chi \to i\bar{\chi}\partial_k \chi + \Lambda_k, \quad \Lambda_k = i\frac{\bar{\Lambda}}{|\Lambda|} \partial_k \left(\frac{\Lambda}{|\Lambda|} \right).$$

We hence find that

$$\tilde{\theta} = \frac{1}{2\pi} \oint (i\bar{\chi}\partial_k \chi + \Lambda_k) dx^k = \theta + \frac{1}{2\pi} \oint \Lambda_k dx^k = \theta,$$

because $\partial_j \Lambda_k - \partial_k \Lambda_j = 0$ and the integral of Λ_k is zero by the Stokes theorem. We now verify that the index can take the values

$$\theta = n, \tag{26}$$

where n is an integer. To verify relation (26), we argue as follows. We regard the complex scalar field $\varphi(x)$ as a map of the reference space to the complex plane. The charged string then maps to the origin of the complex plane. We introduce a contour C in the reference space, setting

$$C: x^k = f^k(\tau), \quad \tau_0 \leq \tau \leq \tau_1.$$

If the contour is closed, then $f^k(\tau_0) = f^k(\tau_1)$. The contour corresponds to a path in the complex plane. If a closed contour in the reference space encircles a line on which the complex scalar field vanishes, then the corresponding closed path on the complex plane encircles the origin. Along the contour, the phase field can be regarded as a function of the phase angle,

$$\varphi/|\varphi| = \cos\alpha + i \sin\alpha.$$

We hence find that

$$(i\bar{\chi}\partial_k\chi)dx^k == -d\alpha.$$

Consequently,

$$\int_C (i\bar{\chi}\partial_k\chi)dx^k = -\alpha(\tau_1) + \alpha(\tau_0).$$

If the considered contour in the reference space is closed, then the path on the complex plane is also closed, and it encircles the origin. Therefore, according to (22),

$$\alpha(\tau_1) - \alpha(\tau_0) = 2\pi n,$$

where the integer n indicates how many times the point following the closed path on the complex plane encircles the origin. Relation (26) is thus verified.

The conclusion is that the theory of a charged string invariant under the local group of turn dilatations (23) can be represented by the Lagrangian

$$L_P = \frac{\hbar c \,\theta}{2\lambda^2}(D_k\bar{\chi})(D^k\chi) - \frac{1}{4}F_{kj}F^{kj}, \tag{27}$$

where θ is the charged-string index and we consider the phase field $\chi = \varphi/|\varphi|$ invariant under local expansions as the fundamental degree of freedom. The constant λ of dimension of length is also necessary because the phase field $\chi = \varphi/|\varphi|$ is dimensionless.

Since a coupling constant q for the phase and electromagnetic fields might not coincide with the coupling constant e for the spinor and electromagnetic fields, $D_k\chi$ and $D_k\bar{\chi}$ in Lagrangian (27) have the forms

$$D_k\chi = \left(\partial_k - \frac{iq}{\hbar c}A_k\right)\chi, \quad D_k\bar{\chi} = \left(\partial_k + \frac{iq}{\hbar c}A_k\right)\bar{\chi}.$$

Lagrangian (27) is invariant under the local group of turn dilatations (23), because it is easy to verify that

$$\left(\partial_k - \frac{iq}{\hbar c}\tilde{A}_k\right)\tilde{\chi}(x) = \frac{\Lambda(x)}{|\Lambda(x)|}\left(\partial_k - \frac{iq}{\hbar c}A_k\right)\chi(x),$$

where

$$\tilde{\chi}(x) = \frac{\Lambda(x)}{|\Lambda(x)|}\chi(x), \quad A_k \to \tilde{A}_k = A_k - \frac{\hbar c}{q}\Lambda_k, \quad \Lambda_k = i\frac{\bar{\Lambda}}{|\Lambda|}\partial_k\left(\frac{\Lambda}{|\Lambda|}\right).$$

Varying Lagrangian (27), we obtain the system of equations

$$\partial_k(i\bar{\chi}D^k\chi) = 0, \tag{28}$$

$$\partial_k F^{kj} + \frac{q\theta}{\lambda^2}(i\bar{\chi}D^j\chi) = 0, \tag{29}$$

which describes the interaction of the electromagnetic and phase fields. The system of equations (28), (29) is consistent, because (28) follows from (29).

We consider one consequence of (29). If there is a closed contour in the reference space that encircles a charged string and for which the relation

$$\oint (\partial^j F_{jk})dx^k = 0,$$

is satisfied, then we have

$$\oint (i\bar{\chi}D_k\chi)dx^k = 0$$

in accordance with (29). It hence follows from formula (26) that

$$\Phi + \frac{\hbar c}{q}2\pi n = \Phi + \frac{hc}{q}n = 0, \tag{30}$$

where by the Stokes theorem,

$$\Phi = \oint A_k dx^k = \frac{1}{2}\int F_{kj}\,dx^k \wedge dx^j \tag{31}$$

is the flux of the electromagnetic field through a surface bounded by the integration contour. It follows from (30) that under the condition $\oint(\partial^j F_{jk})dx^k = 0$, this flux is quantized, and the flux quantum Φ_0 is

$$\Phi_0 = \frac{hc}{q}.$$

Below we establish the relation between the concept of a charged string and the Maxwell-Dirac electrodynamics.

The Maxwell-Dirac electrodynamics is described by the Lagrangian

$$L = -\frac{1}{4}F_{jk}F^{jk} + \frac{i\hbar c}{2}[\bar{\psi}\gamma^k(\partial_k\psi) - (\partial_k\bar{\psi})\gamma^k\psi] - mc^2\bar{\psi}\psi + eA_k(\bar{\psi}\gamma^k\psi),$$

which is invariant under local phase transformations of the spinor field and gauge transformations of the electromagnetic field of the forms

$$\tilde{\psi} = \frac{\Lambda}{|\Lambda|}\psi, \quad \tilde{A}_k = A_k - \frac{\hbar c}{e}\Lambda_k, \quad \Lambda_k = i\frac{\bar{\Lambda}}{|\Lambda|}\partial_k\left(\frac{\Lambda}{|\Lambda|}\right). \tag{32}$$

Comparing the electrodynamics of the phase field with the Maxwell-Dirac electrodynamics, we conclude that the problem of including the concept of a charged string in the Maxwell-Dirac theory admits the following solution. To the Maxwell-Dirac Lagrangian, we add the Lagrangian of the phase field itself and the current-current Lagrangian. For the Lagrangian L_G, solving the considered problem, we thus obtain $L_G = L + L_\chi$, where consequently

$$L_\chi = \frac{\hbar c \theta}{2\lambda^2}(D_k \bar{\chi})(D^k \chi) + \hbar c \theta (i\bar{\chi} D_k \chi)(\bar{\psi}\gamma^k \psi).$$

We consider the equations obtained as a result of varying the Lagrangian $L_G = L + L_\chi$. The Maxwell and Dirac equations in the proposed model become

$$\partial_k F^{kj} + \frac{q\theta}{\lambda^2}(i\bar{\chi}D^j\chi) + (e+g)(\bar{\psi}\gamma^j\psi) = 0, \qquad (33)$$

$$\gamma^k(i\partial_k + \frac{e+g}{\hbar c}A_k + i\theta\,\bar{\chi}\partial_k\chi)\psi = \frac{mc}{\hbar}\psi, \qquad (34)$$

while Eq. (28) remains the same. In (33) and (34), $g = q\theta$.

We note one important circumstance. In the Lorentz gauge $\partial_i A^i = 0$, Eq. (28) reduces to

$$\partial_k(i\bar{\chi}\partial^k\chi) = 0. \qquad (35)$$

It hence follows that the phase field in the Lorentz gauge is external with respect to the electromagnetic and spinor fields. Hence, the phase field uniquely fixes the local gauge and influences the behavior of the electromagnetic and spinor fields without a back reaction if the effects of gravity are neglected, of course. This is the local gauge when $e = q$. In the global gauge $e \neq q$ the Lorentz gauge is impossible and from Eqs. (33) and (34) it follows that the phase field does not allow "switching off the interaction of the electromagnetic field with the spinor field because the effective coupling constant $e + g$ cannot vanish. However, in the local gauge (at $e = q$) this is possible, because we obtain in this case that the effective coupling constant $e + g = e(1 + \theta)$ depends on the orientation of the charged string. Indeed, $e + g = 2e$ at $\theta = 1$ and $e + g = 0$ at $\theta = -1$. In the last case, Eq. (33) becomes Eq. (29), and Eq. (34) describes the states of the spinor field in the field of a charged string. These eqs. can be used to understand the unique electrodynamic properties of the superconducting state. Indeed, at $e+g = 0$ we obtain an explanation of the Meissner effect, which should be observed everywhere except the points occupied by a charged string. The constant λ in this case characterizes the penetration depth of the field. Electrons are then connected with charged strings. The so-called superconducting current S_k, which appears in superconductivity, turns out to be gauge-invariant and is equal to

$$S_k = \frac{q\theta}{\lambda^2}(i\bar{\chi}D_k\chi).$$

The other possible physical applications of the Eqs. of the electrodynamics with charged strings can be found in [4]. We can put forward a conjecture that the superconductivity is the phenomenon of changing the orientations of charged strings in solid.

Newton Equation and Geodesic Motion

In this chapter we consider the motion of a charged massive point particle in the external gravitational and electromagnetic fields via the action principle that is general covariant and reparametrization invariant. Reparametrization is defined as a one-to-one and smooth transformation of the definition region of the parameter along a path in the reference space. We derive the general covariant and reparametrization invariant Newton equations from the Euler-Lagrange equations which describe the geodesic motion and can at first to see the visible changes that the principle of general covariance and reparametrization invariance carries into the classical mechanics. We uncover a hidden physical content of special and general relativity with the concept of natural Time [1] which provides an adequate solution of all unresolved so far problems. Below one can find one more example.

On the basis of the natural concept of Time general covariant definitions of physical momentum, velocity and energy invariant with respect to reparametrization are established. The connection between the change of the orientation of a path and the operation of the charge conjugation is marked.

Let Ω be a domain in R^n. A one-to-one and smooth mapping of Ω into itself is called transformation or local diffeomorphism. The group of local diffeomorphisms is denoted as Diff(Ω). The local diffeomorphism is defined by the smooth functions

$$\alpha^i(x^1, \cdots, x^n) = \alpha^i(x); \quad \alpha^i_{-1}(x^1, \cdots, x^n) = \alpha^i_{-1}(x), \quad i = 1, 2, \cdots n$$

so that

$$y^i = \alpha^i(x), \quad \alpha^i_{-1}(y) = x^i.$$

Let us consider a scalar field $\varphi(x^1, \cdots, x^n) = \varphi(x)$ in the domain Ω. The new scalar field $\widetilde{\varphi}(x)$ in this domain is defined by the transformation of Diff(Ω) as follows:

$$\widetilde{\varphi}(x) = \varphi(\alpha(x)). \tag{36}$$

The transformation (36) is called reparametrization of the scalar field $\varphi(x)$. The reparametrization of the covector field $A_i(x)$ is defined by the equation

$$\widetilde{A}_i(x) = A_k(\alpha(x)) \frac{\partial \alpha^k(x)}{\partial x^i}. \tag{37}$$

A similar formula can be written for any geometrical quantity. We mention only the vector field and symmetrical covariant tensor field g_{ij}.

$$\widetilde{A}^i(x) \frac{\partial \alpha^k(x)}{\partial x^i} = A^k(\alpha(x)), \tag{38}$$

$$\widetilde{g}_{ij}(x) = g_{kl}(\alpha(x)) \frac{\partial \alpha^k(x)}{\partial x^i} \frac{\partial \alpha^l(x)}{\partial x^j}. \tag{39}$$

Now we need to establish an important relation that will be demanded in what follows. The potential of the electromagnetic field $A_i(x) + \partial_i \lambda(x)$ is invariant with respect to reparametrization if

$$A_i(x) + \partial_i \lambda(x) = \widetilde{A}_i(x) + \partial_i \widetilde{\lambda}(x) = A_k(\alpha(x)) \frac{\partial \alpha^k(x)}{\partial x^i} + \partial_i \lambda(\alpha(x)).$$

Writing this condition in the infinitesimal form, we have

$$\xi^l \partial_l A_i + A_l \partial_i \xi^l + \partial_i(\xi^l \partial_l \lambda) = 0.$$

The last equation can be written as follows:

$$\xi^l F_{li} + \partial_i((A_l + \partial_l \lambda)\xi^l) = 0, \tag{40}$$

where F_{li} is the tensor of the electromagnetic field.

By definition, if the potential of the gravitational field is invariant with respect to reparametrization, then

$$\tilde{g}_{ij}(x) = g_{kl}(\alpha(x))\frac{\partial \alpha^k(x)}{\partial x^i}\frac{\partial \alpha^l(x)}{\partial x^j} = g_{ij}(x).$$

Given $g_{ij}(x)$, we can find $\alpha^i(x)$ (or $g_{ij}(x)$ at given $\alpha^i(x)$). This is an infinitesimal form of this equation

$$\xi^l \partial_l g_{ij} + g_{lj}\partial_i \xi^l + g_{il}\partial_j \xi^l = 0. \tag{41}$$

Now we can define the operator of infinitesimal motion $D_{\mathbf{t}}$, which is defined by the gradient of time $t^i = g^{il}\partial_l f = g^{il}t_l$

$$D_{\mathbf{t}}g_{ij} = t^l \partial_l g_{ij} + g_{lj}\partial_i t^l + g_{il}\partial_j t^l = \nabla_i t_j + \nabla_j t_i = 2\nabla_i t_j, \tag{42}$$

where ∇_i is covariant derivative with respect to the connection belonging to g_{ij}. The operator (42) gives a rate of changes of the gravitational field with Time and plays a very important role in the gravidynamics [1]. For example, the linear operator

$$P_j^i = \frac{1}{2}g^{ik}D_{\mathbf{t}}g_{jk} = \frac{1}{2}g^{ik}(\nabla_j t_k + \nabla_k t_j) = g^{ik}\nabla_j t_k = \nabla_j t^i \tag{43}$$

is called the momentum of the gravitational field and the gravitational field is called static if its rate of change is trivial

$$D_{\mathbf{t}}g_{ij} = t^l \partial_l g_{ij} + g_{lj}\partial_i t^l + g_{il}\partial_j t^l = 0.$$

From (42) it follows that if the rate of change of the gravitational field is equal to zero in some system of coordinates, then it will be trivial in any system of coordinates. In accordance with (42) and (43), a momentum of the static gravitational field is trivial and this can be used as an other definition of the static gravitational field. In gravidynamics (general relativity equipped by natural Time and by this put in correspondence with its principles) the notions of the non-static and static gravitational field are invariant concepts in accordance with the fundamental principles of general relativity.

Let us consider the class of functions of one variable $\tau = \alpha(\sigma)$, for which the region of their definition coincides with the region of their values. If $\sigma \in [a, b]$, then $\alpha(\sigma) \in [a, b]$ as well. Every function of this class defines the same path in the reference space R^n in accordance with the equations $x^i = x^i(\alpha(\sigma))$, $i = 1, 2, \cdots n$. For a graphic representation of a set of functions $\tau = \alpha(\sigma)$ one should use the plane σ, τ, with fixed points (a, a), (a, b), (b, b), (b, a). Decreasing functions connect the points (a, b), (b, a)

and increasing ones connect the points (a, a), (b, b). The simplest graphics are defined by the functions $\tau = \sigma$, $\tau = b - (\sigma - a)$. For the functions $\alpha(\sigma)$ in question the following conditions are fulfilled:
$$\frac{d\alpha}{d\sigma} > 0, \quad \frac{d\alpha}{d\sigma} < 0,$$
which define the orientation of the path. In the first case $\alpha(a) = a$, $\alpha(b) = b$, whereas $\alpha(a) = b$, $\alpha(b) = a$ in the case of the path with the opposite orientation. The equations of motion of a point particle are invariant with respect to reparametrization if the functions $x^i = x^i(\alpha(\sigma))$, $i = 1, 2, \cdots n$ are solutions of these equations under any $\alpha(\sigma)$. Under these conditions there is no sense to define causality as in the case of cars on a highway. Let us consider the autonomous system of equations
$$\frac{dx^i}{d\sigma} = v^i(x^1, \cdots x^n)$$
and suppose that $x^i = x^i(\sigma)$, $i = 1, 2, \cdots n$ is its solution. Since
$$\frac{dx^i}{d\sigma} = \frac{dx^i(\alpha(\sigma))}{d\sigma} = \frac{dx^i(\alpha(\sigma))}{d\alpha(\sigma)} \frac{d\alpha(\sigma)}{d\sigma} = v^i(x^1(\alpha(\sigma)), \cdots x^n(\alpha(\sigma))) \frac{d\alpha(\sigma)}{d\sigma}$$
then $x^i = x^i(\alpha(\sigma))$, $i = 1, 2, \cdots n$ is again solution but only under the condition $d\alpha(\sigma)/d\sigma = 1$. Hence, the system of equations in question is invariant only with respect to the transformations $\sigma \to \sigma + a$, where a is constant. It is not invariant with respect to reparametrization. It is clear that the Hamilton equations are not invariant with respect to reparametrization and that is why it is very important from a physical point view to consider in all details the transition from the Euler-Lagrange equations, which are general covariant and invariant with respect to reparametrization, to the Hamilton equations in the case of the motion of a massive charged particle in the external gravitational and electromagnetic fields.

To write Lagrangian for a particle in the gravitational and electromagnetic fields, we give some needed definitions. For any vector ξ^i, η^i we put
$$(\xi|\eta) = g_{ij}\xi^i \eta^j, \quad <\xi|\eta> = \bar{g}_{ij}\xi^i \eta^j,$$
where $\bar{g}_{ij} = 2t_i t_j - g_{ij}$ is auxiliary metric defined by the bilateral symmetry of our world with respect to the gradient of natural Time $t_i = \partial_i f(x)$. From the definition it follows that
$$<\xi|\eta> = 2(t|\xi)(t|\eta) - (\xi|\eta).$$

We investigate the Lagrangian
$$L = -m\sqrt{\bar{g}_{ij}u^i u^j} - eA_i u^i, \tag{44}$$
where $u^i = dx^i/d\sigma$, m and e denote mass and charge of a point particle to which we put in correspondence a path $x^i = x^i(\alpha(\sigma))$, $i = 1, 2, \cdots n$. The coordinates and parameter σ have the dimension of length. The action is dimensionless and hence the mass m and components of the electromagnetic potential have the dimension of inverse length and the charge e is dimensionless.

We verify that a shortened action ($e = 0$) is invariant with respect to the reparametrization. Since $\sigma \subset [a, b]$, we write the action in the form of definite integral

$$A = \int_a^b L(\sigma)d\sigma = F(b) - F(a),$$

where $F(\sigma)$ is the primitive of $L(\sigma)$ and $x^i = x^i(\sigma)$. For $x^i = x^i(\alpha(\sigma))$ we write the action in the following form:

$$\widetilde{A} = \int_a^b \widetilde{L}(\sigma)d\sigma.$$

Since

$$\widetilde{L}(\sigma) = L(\alpha(\sigma))|d\alpha/d\sigma|,$$

we will distinguish two cases. If $d\alpha/d\sigma > 0$, then $\alpha(a) = a$, $\alpha(b) = b$ and hence $\widetilde{A} = F(\alpha(b)) - F(\alpha(a)) = F(b) - F(a) = A$. For $d\alpha/d\sigma < 0$, $\alpha(a) = b$, $\alpha(b) = a$ and hence $\widetilde{A} = -F(\alpha(b)) + F(\alpha(a)) = F(b) - F(a) = A$. Our statement is verified.

The Euler-Lagrange equations $\delta A = 0$ can be written as follows:

$$\frac{d}{d\sigma}\left(-m\frac{\bar{g}_{ij}u^j}{\sqrt{<u|u>}} - eA_i\right) = -\frac{m}{2}\frac{\partial \bar{g}_{jk}}{\partial x^i}\frac{u^j}{\sqrt{<u|u>}}u^k - e\frac{\partial A_k}{\partial x^i}u^k,$$

where $<u|u> = \bar{g}_{ij}u^i u^j$. After some transformations these equations can be written in the following form:

$$\frac{dQ^i}{d\sigma} + \bar{\Gamma}^i_{kj}u^k Q^j = \frac{e}{c}\bar{F}^i_k u^k, \qquad (45)$$

where

$$Q^i = \frac{mu^i}{\sqrt{<u|u>}}, \quad \bar{F}^i_k = \bar{g}^{il}F_{lk},$$

and $\bar{\Gamma}^i_{kj}$ are the Christoffel symbols of the auxiliary metric $\bar{g}_{ij}$. If Γ^i_{kj} are the Christoffel symbols of the metric g_{ij}, then

$$\bar{\Gamma}^i_{kj} = \Gamma^i_{kj} + 2t^i \nabla_k t_j.$$

The last relation and concept of natural or gradient Time is the starting point to derive from equations (45) the general covariant Newton equations (Newton's second law) invariant with respect to path reparametrization

$$x^i = x^i(\sigma) \to x^i = x^i(\alpha(\sigma)).$$

We define that some quantity (which is defined along a path $x^i = x^i(\sigma)$) is invariant with respect to the path reparametrization if it does not depend from the factor $d\alpha(\sigma)/d\sigma$. It is evident that all physical fields are invariant with respect to the path reparametrization if a path goes through the region of definition of these fields. It is easy to see that the vector Q^i is invariant with respect to the path reparametrization but u^i is not. It is necessary to strictly distinguish the path reparametrization and the change of parametrization of a path $\tau = \varphi(\sigma)$.

The path reparametrization changes the orientation of the path, if $d\alpha/d\sigma < 0$. Let $x^i = x^i(\sigma)$ be a solution of equations (45), then $x^i = x^i(\alpha(\sigma))$ is again a solution of the same equations, if $d\alpha/d\sigma > 0$. In the opposite case $d\alpha/d\sigma < 0$ and the functions $x^i = x^i(\alpha(\sigma))$ will be a solution of equations (45) only after changing the sign of the electric charge e. Really, for $x^i = x^i(\sigma)$ we have

$$Q^i(\sigma) = \frac{dx^i(\sigma)}{\sqrt{\bar{g}_{ij}(x(\sigma))dx^i(\sigma)dx^j(\sigma)}}$$

and hence, for $x^i = x^i(\alpha(\sigma))$ we obtain

$$\widetilde{Q}^i(\sigma) = \frac{dx^i(\alpha(\sigma))}{\sqrt{\bar{g}_{ij}(x(\alpha(\sigma)))dx^i(\alpha(\sigma))dx^j(\alpha(\sigma))}} = Q^i(\alpha(\sigma))\varepsilon,$$

$$\varepsilon = \frac{d\alpha(\sigma)/d\sigma}{|d\alpha(\sigma)/d\sigma|} = \pm 1.$$

We see the direct connection between the charge conjugation and the orientation of a path of a charged massive particle. The symmetry of this kind attracts attention since there is evident but unclear discrepancy between matter and antimatter in the Universe.

Now we prove that if the vector field ξ^i is a solution to equations (40) and besides

$$\nabla_i \xi_j + \nabla_j \xi_i = 0, \quad \xi_i = g_{ij}\xi^j, \quad t_i\xi^i = 0,$$

then the invariant

$$I = \xi^i(Q_i - eA_i - e\partial_i\lambda)$$

is a first integral of equations (45); $dI/d\sigma = 0$. We have

$$\frac{d}{d\sigma}(\xi_i(Q^i)) = Q^i\frac{d\xi_i}{d\sigma} + \xi_i\frac{dQ^i}{d\sigma},$$

$$Q^i\frac{d\xi_i}{d\sigma} = Q^i(\partial_k\xi_i - \Gamma^l_{ki}\xi_l + \Gamma^l_{ki}\xi_l)u^k = Q^i(\nabla_k\xi_i + \Gamma^l_{ki}\xi_l)u^k.$$

Since $Q^i(\nabla_k\xi_i)u^k = \frac{1}{2}(\nabla_k\xi_i + \nabla_i\xi_k)Q^iu^k = 0$, then

$$\frac{d}{d\sigma}(\xi_iQ^i) = \xi_i(\frac{dQ^i}{d\sigma} + \Gamma^i_{kj}u^kQ^j) = \xi_i(\frac{dQ^i}{d\sigma} + \Gamma^i_{kj}u^kQ^j) = \xi_i(\frac{dQ^i}{d\sigma} + \overline{\Gamma}^i_{kj}u^kQ^j),$$

because $\xi_i\overline{\Gamma}^i_{kj} = \xi_i(\Gamma^i_{kj} + 2t^i\nabla_k t_j) = \xi_i\Gamma^i_{kj}$. Now we take into account that in accordance with (40)

$$\xi_i\overline{F}^i_k u^k = \xi_i\bar{g}^{il}F_{lk}u^k = \xi_i(2t^it^l - g^{il})F_{lk}u^k = -\xi^l F_{lk}u^k = \frac{d}{d\sigma}(\xi^l(A_l + \partial_l\lambda))$$

and hence our statement is proven.

Now we are strongly motivated to consider the decomposition

$$Q^i = Wt^i + \pi^i, \quad (t|\pi) = t_i\pi^i = 0, \qquad (46)$$

and derive equations for W and π^i from equations (45). We have

$$\frac{dQ^i}{d\sigma} + \overline{\Gamma}^i_{kj} u^k Q^j = \frac{d\pi^i}{d\sigma} + \overline{\Gamma}^i_{kj} \pi^j u^k + W(\frac{dt^i}{d\sigma} + \overline{\Gamma}^i_{kj} t^j u^k) + \frac{dW}{d\sigma} t^i.$$

Taking into account the relations $t^k \nabla_k t^i = t^k \nabla_i t^k = 0$, $W u^k = (t_l u^l) Q^k$ it is easy to check that

$$W(\frac{dt^i}{d\sigma} + \overline{\Gamma}^i_{kj} t^j u^k) = (t_l u^l) \pi^k \nabla_k t^i. \tag{47}$$

We put

$$\frac{d\pi^i}{d\sigma} + \overline{\Gamma}^i_{kj} \pi^j u^k = \frac{d\pi^i}{d\sigma} + \widetilde{\Gamma}^i_{kj} \pi^j u^k + (\pi^j \nabla_j t_k u^k) t^i = \frac{D\pi^i}{d\sigma} + (\pi^j \nabla_j t_k u^k) t^i$$

and have

$$t_i \frac{D\pi^i}{d\sigma} = 0,$$

where

$$\widetilde{\Gamma}^i_{kj} = \Gamma^i_{kj} + t^i \nabla_k t_j.$$

We conclude that $D\pi^i/d\sigma$ can be considered as a rate of change of the physical momentum with respect to the parameter σ.

To complete our investigation, we transform the right-hand side of equations (45) as well. The general covariant definition of the electric and magnetic fields is given by the relations [1]

$$E_k = t^i F_{ki}, \quad H_k = t^i \widetilde{F}_{ki}, \quad \widetilde{F}_{ki} = \frac{1}{2} e_{kijl} F^{jl}.$$

Since $t^i E_i = t^i H_i = 0$, then

$$F_{ki} = -t_k E_i + t_i E_k - e_{kijl} t^j H^l$$

and hence,

$$\overline{F}^i_k u^k = -t^i E_k u^k - (t_k u^k) E^i - [u \times H]^i, \quad [u \times H]_i = e_{ijkl} t^j u^k H^l.$$

We conclude that equation (45) can be written as a system of two equations

$$\frac{dW}{d\sigma} + \pi^i \nabla_i t_k u^k + e E_k u^k = 0, \tag{48}$$

$$\frac{D\pi^i}{d\sigma} + (t_l u^l) \pi^j \nabla_j t^i + e(t_l u^l) E^i + e[u \times H]^i = 0. \tag{49}$$

We see that there is small asymmetry in the expression for the Lorentz force F^i due to the factor $t_i u^i$ and this is a very important argument to introduce the fundamental concept of physical velocity. We write the sequence

$$u^i = (t_l u^l) \frac{u^i}{t_l u^l} = (t_l u^l)(\frac{u^i}{t_l u^l} - t^i + t^i) = (t_l u^l)(v^i + t^i)$$

and have the beautiful expression the the Lorentz force

$$F^i = (t_l u^l)\Big(eE^i + e[v \times H]^i\Big),$$

since $[t \times H] = 0$. Thus, the definition of physical velocity with respect to the parameter σ is given by the expression

$$v^i = \frac{u^i}{t_l u^l} - t^i, \quad t_i v^i = 0 \qquad (50)$$

which is general covariant and invariant with respect to the path reparametrization. The other evidences in support of this definition of the physical velocity can be presented as follows. Since $u^i = (t_l u^l)(v^i + t^i)$, then

$$<u|u> = |t_l u^l|\sqrt{1 - v^2}$$

and hence,

$$Q^i = \epsilon\left(\frac{mv^i}{\sqrt{1-v^2}} + \frac{mt^i}{\sqrt{1-v^2}}\right), \quad \epsilon = \frac{|t_l u^l|}{t_l u^l} = \pm 1.$$

We conclude that

$$\pi^i = \frac{\epsilon m v^i}{\sqrt{1-v^2}}, \qquad (51)$$

$$W = \frac{\epsilon m}{\sqrt{1-v^2}}. \qquad (52)$$

At last, we have the following expression for the Lagrangian (44):

$$L = (t_l u^l)\Big(-\epsilon m \sqrt{1-v^2} - e\Phi_i v^i - e\varphi\Big), \qquad (53)$$

where $\varphi = (t|A) = t^l A_l$ is the scalar potential of the electromagnetic field and $\Phi_i = A_i - t_i \varphi$ is its vector potential $(t|\Phi) = t^i \Phi_i = 0$.

Equations (48) and (49) read

$$\frac{1}{t_l u^l}\frac{dW}{d\sigma} + \pi^i \nabla_i t_k v^k + eE_k v^k = 0, \qquad (54)$$

$$\frac{1}{t_l u^l}\frac{D\pi^i}{d\sigma} + \pi^k \nabla_k t^i + eE^i + e[v \times H]^i = 0. \qquad (55)$$

Factor $1/t_l u^l$ provides the invariance of equations (54) and (55) with respect to the path reparametrization. Equation (54) expresses the law of energy conservation and equations (55) represent the generalization of the Newton second law defined by the principles of general relativity.

From (54) and (55) we have the following expression for the gravitational force:

$$F_g^i = \pi^k \nabla_k t^i = \pi^k P_k^i, \qquad (56)$$

which is general covariant (if F_g^i is equal to zero in some system of coordinates, then it will be trivial in any system of coordinates.) We conclude that the gravitational force is not trivial only for the non-static gravitational fields. Remind that for the static gravitational field

$$D_{\mathbf{t}}g_{ij} = t^l \partial_l g_{ij} + g_{lj} \partial_i t^l + g_{il} \partial_j t^l = \nabla_i t_j + \nabla_j t_i = 2\nabla_i t_j = 0,$$

where ∇_i is covariant derivative with respect to the connection belonging to g_{ij}.

We have obtained the general covariant and invariant with respect to the path reparametrization expressions for the physical velocity, momentum and energy of massive charged particles moving in the external gravitational and electromagnetic fields.

Now we see that transformation which defines the physical velocity through the component of tangential vector u^i is not reversible, because v^i has three independent components and u^i has four independent components. It is evident that we can put $t_l u^l = 1$ and the transition from the Euler-Lagrange equations to the Hamilton equations will be trivial. But we need to estimate the consequences of this condition. We see that the equation $t_l u^l = 1$ is general covariant but not invariant with respect to the path reparametrization. A reason is as follows. For the natural Time we have $t = f(x^1, x^2, x^3, x^4)$. Hence, along a path $x^i = x^i(\sigma)$ $t = f(x^1(\sigma), x^2(\sigma), x^3(\sigma), x^4(\sigma))$ and $dt = t_i u^i d\sigma = d\sigma$. Thus, we see that under the condition $t_l u^l = 1$ a path is automatically parametrized by the moments of the natural Time. We can speak about the trajectory of motion is this case, which is defined by the equations

$$\frac{dW}{dt} + \pi^i \nabla_i t_k v^k + eE_k v^k = 0, \tag{57}$$

$$\frac{D\pi^i}{dt} + \pi^k \nabla_k t^i + eE^i + e[v \times H]^i = 0, \tag{58}$$

where

$$\frac{D\pi^i}{dt} = \frac{d\pi^i}{dt} + \widetilde{\Gamma}^i_{kj} \pi^j \frac{dx^k}{dt}$$

is the general covariant definition of rate of the physical momentum change with respect to the natural Time. For the physical momentum and energy we have the same equations (51) and (52) but with the following definition of the physical velocity v^i with respect to the natural Time

$$v^i = \frac{dx^i}{dt} - t^i \tag{59}$$

and $\varepsilon = 1$,

$$\pi^i = \frac{mv^i}{\sqrt{1-v^2}}, \tag{60}$$

$$W = \frac{m}{\sqrt{1-v^2}}. \tag{61}$$

Thus, the concept of natural Time on the fundamental level provides full correspondence of general relativity with the classical mechanics and special relativity and uncover its hidden so far deep physical content.

If $v^i = 0$, then

$$\frac{dx^i}{dt} = t^i.$$

This system of equations defines the congruence of curves called the lines of Time. It is easy to show that any solution of the system in question is a solution to equations

$$\frac{d^2 x^i}{dt^2} + \Gamma^i_{kj} \frac{dx^k}{dt} \frac{dx^j}{dt} = 0,$$

which define the geodesic motion with respect to the metric g_{ij}. A charged massive particle can move along a line of time, if $E^i = 0$.

Thus, it is shown that the concept of natural Time uncovers the physical content of general relativity which in the case in question is represented by the Newton equations (54)-(58). The prediction of the gravitational force (56), which is defined by the momentum of the gravitational field and is not trivial only for the non-static gravitational fields, may have practical meaning as well. We should also like to emphasize the connection between the orientation of path and charge conjugation which may have applications in physics of elementary particles and statistical physics. The adequate solution of the well-known problem of zero Hamiltonian should be mentioned as well.

Conclusion

We briefly formulate the obtained results and note why they can be interesting in modern theoretical physics. The electromagnetic field is represented from different and unknown points of view. The duality of natural Time is considered. It is established that there are two different Times in the nature and dual Time is tightly connected with the natural rotation. It is important to emphasize here a deep connection between the symmetry group of the quantum Spherical Top and the duality of Time since this connection demonstrates that in a certain sense the well-known idea of "rotating rigid body" (also mentioned as the Top) of classical mechanics is as fundamental as the idea of "mass point", i. e. the first concept can be reduced to the second one at the fundamental (field -theoretical) level and dual Time plays a key role in this consideration. On this ground, the equations of dual electrodynamics are established and the new theoretical approach to the world of leptons and dual particles (quarks) is formulated. At the present time, quantum chromodynamics has no alternative, but in the framework of this theory we have no answer to the set of principle questions. To explain the confinement mechanism is a very important but still unresolved problem in particle physics. Hence, new approaches and concepts are desirable in any case. From this point of view our suggestion to consider leptons on the ground of the one space-time structure and connect quarks (dual particles) with the dual space-time structure on the same four-dimension Euclidian space looks like quite timely. We put forward a conjecture that this beautiful duality is adequate to the nature of things. The problem of time and everything connected with this topic are up to now significant. The results obtained are significant because they give a simple and evident explanation of lepton-quark symmetry, quark confinement and baryon number conservation. The confinement and the baryon number conservation simply mean that the quarks (dual particles)cannot change their space-time structure because they are doomed for the eternal natural rotation. At last, we should like to stress that the dual space-time structure is an evident guide in the attempts to find Lorentz violation (I mean a research program on this topic proposed by Alan Kostelesky). We mention Eqs. (9)-(12) as the main result of this consideration.

From classical mechanics it follows that for solution of a certain range of problems it is possible not to consider the internal structure of the object in the question and accept that it is at the same internal state (concept of a "point particle"). With the discovery of quantum mechanics we know that this internal structure exists in the form of a wave field and at this level the internal symmetry plays the same fundamental role as the external symmetry (a transformation of internal symmetry affects functions of the field and does not touch upon the coordinates). As a new evidence of significance of this discovery it is shown that the electromagnetic field emerges as a ground state of a new field (called the generalized electromagnetic field) with the general linear group as group of local internal symmetry. The formulated theory of this field predicts the existence of the shadow world which consists from the particles (heavy photons) which do not interact at all with any detectors and the influence of which becomes apparent only in the form of gravity and a simplest form of energy (because its internal symmetry is very wide). The generalized Maxwell equations (14) and (15) represent the main result of this investigation.

We show that the ground state of the electromagnetic field itself has a hidden structure which takes a form of a complex scalar field associated with a one-dimensionally extended object, called a charged string. The theory which predicts existence of these objects is formulated and this gives rise to put forward the conjecture that the superconductivity is tightly connected with the orientations of the charged strings in solid. The main result of this consideration is presented by the equations (28), (29) and (33),(34).

There is no general relativistic, that is real Newton equations. These general covariant and reparametrization invariant equations are derived (on the basis of the natural concept of Time) from the equations of geodesic motion with the general covariant and reparametrization invariant expressions for the physical momentum, velocity and energy of a massive and charged point particle. It is marked the connection between the change of the orientation of a path and operation of the charge conjugation. The prediction of the gravitational force (56), which is defined by the momentum of the gravitational field and is not trivial only for the non-static gravitational fields may have practical meaning as well. We also should like to mention the connection between the orientation of a path and the charge conjugation which may have applications in physics of elementary particles and statistical physics. The adequate solution of the well known problem of zero Hamiltonian should be mentioned as well. Thus, it is shown that concept of natural Time uncover the physical content of general relativity which in the case in question is represented by the Newton equations (54)-(58) as the main result of this consideration.

References

[1] Pestov, Ivanhoe. 2005. "Field Theory and the Essence of Time." In *Horizons in World Physics, Volume 248. Spacetime Physics Research Trends,* edited by Albert Reimer, 1-29. New York: Nova Science Publishers, Inc.

[2] Pestov, Ivanhoe. 2006. "Dark Matter and Potential Fields." In *Dark Matter: New Research,* edited by J.Val Blain, 73-90. New York: Nova Science Publishers, Inc.

[3] Biedenharn, L.C., and Louck J.D. 1981. *Angular Momentum in Quantum Physics.* Massachusetts: Addison-Wesley Publishing Company Reading.

[4] Pestov, I.B. 2017. "Electrodynamics with Charged Strings." *Theoretical and Mathematical Physics* 191(2): 665-72.

In: An Essential Guide to Electrodynamics
Editor: Norma Brewer

ISBN: 978-1-53615-705-5
© 2019 Nova Science Publishers, Inc.

Chapter 5

RADIATION OF ELECTROMAGNETIC WAVES INDUCED BY ELECTRON BEAM PASSAGE OVER ARTIFICIAL MATERIAL PERIODIC INTERFACES

Yuriy Sirenko[1,2,*], *Petro Melezhik*[1],
Anatoliy Poyedinchuk[1], *Seil Sautbekov*[3],
Alexandr Shmat'ko[4], *Kostyantyn Sirenko*[1],
Alexey Vertiy[1] *and Nataliya Yashina*[1]

[1]Department of Diffraction Theory and Electronics,
O.Ya. Usikov Institute for Radiophysics and Electronics, Kharkiv, Ukraine
[2]Department of Applied Mathematics,
V.N. Karazin Kharkiv National University, Kharkiv, Ukraine
[3]Department of Physics and Technology,
Al-Farabi Kazakh National University, Almaty, Republic of Kazakhstan
[4]Department of Radiophysics, Biomedical Electronics and Computer Systems,
V.N. Karazin Kharkiv National University, Kharkiv, Ukraine

Abstract

The chapter is focused at accurate and profound investigation, interpretation and explanation of resonant and anomalous phenomena in radiated electromagnetic field that arises due to the passage of charged particles beams over arbitrary shaped periodic interface of natural or artificial material including smartmaterials and metamaterials. Reliability of the results is assured by the fact that the study is based on rigorous accurate solutions to electromagnetic boundary and initial boundary value problems and corresponding robust numerical algorithms.

Two types of structures are considered in theory: (i) infinite arbitrary profiled periodic interfaces of conventional or artificial materials with a priori given dispersion law, their consideration is based on frequency domain (FD) methods of analytical regularization; and (ii)

---

[*] Corresponding Author's E-mail: yks2002sky@gmail.com ().

infinite structures constructed of periodic arrays of various materials, their consideration is based on solutions to the corresponding electrodynamic problems, which are developed with a help of the method of exact absorbing conditions (EAC) enabling the consideration of the problems both in time domain (TD) and FD.

Keywords: analytical regularization method, method of exact absorbing conditions, density-modulated electron flow, periodic interface, artificial materials, Smith-Purcell and Vavilov-Cherenkov radiation phenomena

Introduction

A plane, density-modulated electron beam, moving at a constant speed over an infinite one-dimensional periodic grating, generates homogeneous plane electromagnetic waves in the environment space. The number of waves, their wavelength and direction of propagation are determined by the speed of the electron beam, its period of modulation, and by the length of the period of grating. The same beam moving at a constant speed in a medium, where the speed of light is less than the speed of charged particles, generates in it two homogeneous plane electromagnetic waves, diverging from the direction of flow. The length of these waves and the direction of their propagation are determined by the period of beam modulation, the ratio of its velocity to the speed of light in the medium, and by the sign of refractive index of the medium. The field of plane waves propagating above or below the grating or in a sufficiently optically dense medium is generated by the field of the charged particles flow.

That may serve as a concise representation of the well-known effects of Smith-Purcell and Vavilov-Cherenkov radiation [1–5]. Discovered in the first half of the 20th century, they are still of a great interest, and quite often are considered in various kinds of basic and applied research. Adequate modeling of these effects, the analysis of physical features observed in their implementation at periodic interfaces between media (conventional and artificial with non-standard properties) is the main topic of this work.

In electromagnetic modeling, the field of a plane, density-modulated electron flow is identified with the field of an inhomogeneous plane wave arriving at an infinite periodic grating, or with the field of a surface (slow) wave of a dielectric waveguide located near a finite periodic structure. Within the frames of these models, only the wave analogs of the Smith-Purcell and Vavilov-Cherenkov effects are simulated. Namely, the models describe the diffraction effects of the classical grating theory and surface-to-spatial mode conversion effects: an inhomogeneous plane wave or a surface wave of an open guiding structure, whose exponentially decaying part sweeps the surface of a grating or an interface of sufficiently optically dense medium, creates in this medium or in the radiation zones of a periodic structure a wave that can propagate (if there is no attenuation) infinitely far [2,6–12]. The differences in the adequacy of the corresponding models are mainly due to the fact that the first of them implements the so-called 'approximation of a given current' or 'approximation of a given field' approach when the amplitude of an inhomogeneous plane wave, which gives energy to outgoing homogeneous plane waves, remains constant along infinite interaction space with periodic structure. The model of the system 'dielectric waveguide – finite grating' is free from this drawback. But it is important, that together they can effectively and accurately solve all the theoretical and practical problems related to the study and practical application of the Smith-Purcell and Vavilov-Cherenkov effects, and their wave analogs. In the theoretical part of this chapter, we consider only those models that implement the approximation of a given field.

For periodic structures made of conventional materials, such effects (diffraction radiation effects) have been consistently and thoroughly, theoretically and experimentally studied in the last three decades of the 20th century in *O.Ya. Usikov* Institute for Radiophysics and Electronics, Kharkiv, Ukraine (IRE NASU) in the department headed by academician Viktor Shestopalov. In numerical simulations, the models of the method of analytical regularization [8, 13–19] implementing the approximation of a given current have been used, and the radiation field of an infinite one-dimensionally periodic grating placed in the field of an inhomogeneous plane wave was studied. The practical outputs resulted in the creation of the diffraction radiation generators – stable, coherent sources of the millimeter range electromagnetic waves, operating on the Smith-Purcell effect [2, 20, 21], in the construction of planar and linear diffraction antennas [11, 12, 22–28], unique in their characteristics, for radar and radiometric ground, airborne and satellite-based complexes of different purpose. The method of exact absorbing conditions developed in the last two decades for solving initial boundary value problems of computational electrodynamics [11, 12, 29–38] allowed significant progress in the directions indicated above; a computational experiment, carried out for models 'dielectric waveguide – finite grating', now gives the same reliable results as a full-scale experiment. But much faster and, that is the most important, at a much lower costs.

The resource, not yet actively involved by neither theoreticians nor applied scientists studying and exploiting the effects of diffraction radiation when solving actual practical problems, is associated with the use of artificial and smart materials. These are materials whose properties could be adjusted and modified under the influence of external factors (temperature, light, pressure), and metamaterials whose unusual electrodynamic characteristics (band gaps, negative refraction, and so on) are due to their structure (most often periodic), but not to the properties of individual substances they composed of. The scientists aiming at a significant progress in this direction will face many different problems. In this chapter we only demonstrate the possibilities of the theoretical approaches we have developed, namely the method of analytical regularization and the method of exact absorbing conditions, and their ability to effectively solve traditional and new problems arising in the study and practical application of diffraction radiation effects.

We use SI, the International System of Units, for all physical parameters except the time t that is the product of the natural time and the velocity of light in vacuum, thus t, is measured in meters. According to SI, all lineargeometrical parameters (a, b, etc.) are given in meters. However, this is obviously not a serious obstacle to extend the results to any other geometrically similar structure. As a rule, the dimensions in the text are omitted, and the majority of results are presented in the form of dimensionless parameters.

Models of the Method of Analytical Regularization

In free space, the field $\vec{U}^i(g,k) = \{\vec{E}^i(g,k), \vec{H}^i(g,k)\}$ of a plane density modulated electron beam, whose instantaneous charge density is $\rho\delta(z-a)\exp[i((k/\beta)y - kt)]$, corresponds to the H-polarized electromagnetic field ($\partial/\partial x \equiv 0$, $E_x^i = H_y^i = H_z^i = 0$) and [2, 39]

$$H_x^i(g,k) = 2\pi\rho\beta\exp\left\{i\left[\sqrt{k^2-(k/\beta)^2}\,|z-a|+(k/\beta)y\right]\right\}[|z-a|/(z-a)]; \quad z \neq a, \qquad (1)$$

$$E_y^i(g,k) = -(\eta_0/ik)\partial_z H_x^i(g,k), \quad E_z^i(g,k) = (\eta_0/ik)\partial_y H_x^i(g,k).$$

Here, $\delta(...)$ is the Dirac δ-function; ρ and k are the amplitude and frequency of the flow modulation, and $0 < \beta < 1$ is its relative velocity; $\eta_0 = (\mu_0/\varepsilon_0)^{1/2}$ is the free space impedance, ε_0 and μ_0 are vacuum's permittivity and permeability; $g = \{y,z\}$ is a point in space R^2; $\exp(-ikt)$ is the time factor of harmonic fields.

Let $-2\pi\rho\beta\sqrt{l}\exp\left[-ka\sqrt{(1/\beta)^2-1}\right] = 1$. Then the field (1) of the electrons moving over the periodic boundary S, separating the conventional medium (vacuum) and the dispersive medium (see Figure 1a), generates in the regions $z \geq 0$ and $z \leq -b$ H-polarized field $\vec{U}^s(g,k) = \{\vec{E}^s(g,k), \vec{H}^s(g,k)\}$ with nonzero components [8, 17]:

$$H_x^s(g,k) = \sum_{n=-\infty}^{\infty} \varphi_n(y) \begin{cases} R_{n1}(k)\exp(i\Gamma_n z); & z \geq 0 \\ T_{n1}(k)\exp\left[-i\Gamma_n^{\varepsilon,\mu}(z+b)\right]; & z \leq -b \end{cases} \qquad (2)$$

and

$$E_y^s(g,k) = -\frac{\eta_0}{ik\varepsilon(g)}\partial_z H_x^s(g,k), \quad E_z^s(g,k) = \frac{\eta_0}{ik\varepsilon(g)}\partial_y H_x^s(g,k). \qquad (3)$$

Here, l and b are the period and height of grating mounts $S = \{g: z = f(y), -b \leq f(y) \leq 0\}$; $k = 2\pi/\lambda$, λ is the wavelength of electromagnetic waves in free space; $\varepsilon(g)$ and $\mu(g)$ ($\mathrm{Im}\,\varepsilon(g) = \mathrm{Im}\,\mu(g) = 0$) are relative dielectric and magnetic permeability of wave propagation medium ($\varepsilon(g) = \mu(g) \equiv 1$ for $z > f(y)$ and $\varepsilon(g) = \varepsilon(k)$, $\mu(g) = \mu(k)$ for $z < f(y)$); $\varphi_n(y) = l^{-1/2}\exp(i\Phi_n y)$, $\Phi_n = 2\pi(n+\Phi)/l$, $\mathrm{Im}\,\Phi = 0$, $|\Phi| < 0.5$ and $\Phi_1 = k/\beta$; $\Gamma_n = \Gamma_n(\Phi) = \sqrt{k^2 - \Phi_n^2}$ and $\mathrm{Re}\,\Gamma_n \geq 0$, $\mathrm{Im}\,\Gamma_n \geq 0$. Signs of the real and imaginary parts of the root $\Gamma_n^{\varepsilon,\mu} = \Gamma_n^{\varepsilon,\mu}(\Phi) = \sqrt{k^2\varepsilon(k)\mu(k) - \Phi_n^2}$ are set so that all partial components $\vec{U}_n^T(g,k): H_x(g,k) = T_{n1}(k)\exp\left[-i\Gamma_n^{\varepsilon,\mu}(z+b)\right]\varphi_n(y)$ of the field $\vec{U}^s(g,k)$ in the region $z < -b$ representing fields of outgoing plane waves, i.e., homogeneous waves ($\mathrm{Im}\,\Gamma_n^{\varepsilon,\mu} = 0$), transferring energy in the direction $z = -\infty$, or inhomogeneous waves ($\mathrm{Re}\,\Gamma_n^{\varepsilon,\mu} = 0$) exponentially decaying when moving in the same direction.

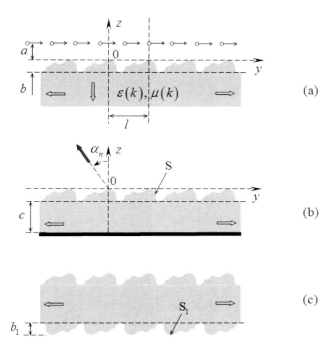

Figure 1. Structures with periodic boundary between two media: (a) half-space filled with dispersive material; (b) layer of dispersive material on metal substrate; (c) layer of dispersive material in free space.

Assuming $\Phi_1 = k/\beta$, we identify the own field of the electron flow, whose exponentially decaying part sweeps the boundary S, with the field of H-polarized inhomogeneous plane wave $\vec{U}_1^i(g,k): H_x^i(g,k) = \exp(-i\Gamma_1 z)\varphi_1(y)$, $\Gamma_1 = \sqrt{k^2 - \Phi_1^2} = ik\sqrt{\beta^{-1} - 1}$, which falls onto this boundary and generates in the reflection ($z > 0$) and transition zones ($z < -b$) homogenous and inhomogeneous plane waves $\vec{U}_n^R(g,k): H_x(g,k) = R_{n1}(k)\exp(i\Gamma_n z)\varphi_n(y)$ and $\vec{U}_n^T(g,k)$. In the electrodynamic theory of gratings, these waves are called spatial harmonics of periodic structure [8, 15, 17]. The ones whose numbers n correspond to real propagation constants Γ_n and $\Gamma_n^{\varepsilon,\mu}$ are able to propagate infinitely far from the boundary S. In the reflection zone $z > 0$, they leave boundary at the angles $\alpha_n(k) = -\arcsin[\Phi_n(k)/k]$ that are counted anti-clockwise from the axis z. Obviously, for any fixed values k and β the number of such waves, N, is also finite $n = 0, -1, -2, ..., -N+1$.

Complex amplitude coefficients $R_{n1}(k)$ and $T_{n1}(k)$ are intricate functions of the frequency k, and geometric and constitutive parameters of wave propagation media. We find them by solving numerically in the interval $0 \leq y \leq l$ (in the Floquet channel) the standard boundary value problem of the electrodynamic theory of gratings [8, 17]

$$\begin{cases} \left[\partial_y^2 + \partial_z^2 + k^2\varepsilon(g)\mu(g)\right]H_x(g,k) = 0; \quad -b \leq z \leq 0 \\ H_x\{\partial_y H_x\}(l,z,k) = \exp(i2\pi\Phi)H_x\{\partial_y H_x\}(0,z,k); \quad -b \leq z \leq 0 \\ H_x(g,k) = \begin{cases} H_x^i(g,k) \\ 0 \end{cases} + \sum_{n=-\infty}^{\infty}\varphi_n(y)\begin{cases} R_{n1}(k)\exp(i\Gamma_n z); \quad z \geq 0 \\ T_{n1}(k)\exp\left[-i\Gamma_n^{\varepsilon,\mu}(z+b)\right]; \quad z \leq -b \end{cases} \\ H_x(g,k) \text{ and } \vec{E}_{tg}(q,k), q = \{x,y,z\} \text{ are continuous when crossing} \\ S \text{ and virtual boundaries } z = 0 \text{ and } z = -b \end{cases} \qquad (4)$$

relatively to the $H_x(g,k)$ component of the full field

$$\vec{U}(g,k) = \{\vec{E}(g,k), \vec{H}(g,k)\} = \begin{cases} \vec{U}_1^i(g,k) + \vec{U}^s(g,k); \quad z \geq 0 \\ \vec{U}^s(g,k); \quad z < 0 \end{cases}.$$

This problem allows to determine (in the approximation of a given current) the electromagnetic field (the field of diffraction radiation) generated by the density-modulated electron flow. In numerical solution, we use the method of analytical regularization [8, 13–15, 18, 19], which provided most of physical and applied results of the electrodynamic theory of gratings associated with resonant and anomalous spatial-frequency and spatial-time transformations of electromagnetic fields in periodic structures [8–10, 15, 17, 19, 40].

The approach used here can be described briefly as follows [18, 41]. The system of orthonormal functions $\varphi_n(y) = l^{-1/2}\exp(i\Phi_n y)$, $n = 0, \pm 1, \pm 2,...$ is complete in the space $L_2(0,l)$ of functions with the integrable on the interval $0 \leq y \leq l$ module squared. This allows us to write down the conditions (4), which is related to the continuity of the tangential field components $\vec{U}(g,k)$ on the boundary S, in the form of an infinite system of linear algebraic equations

$$\begin{cases} \sum_{n=-\infty}^{\infty} F_{mn}^+ x_n^+ - \sum_{n=-\infty}^{\infty} F_{mn}^- x_n^- = b_m^+; \quad m = 0, \pm 1, \pm 2,... \\ \sum_{n=-\infty}^{\infty} K_{mn}^+ x_n^+ - \varepsilon^{-1}(k)\sum_{n=-\infty}^{\infty} K_{mn}^- x_n^- = b_m^+; \quad m = 0, \pm 1, \pm 2,... \end{cases} \qquad (5)$$

Here, $x_n^+ = R_{n1}\exp(-i\Gamma_n b)$ and $x_n^- = T_{n1}\exp(-i\Gamma_n^{\varepsilon,\mu}b)$ are new unknowns and

$$F_{mn}^+ = L_{m-n}^+(\Gamma_n), F_{mn}^- = L_{m-n}^-(\Gamma_n^{\varepsilon,\mu}), K_{mn}^+ = F_{mn}^+\left[\frac{\Gamma_n l}{2\pi} - \frac{(m-n)\Phi_n}{\Gamma_n}\right],$$

$$K_{mn}^- = -F_{mn}^-\left[\frac{\Gamma_n^{\varepsilon,\mu}l}{2\pi} - \frac{(m-n)\Phi_n}{\Gamma_n^{\varepsilon,\mu}}\right], b_m^+ = L_m^+(\Gamma_1), b_m^- = b_m^+\left(\frac{\Gamma_1 l}{2\pi} - \frac{m\Phi_1}{\Gamma_1}\right),$$

$$L_n^+(q) = \frac{1}{l}\int_0^l e^{i\left[q(f(y)+b)-\frac{2\pi n}{l}y\right]}dy, \quad L_n^-(q) = \frac{1}{l}\int_0^l e^{-i\left[qf(y)+\frac{2\pi n}{l}y\right]}dy.$$

The conversion to finite-dimensional analogues in (5) leads to ill-conditioned systems of equations, so the problem needs regularization. We begin the corresponding procedure by introducing a periodic function $\overline{f}(\overline{y}) = -f(\overline{y}l/2\pi)b^{-1}$. Assume also that the function $\overline{f}(\overline{y})$ is twice continuously differentiable. Suppose that in a finite number of points $\overline{y}_s^+$, $s = 1, 2, ..., S^+$ and $\overline{y}_s^-$, $s = 1, 2, ..., S^-$ within the period $0 \leq \overline{y} \leq 2\pi$, this function satisfies the following conditions:

$$\overline{f}(\overline{y}_s^+) = 0, \quad d_{\overline{y}}\overline{f}(\overline{y}_s^+) = 0, \quad d_{\overline{y}}^2\overline{f}(\overline{y}_s^+) > 0,$$
$$\overline{f}(\overline{y}_s^-) = 1, \quad d_{\overline{y}}\overline{f}(\overline{y}_s^-) = 0, \quad d_{\overline{y}}^2\overline{f}(\overline{y}_s^-) < 0.$$

Under such assumptions, we can obtain the following asymptotic estimates for the matrix elements of the system (5):

$$F_{mn}^\pm \approx \sum_{s=1}^{S^\pm} \frac{\exp\left[-i(m-n)\overline{y}_s^\pm\right]\exp\left[-(m-n)^2/2b|\Phi_n d_{\overline{y}}^2\overline{f}(\overline{y}_s^\pm)|\right]}{\sqrt{2\pi b|\Phi_n d_{\overline{y}}^2\overline{f}(\overline{y}_s^\pm)|}}, \quad (6)$$

$$K_{mn}^\pm \approx \mp F_{mn}^\pm \frac{i2\pi mn}{l|\Phi_n|}; \quad |n|,|m| \to \infty.$$

From (6), in particular, it follows that (5) is an operator equation of the first kind, and this is precisely what makes impossible the application of the truncation method to solve it numerically. Using the representations (6), we introduce matrix operators

$$J^\pm = \left\{\delta_n^m \sqrt{\tilde{\Phi}_n 2\pi b/l}\left[\sum_{s=1}^{S^\pm} 1/\sqrt{d_{\overline{y}}^2\overline{f}(\overline{y}_s^\pm)}\right]^{-1}\right\}_{m,n=-\infty}^\infty, \quad M = \{\delta_n^m \tau_n\}_{m,n=-\infty}^\infty, \quad (7)$$

and new unknowns $y^\pm = \{y_n^\pm\}$ such that $x^\pm = \{x_n^\pm\}_{n=-\infty}^\infty = J^\pm y^\pm$. Here, δ_n^m is the Kronecker symbol, $\tau_0 = 1$, $\tau_{n\neq 0} = (i|n|)^{-1}$, $\tilde{\Phi}_n = n + \Phi_1 l/2\pi$. Now we are able to carry out the right-side regularization of the system (5):

$$\begin{cases} F^+J^+y^+ - F^-J^-y^- = b^+ \\ K^+J^+y^+ - \varepsilon^{-1}(k)F^-K^-y^- = b^- \end{cases}; \quad F^\pm = \{F_{mn}^\pm\}_{m,n=-\infty}^\infty, \quad K^\pm = \{K_{mn}^\pm\}_{m,n=-\infty}^\infty. \quad (8)$$

Applying the operator M to the second equation of system (8), we perform the left-side regularization of the problem. After several simple transformations, we arrive at the system of operator equations

$$\begin{cases} \varepsilon(k)\left[1+\varepsilon(k)\right]^{-1} y^+ + P^+ y^+ + P^- y^- = a^+ \\ \varepsilon(k)\left[1+\varepsilon(k)\right]^{-1} y^- + Q^+ y^+ + Q^- y^- = a^- \end{cases}, \quad (9)$$

equivalent to (5). Here,

$$P^+ = \left[\varepsilon^{-1}(k)F^+ + MK^+\right]J^+, \quad P^- = -\varepsilon^{-1}(k)\left[F^- + MK^-\right]J^-,$$
$$Q^+ = \left[MK^+ - F^+\right]J^+, \quad Q^- = \left[F^- - \varepsilon^{-1}(k)MK^-\right]J^-,$$

$a^+ = \varepsilon^{-1}(k)b^+ + Mb^-$, $a^- = Mb^- - b^+$. The estimates (6) allow us to prove the compactness of operators $P^\pm$ and $Q^\pm$ in the space l_2 of infinite sequences $a = \{a_n\}_{n=-\infty}^{\infty}$ such that $\sum_n |a_n|^2 < \infty$. This means that the original problem (4) is reduced to a system of operator equations of the second kind (9) (to a system of Fredholm operator equations), which numerical solution can be obtained by the truncation method converging in the norm of the space l_2. The regularization of problem (4) is completed. Let us now analyze the most general properties of its solution $\vec{U}(g,k)$.

The Pointing's complex power theorem for field $\vec{U}(g,k)$ in the volume $[0 \leq x \leq 1] \times [0 \leq y \leq l] \times [-b \leq z \leq 0]$ implies the fundamental relation

$$\sum_{n=-\infty}^{\infty} \left[|R_{n1}|^2 \operatorname{Re}\Gamma_n + |T_{n1}|^2 \operatorname{Re}\Gamma_n^{\varepsilon,\mu} \varepsilon^{-1}(k)\right] (\operatorname{Im}\Gamma_1)^{-1} = 2\operatorname{Im} R_{11}. \quad (10)$$

This relation determines all energy characteristics of the diffraction radiation processes considered in the approximation of a given current [8, 40]. The value in the left of (10) is the total electromagnetic energy $W = W^\uparrow + W^\downarrow$ radiated into the half-spaces $z \geq 0$ and $z \leq -b$. The values $W_{n1}^R = |R_{n1}|^2 \operatorname{Re}\Gamma_n (\operatorname{Im}\Gamma_1)^{-1}$ and $W_{n1}^T = |T_{n1}|^2 \operatorname{Re}\Gamma_n^{\varepsilon,\mu} \varepsilon^{-1}(k)(\operatorname{Im}\Gamma_1)^{-1}$, composing $W^\uparrow$ and $W^\downarrow$, characterize the distribution of energy lost by the flow of electrons in the channels open to radiation, i.e., between harmonics of the spatial spectrum $\vec{U}_n^R(g,k)$ and $\vec{U}_n^T(g,k)$ such that $\operatorname{Re}\Gamma_n \geq 0$ and/or $\operatorname{Re}\Gamma_n^{\varepsilon,\mu}\varepsilon^{-1}(k) > 0$. The last inequality and the relation $\operatorname{Re} P_y(k) = \varepsilon^{-1}(k)\sum_{n:\operatorname{Im}\Gamma_n^{\varepsilon,\mu}=0} |T_{n1}|^2 \Phi_n$, which is the real part of (averaged over the period l of the boundary S) component $P_y(k)$ of the complex Poynting vector $\vec{P}(k)$ of the field $\vec{U}(g,k)$ in the plane $z = -b$, allow to determine unambiguously and completely strictly the direction of phase velocity of the harmonics $\vec{U}_n^T(g,k)$ propagating in the region $z<-b$ and the energy transfer direction of this harmonic. For the conventional medium, these directions coincide and are set by the vector $\Phi_n \vec{y} - \Gamma_n^{\varepsilon,\mu} \vec{z}$, $\Gamma_n^{\varepsilon,\mu} > 0$. For a bi-negative medium, $\Gamma_n^{\varepsilon,\mu} < 0$, the phase velocity is oriented along the vector $\Phi_n \vec{y} - \Gamma_n^{\varepsilon,\mu} \vec{z}$, and the energy transfer direction is oriented along the vector $-\Phi_n \vec{y} + \Gamma_n^{\varepsilon,\mu} \vec{z}$. In a medium with only one negative constitutive parameter, the harmonics $\vec{U}_n^T(g,k)$ transferring energy in the direction $z=-\infty$ are not excited.

When $H_x^i(g,k) \equiv 0$ and k is fixed, we obtain from (4) a homogeneous (spectral) problem with nontrivial solutions $H_x(g,\bar{\Phi})$ existing for no more than a countable set of eigenvalues $\{\Phi = \bar{\Phi}\} \in F$ and determining the fields of the eigen waves $\vec{U}(g,\bar{\Phi}) = \{\vec{E}(g,\bar{\Phi}), \vec{H}(g,\bar{\Phi})\}$ of a periodic media interface [8, 40]. If any eigen value $\bar{\Phi}$ belongs to the axis $\mathrm{Re}\,\Phi$ of the first (physical) sheet of the surface F (this is the Riemann surface onto which the solution of the problem (4) is analytically continued from the real values of the spectral parameter Φ) and $\mathrm{Im}\,\Gamma_n(\bar{\Phi}) > 0$, $\mathrm{Im}\,\Gamma_n^{\varepsilon,\mu}(\bar{\Phi})\varepsilon^{-1}(k) > 0$ for all $n=0,\pm 1,\pm 2,\ldots$, then we are dealing with ordinary (or correct) surface waves propagating near a media boundary without attenuation.

Above we briefly described the main points related to the use of the analytical regularization method for analyzing the effects of diffraction radiation in the system 'flat, density-modulated electron flow – periodic boundary of a conventional medium and a dispersive medium'. Obviously, the method of generalized scattering matrices provides an accurate solution to the model boundary value problem (4) arising in the case of more complex objects placed in the field of a beam of charged particles (see Figures 1b and 1c).

Interface 'Vacuum – Plasma-Like Medium'. Numerical Results

Suppose that in the problem (4), the constitutive parameters of the medium filling the half-space $z<f(y)$ are given by the relations

$$\varepsilon(k) = 1 - k_\varepsilon^2/k^2 \quad \text{and} \quad \mu(k) = 1 - k_\mu^2/k^2. \quad (11)$$

Such a medium can be called plasma-like, and the real numbers k_ε and k_μ are its characteristic frequencies. Let us also set $l=2\pi$ for all simulated here periodic boundaries in order to simplify all analytical and numerical results in terms of dimensionless parameters that are commonly used in the theory of periodic structures. They are: coordinates $\{Y,Z\} = \{2\pi y/l, 2\pi z/l\}$, time $\tau = 2\pi t/l$, and frequency $\kappa = l/\lambda = kl/2\pi$.

The characteristic feature of periodic boundaries discussed in this section is the ability to control material parameters of a medium filling the area $z<f(y)$ with the help of external

influences that change the frequencies k_ε and k_μ. Among various unusual properties of such periodic boundaries swept by the field of a density-modulated electron flow, we can distinguish resonant regimes appearing in generation of plane waves propagating into the half-spaces $z > 0$ and $z < -b$, in the case when the flow velocity is close to the phase velocity of an eigen surface wave of the boundary S.

For small values k and b discussed below, the periodic boundary separating usual medium (vacuum) and dispersive medium with parameters (11) is capable of supporting forward (or direct) surface waves in the frequency range [42]

$$K_2 = k_\varepsilon k_\mu \Big/ \sqrt{k_\varepsilon^2 + k_\mu^2} < k < k_\varepsilon / \sqrt{2} = K_1; \quad k_\varepsilon > k_\mu. \tag{12}$$

The eigen values $\bar{\Phi}^{direct,\pm,m}$ corresponding to these waves have form

$$\frac{2\pi}{l}\left(\bar{\Phi}^{direct,\pm,m} + m\right) \approx \pm \frac{k}{k_\varepsilon} \sqrt{\frac{(k_\varepsilon^2 - k^2)(k_\mu^2 - k^2)}{(k_\varepsilon^2 - 2k^2)}}; \quad m = 0, \pm 1, \pm 2, \ldots$$

In the range

$$k_\varepsilon/\sqrt{2} < k < k_\varepsilon k_\mu \Big/ \sqrt{k_\varepsilon^2 + k_\mu^2}; \quad k_\varepsilon < k_\mu \tag{13}$$

the same boundary can support backward waves, which have oppositely directed phase and group velocities. For such waves

$$\frac{2\pi}{l}\left(\bar{\Phi}^{back,\pm,m} + m\right) \approx \pm \frac{k}{k_\varepsilon} \sqrt{\frac{(k_\varepsilon^2 - k^2)(k_\mu^2 - k^2)}{(2k^2 - k_\varepsilon^2)}}; \quad m = 0, \pm 1, \pm 2, \ldots$$

The velocity of electrons moving synchronically with surface waves is determined by the relations (synchronism conditions) $\beta = kl/2\pi\bar{\Phi}^{direct,\pm,m}$ or $\beta = kl/2\pi\bar{\Phi}^{back,\pm,m}$.

Vavilov-Cherenkov radiation into the lower half-space occupied by a dispersive medium (radiation of the harmonic $\vec{U}_1^T(g,k)$) is possible if the condition $\operatorname{Re}\Gamma_1^{\varepsilon,\mu}\varepsilon^{-1}(k) > 0$ holds or, equivalently, the condition $\beta^2 > \left[\varepsilon(k)\mu(k)\right]^{-1}$ holds, i.e., only in the case of a bi-negative or bi-positive (conventional environment) medium. Hence, for the dispersion law (11) we obtain the following restriction on the frequency range where Vavilov-Cherenkov radiation can be observed:

$$k < K_0 = \sqrt{2} k_\varepsilon k_\mu \Big/ \sqrt{k_\varepsilon^2 + k_\mu^2 + \sqrt{(k_\varepsilon^2 - k_\mu^2)^2 + 4k_\varepsilon^2 k_\mu^2/\beta}}. \tag{14}$$

For all β from the interval $0 < \beta < 1$, the value $K_0 < k_\varepsilon k_\mu / \sqrt{k_\varepsilon^2 + k_\mu^2}$, $K_0 \to 0$ for $\beta \to 0$ and $K_0 \to k_\varepsilon k_\mu / \sqrt{k_\varepsilon^2 + k_\mu^2}$ when $\beta \to 1$. The parameter intervals (13) and (14), which provide the possibility of existence of backward surface waves, partially overlap. The frequency band where forward surface waves can exist does not overlap the frequency interval (14).

One more special feature of the periodic boundary of the medium with parameters (11) should be discussed before proceeding to the analysis of the results of computational experiments. In [41], the existence of a finite accumulation point $\bar{k}^{accum} = k_\varepsilon / \sqrt{2}$ in the spectrum of an operator of the problem (4) corresponding to such a boundary was proved. The set of complex frequencies $k = \bar{k} \in K$ corresponding to non-trivial solutions $\vec{U}(g, \bar{k}) = \{\vec{E}(g, \bar{k}), \vec{H}(g, \bar{k})\}$ of the homogeneous ($H_x^i(g, k) \equiv 0$) problem (4) is called its spectrum. Functions $\vec{U}(g, \bar{k})$ are eigen oscillations of the field of the structure, relevant to eigen frequencies $\bar{k}$; K is the Riemannian surface, which defines the natural limits of the analytic continuation of the problem (4) to the domain of complex parameter values k [8, 40]. The existence of a real point of accumulation $\bar{k}^{accum}$ is manifested in the thickening of resonant peaks in the amplitude-frequency characteristics of the structure in the close vicinity of the frequency $k = \bar{k}^{accum}$.

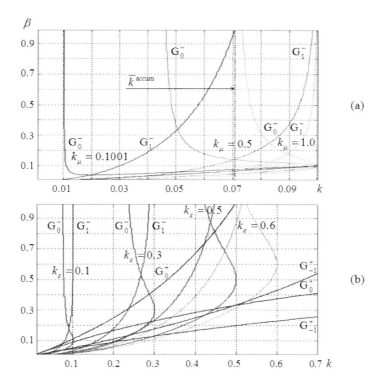

Figure 2. To the determination of the parameters variations domains where the Vavilov-Cherenkov radiation can be observed ($l = 2\pi$): (a) $k_\varepsilon = 0.1$; (b) $k_\mu = 1.0$.

It is extremely important in the study of diffraction radiation to determine correctly the limits of parameter variation where the given regime of the electron flow field transformation into the field of waves outgoing infinitely far from the periodic interface between the media is implemented. The regime identifier $\{N^+, N^-\}$ is given by the number of harmonics N^+ and N^-, which propagate without attenuation in the reflection and transition zones of the periodic structure, and the boundaries of the domains corresponding to this regime in the plane of the variables k and β are given by the curves $G_n^+ : \Gamma_n(k,\beta) = 0$ and $G_n^- : \Gamma_n^{\varepsilon,\mu}(k,\beta) = 0$. Here, n are the numbers of harmonics $\vec{U}_n^R(g,k)$ and $\vec{U}_n^T(g,k)$, included in the numbers N^+ and N^-.

In Figure 2, for the parameters k and β from the region situated to the left from the curve G_n^-, the wave $\vec{U}_n^T(g,k)$ propagates in the structure's transition zone without attenuation. In the reflection zone, the wave $\vec{U}_n^R(g,k)$ propagates without attenuation for the parameters k and β from the domain located between the branches of V-shaped curve G_n^+. With this in mind, we can easily determine the domains of parameters variation for which, in the situation under consideration, Vavilov-Cherenkov radiation is implemented, and radiation of harmonics $\vec{U}_n^R(g,k)$ for all n and of harmonics $\vec{U}_n^T(g,k)$ for $n \neq 1$ (Smith-Purcell radiation) is absent. On the fragments of Figures 3, 4, and 5, all the points $\{k,\beta\}$ lying between the curves G_1^- and G_0^- (the curve G_0^- is passing to the left of G_1^-) can be attributed to these domains.

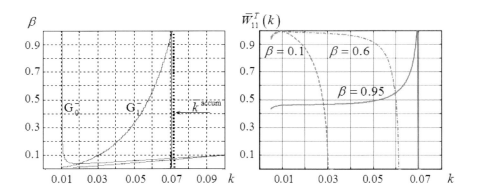

Figure 3. Efficiency of the reverse Vavilov-Cherenkov radiation (diffraction radiation on harmonic $\vec{U}_1^T(g,k)$) into dispersive medium with characteristic frequencies $k_\varepsilon = 0.1$, $k_\mu = 0.10001$ and sinusoidal boundary $z = 0.5b\left[\cos(2\pi y/l) - 1\right]$, $l = 2\pi$, $b = 0.4$.

Figure 3 shows the results of one of the computational experiments related to determining the energy characteristics $\overline{W}_{11}^T(k) = W_{11}^T(k) / \max_k W_{11}^T(k)$ ($\max_k W_{11}^T(k)$ is equal to 2.5288 for $\beta = 0.95$, 1.9148 for $\beta = 0.6$, and 0.3876 for $\beta = 0.2$) of the Vavilov-Cherenkov reverse radiation – the permittivity and permeability of the dispersive medium in the frequency range considered $0.01 \leq k \leq 0.09$ are negative. When $\beta \to 1$, the boundary G_1^- of the Vavilov-

Cherenkov radiation region approaches the straight line $k = \bar{k}^{\,\text{accum}} \approx 0.07$, but does not intersect it. On the curve located slightly to the left of the straight line $k = \bar{k}^{\,\text{accum}}$ and intersecting for large β the boundary G_1^-, one of the synchronism conditions is fulfilled. Therefore, for $\beta = 0.95$, the characteristic $\overline{W}_{11}^T(k)$ on the right fragment of Figure 3 changes much more dynamically than the similar characteristics for $\beta = 0.1$ and $\beta = 0.6$.

Figure 4 illustrates the implementation of reverse Vavilov-Cherenkov radiation for the periodic boundary of the dispersive medium with the parameters $k_\varepsilon = 0.1$ and $k_\mu = 1.0$, $-1.0 \leq \varepsilon(k) \leq 0$, $\mu(k) \ll \varepsilon(k)$ in the frequency band $0.07 \leq k \leq 0.1$. As the flow velocity β increases, the frequency band where these effects can be observed (the interval between the boundaries G_0^- and G_1^-) becomes wider and can contain from two to three points $\{k, \beta\}^{\text{synchr}}$ which fulfill one of the synchronism conditions. As a result, on the curve $\overline{W}_{11}^T(k, \beta)$ ($\max_k W_{11}^T(k)$ is equal to 0.6538 for $\beta = 0.95$, 2.6676 for $\beta = 0.6$, and 1.4276 for $\beta = 0.2$), we observe two (for $\beta = 0.2$) or even three (for $\beta = 0.6$ and $\beta = 0.95$) resonant peaks. The accumulation point $k = \bar{k}^{\,\text{accum}} \approx 0.07$ is out of the frequency band with pure Vavilov-Cherenkov radiation. Therefore, all resonant bursts of characteristics are easily predicted using the synchronism conditions and information about the approximate values of the eigen values $\overline{\Phi}^{\text{direct},\pm,m}$, $\overline{\Phi}^{\text{back},\pm,m}$.

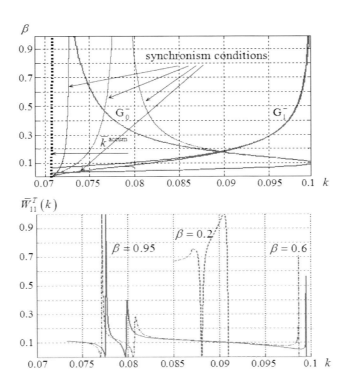

Figure 4. Same as in Figure 3, but for dispersive medium with parameters $k_\varepsilon = 0.1$, $k_\mu = 1.0$.

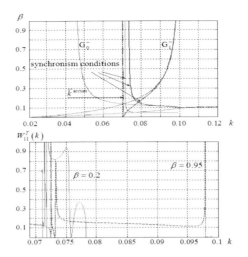

Figure 5. Same as in Figure 3, but for dispersive medium with parameters $k_\varepsilon = 0.1$, $k_\mu = 0.5$.

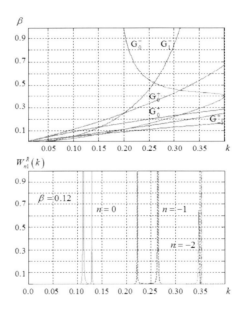

Figure 6. Smith-Purcell radiation into free space over dispersive medium with parameters $k_\varepsilon = 0.5$, $k_\mu = 0.4$ and sinusoidal boundary $z = f(y)$, $l = 2\pi$, $b = 0.4$.

With the values $k_\varepsilon = 0.1$ and $k_\mu = 0.5$ comprising constitutive parameters of the dispersive medium, the range including the frequency intervals where Vavilov-Cherenkov radiation can be observed for all $\beta \geq 0.14$ contains an accumulation point and from two to three points $\{k, \beta\}^{\text{synchr}}$ providing the fulfilment of one of the synchronism conditions (see Figure 5). In this parameter domain, the medium located below the periodic boundary is bi-negative. The values of the functions determining the radiation efficiency change especially sharply in the near vicinity of the point $k = \bar{k}^{\text{accum}}$ where for $\beta \geq 0.2$ two points $\{k, \beta\}^{\text{synchr}}$

fall into this vicinity. In the corresponding fragment of Figure 5, the functions $W_{11}^T(k)$ are truncated at the level $W_{11}^T(k)=1$. Their maximum values, determined when observing the frequency range in increments of 0.00001, are 233.5004 for $\beta = 0.2$ and 6.426 for $\beta = 0.95$.

The last of the results discussed in this section is related to the study of Smith-Purcell radiation on harmonics propagating in free space (in the zone $z>0$). In Figure 6, the boundaries $G_n^\pm$ in coordinates k, β and functions $W_{n1}^R(k,\beta)$ (also truncated at $W_{n1}^R(k)=1$) are presented for the case $k_\varepsilon = 0.5$, $k_\mu = 0.4$, $\beta = 0.12$, $n = 0, -1, -2$. Functions $W_{n1}^R(k,\beta)$ characterize the conversion efficiency of the electron flow field into the radiation field of the harmonic $\vec{U}_0^R(g,k)$ ($\max_k W_{01}^R(k) \approx 1643$), and then, as k grows, into the field of the harmonic $\vec{U}_{-1}^R(g,k)$ ($\max_k W_{-11}^R(k) \approx 50.5$), and further into the field of the harmonic $\vec{U}_{-2}^R(g,k)$ ($\max_k W_{-21}^R(k) \approx 14353$). Such a sequence appears because the frequency intervals, where each of these harmonics propagates without attenuation, do not overlap each other when $\beta = 0.12$. The corresponding points of the k, β plane do not fall into the Vavilov-Cherenkov radiation region. All the resonant bursts of functions $W_{n1}^R(k)$ occur in a small vicinity of points $\{k,\beta\}^{synchr}$ (in the viewed part of the plane of variables k and β, they are located near the curves G_n^+), and the bursts are especially strong (in the case of $W_{-21}^R(k)$) in the area where one of the points $\{k,\beta\}^{synchr}$ is located near the accumulation frequency $k = \bar{k}^{accum} \approx 0.354$.

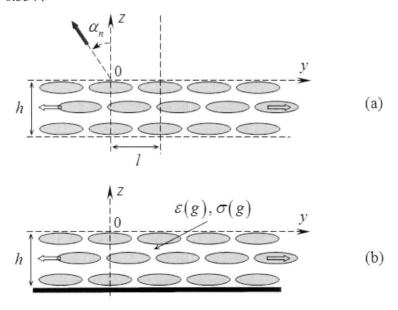

Figure 7. Structures with periodic interfaces: (a) dielectric layer of finite thickness, $\varepsilon(y,z) = \varepsilon(y+l,z)$ and $\sigma(y,z) = \sigma(y+l,z)$; (b) 2-D photonic crystal of limited thickness on metal substrate.

Models of the Method of Exact Absorbing Conditions

Consider a dielectric layer of finite thickness which is periodic along the y-axis and homogenous along the x-axis, Figure 7a. The field $\vec{U}^s(g,k) = \{\vec{E}^s(g,k), \vec{H}^s(g,k)\}$ is generated in the reflection ($z > 0$) and transition ($z < -h$) zones of this dielectric layer when excited by the homogenous ($\operatorname{Im}\Gamma_p = 0$) or inhomogeneous ($\operatorname{Im}\Gamma_p > 0$) plane wave $\vec{U}^i_p(g,k) : H^i_x(g,k) = \exp(-i\Gamma_p z)\varphi_p(y)$. In particular, when $p = 1$ and $\Phi_1 = k/\beta$, it is excited by the field (1) of an electron beam with $-2\pi\rho\beta\sqrt{l}\exp\left[-ka\sqrt{(1/\beta)^2 - 1}\right] = 1$. According to the method of exact absorbing conditions, the amplitude coefficients $R_{np}(k)$ and $T_{np}(k)$ of the $\vec{U}^s(g,k)$ field's component

$$H^s_x(g,k) = \sum_{n=-\infty}^{\infty} \varphi_n(y) \begin{cases} R_{np}(k)\exp(i\Gamma_n z); & z \geq 0 \\ T_{np}(k)\exp[-i\Gamma_n(z+h)]; & z \leq -h \end{cases} \quad (15)$$

are determined from solution of the following correctly formulated [35, 36] in the closure $\overline{\Omega}$ of the domain $\Omega = \{g = \{y,z\} : 0 < y < l, -h < z < 0\}$ initial boundary value problem:

$$\begin{cases} \left[-\varepsilon(g)\partial_t^2 - \sigma(g)\eta_0\partial_t + \partial_y^2 + \partial_z^2\right]H_x(g,t) = 0; & g \in \Omega, \ t > 0 \\ H_x(g,0) = 0, \ \partial_t H_x(g,t)\big|_{t=0} = 0; & g = \{y,z\} \in \overline{\Omega} \\ H_x(g,t) \text{ and } \vec{E}_{tg}(q,t) \text{ are continuous when crossing } \Sigma^{\varepsilon,\sigma}, \\ H_x\{\partial_y H_x\}(l,z,t) = \exp(i2\pi\Phi)H_x\{\partial_y H_x\}(0,z,t) \text{ for } -h < z < 0, \\ D^+\left[H_x(g,t) - H^i_x(g,t)\right]\big|_{g\in L^+} = 0 \text{ and } D^-\left[H_x(g,t)\right]\big|_{g\in L^-} = 0; \ t \geq 0. \end{cases} \quad (16)$$

Here, $H_x(g,t)$ is one of the three non-zero components of the total field

$$\vec{U}(g,t) = \{\vec{E}(g,t), \vec{H}(g,t)\} = \begin{cases} \vec{U}^i_p(g,t) + \vec{U}^s(g,t); & z \geq 0 \\ \vec{U}^s(g,t); & z < 0, \end{cases}$$

generated by the pulse wave $\vec{U}^i_p(g,t) = \{\vec{E}^i(g,t), \vec{H}^i(g,t)\} :: H^i_x(g,t) = v_p(z,t)\varphi_p(y)$, $z \geq 0$ (with its spectral components, we associate the fields of plane monochromatic waves $\vec{U}^i_p(g,k)$), and $\left[\varepsilon(g)\eta_0^{-1}\partial_t + \sigma(g)\right]E_y(g,t) = \partial_z H_x(g,t)$, $\left[\varepsilon(g)\eta_0^{-1}\partial_t + \sigma(g)\right]E_z(g,t) = -\partial_y H_x(g,t)$. Real-valued piecewise-constant functions $\varepsilon(g) : \varepsilon(g) > 0$, $\varepsilon(y,z) = \varepsilon(y+l,z)$

and $\sigma(g): \sigma(g) > 0$, $\sigma(y,z) = \sigma(y+l,z)$ set the relative permittivity and conductivity of the layer $-h \leq z \leq 0$ of material; $\Sigma^{\varepsilon,\sigma}$ are the surfaces on which these functions have discontinuities. $D^+ \left[H_x(g,t) - H_x^i(g,t) \right]\Big|_{g \in L^+} = 0$ and $D^- \left[H_x(g,t) \right]\Big|_{g \in L^-} = 0$ are the exact absorbing conditions for pulse waves $\bar{U}^s(g,t)$ generated by the wave $\bar{U}_p^i(g,t)$ and propagating into the reflection and transition zones of periodic dielectric layer across the boundaries L^+ and L^- of the domain Ω in the planes $z=0$ and $z=-h$. The analytical form of these conditions used in this work is determined by the relations [29, 31, 33, 35]

$$\begin{cases} H_x(y,0,t) - H_x^i(y,0,t) = -\sum_{n=-\infty}^{\infty} \left\{ \int_0^t J_0 \left[\Phi_0(t-\tau) \right] \times \right. \\ \left. \times \left[\int_0^l \partial_{\tilde{z}} \left[H_x(\tilde{y},\tilde{z},\tau) - H_x^i(\tilde{y},\tilde{z},\tau) \right]\Big|_{\tilde{z}=0} \varphi_n^*(\tilde{y}) d\tilde{y} \right] d\tau \right\} \varphi_n(y), \\ H_x(y,-h,t) = \sum_{n=-\infty}^{\infty} \left\{ \int_0^t J_0 \left[\Phi_0(t-\tau) \right] \left[\int_0^l \partial_{\tilde{z}} H_x(\tilde{y},\tilde{z},\tau) \Big|_{\tilde{z}=-h} \varphi_n^*(\tilde{y}) d\tilde{y} \right] d\tau \right\} \varphi_n(y). \end{cases} \quad (17)$$

Here, $J_0(...)$ is the Bessel cylindrical function, $0 \leq y \leq l$, $t \geq 0$, and the asterisk $*$ stands for the complex conjugation.

Computational schemes for initial boundary value problems equipped with absorbing conditions of this type are stable. They quickly converge and lead to reliable and credible physical results in the numerical analysis of anomalous and resonant spatial-temporal and spatial-frequency transformations of electromagnetic waves [38, 43]. The construction of exact absorbing conditions is the most difficult stage in the implementation of corresponding method for diverse and complex problems of computational electrodynamics. A brief history of this method and main analytical and physical results are presented in [12, 23, 29, 31–38].

At the boundaries L^+ and L^- of the region Ω, the function $H_x^s(g,t)$ is represented by the following series of complete on the interval $0 \leq y \leq l$ orthonormal system of functions $\{\varphi_n(y)\}_{n=-\infty}^{\infty}$:

$$H_x^s(y,0,t) = \sum_{n=-\infty}^{\infty} u_{np}^+(t) \varphi_n(y), \quad H_x^s(y,-h,t) = \sum_{n=-\infty}^{\infty} u_{np}^-(t) \varphi_n(y).$$

The amplitude coefficients $R_{np}(k)$ and $T_{np}(k)$, which define all electrodynamic characteristics of the layer, are found from the relations [34, 35]

$$R_{np}(k) = \tilde{u}_{np}^+(k) / \tilde{v}_1(0,k) \quad \text{and} \quad T_{np}(k) = \tilde{u}_{np}^-(k) / \tilde{v}_1(0,k). \quad (18)$$

Here, $\tilde{f}(k) = \int_0^T f(t) \exp(ikt) dt$, and T is the upper limit of the observation interval $0 \leq t \leq T$ in the numerical solution of the initial boundary value problem (16).

The elements $R_{np}(k)$ and $T_{np}(k)$ of the generalized scattering matrices $\{R_{np}(k)\}_{n,p=-\infty}^{\infty}$ and $\{T_{np}(k)\}_{n,p=-\infty}^{\infty}$ are related by the energy balance equations

$$\sum_{n=-\infty}^{\infty} \left[|R_{np}|^2 + |T_{np}|^2 \right] \begin{Bmatrix} \operatorname{Re}\Gamma_n \\ \operatorname{Im}\Gamma_n \end{Bmatrix} = \begin{Bmatrix} \operatorname{Re}\Gamma_p + 2\operatorname{Im} R_{pp} \operatorname{Im}\Gamma_p \\ \operatorname{Im}\Gamma_p - 2\operatorname{Im} R_{pp} \operatorname{Re}\Gamma_p \end{Bmatrix} - \frac{k^2}{\mu_0} \begin{Bmatrix} W_1 \\ W_2 \end{Bmatrix} \quad (19)$$

and by the reciprocity relations

$$R_{np}(\Phi)/\Gamma_p(\Phi) = R_{-p,-n}(-\Phi)/\Gamma_{-n}(-\Phi), \quad (20)$$

which are the corollaries from the Pointing's complex power theorem and the Lorentz lemma [8, 40]. In (19), we have used the following designations:

$$W_1 = \frac{\eta_0 \varepsilon_0}{k} \int_\Omega \sigma(g) |\vec{E}(g,k)|^2 dg \quad \text{and}$$

$$W_2 = \int_\Omega \left[\varepsilon(g) \varepsilon_0 |\vec{E}(g,k)|^2 - \mu_0 |\vec{H}(g,k)|^2 \right] dg.$$

Every harmonic $\vec{U}_n^R(g,k)$ or $\vec{U}_n^T(g,k)$ of the field $\vec{U}^s(g,k)$, for which $\operatorname{Im}\Gamma_n = 0$ and $\operatorname{Re}\Gamma_n > 0$, is a homogeneous plane wave propagating away from a grating at the angle $\alpha_n = -\arcsin(\Phi_n/k)$ into the reflection zone $z > 0$, and at the angle $\alpha_n = \pi + \arcsin(\Phi_n/k)$ into the transmission zone $z < -h$. All angles are measured anticlockwise from the z-axis in the $y0z$-plane, Figure 7a. For $\operatorname{Re}\Gamma_p > 0$, the angle $\alpha_p^i = \arcsin(\Phi_p/k)$ is the angle of incidence of the wave $\vec{U}_p^i(g,k)$ onto a grating. According to (19), the values

$$W_{abs}(k) = \frac{k^2}{\mu_0 |\Gamma_p|} W_1, \quad W_{np}^R(k) = |R_{np}|^2 \frac{\operatorname{Re}\Gamma_n}{|\Gamma_p|}, \quad W_{np}^T(k) = |T_{np}|^2 \frac{\operatorname{Re}\Gamma_n}{|\Gamma_p|} \quad (21)$$

determine the relative part of energy lost to absorption and directed by a grating into the relevant spatial harmonic.

If a grating is excited by an inhomogeneous plane wave ($\operatorname{Im}\Gamma_p > 0$), the near-field to far-field conversion efficiency (diffraction radiation efficiency) is determined by the value of $\operatorname{Im} R_{pp}$ (see (19)), which in this case is non-negative and

$$2\operatorname{Im} R_{pp} = \sum_{n}\left(W_{np}^{R} + W_{np}^{T}\right) + W_{abs}. \tag{22}$$

As follows from (20) and the equalities $\Phi_n(\Phi) = -\Phi_{-n}(-\Phi)$ and $\Gamma_n(\Phi) = \Gamma_{-n}(-\Phi)$, one can study the excitation of a reflecting grating by an inhomogeneous plane wave in the context of conventional for the gratings theory diffraction problem: a structure is excited by a homogeneous plane wave $\vec{U}_{-n}^i(g,k,-\Phi)$ and the coefficient $R_{-p,-n}(-\Phi)$ of conversion into damped spatial harmonic $\vec{U}_{-p}^R(g,k,-\Phi)$ is calculated.

We have described briefly the main points related to the application of the method of exact absorbing conditions for solving problems of diffraction of plane homogeneous and inhomogeneous waves $\vec{U}_p^i(g,k): H_x^i(g,k) = \exp(-i\Gamma_p z)\varphi_p(y)$ by a periodic dielectric layer of finite thickness. The case of inhomogeneous waves ($\operatorname{Im}\Gamma_p > 0$) and, in particular, the case with $p=1$, $\operatorname{Im}\Gamma_1 > 0$, ($\Phi_1 = k/\beta$) allows us to analyze the effects of diffraction radiation in a system 'density-modulated electron flow – periodic dielectric layer' using this method. Modifications that need to be made to allow the analysis of electrodynamic characteristics of the same layer on a perfect metal substrate (see Figure 7b) are obvious, and we will not dwell on them here.

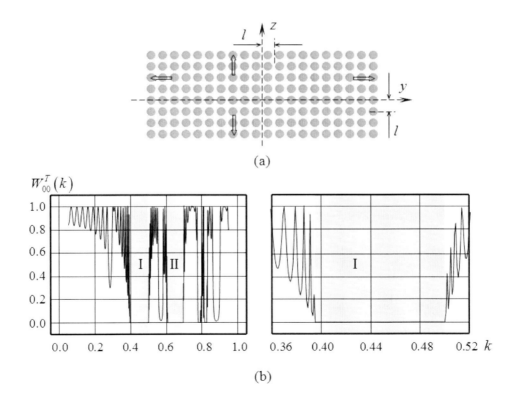

Figure 8. (a) 2-D photonic crystal. (b) Bandgaps (BGs) for the crystal of finite thickness: $h = 10l$; $\Phi = 0$.

Interface 'Vacuum – 2-D Photonic Crystal'. Numerical Results

Consider a 2-D photonic crystal made of circular dielectric cylinders ($\varepsilon = 8.9$, $\sigma = 0$) with directrices parallel to the x-axis, Figure 8a. The axes of the cylinders at the intersection with the planes $x = const$ set the nodes of rectangular grid, which is infinite in the directions y and z, its cells' size is $l \times l$. The cylinder's radius is $r = 0.38l$. The paper [37] is devoted to the determination of electrodynamic characteristics of such spatially bounded crystals when placed in a field of E-polarized waves. Below we discuss several issues related to the excitation of such structures (see Figure 7) by the field of a density-modulated electron flow (plane inhomogeneous H-polarized wave). First, we present some auxiliary results confirming that a layer cut from an ideal (infinite in all directions) photonic crystal retains basic properties of that crystal if layer's thickness is sufficient (but finite).

Let us cut out (by planes $z = const$) a grating from a photonic crystal. The grating's thickness h varies from $3l$ to $10l$ ($l = 2\pi$). The grating is excited with a normally incident ($\Phi = 0$) ultra-wideband H-polarized pulse

$$\vec{U}_0^i(g,t): H_x^i = v_0(z,t)\varphi_0(y); \quad v_0(0,t) = 4\frac{\sin\left[\Delta k\left(t - \tilde{T}\right)\right]}{\left(t - \tilde{T}\right)}\cos\left[\tilde{k}\left(t - \tilde{T}\right)\right] \times$$
$$\times \chi\left(\bar{T} - t\right) = F_1(t), \quad \tilde{k} = 0.5, \quad \Delta k = 0.45, \quad \tilde{T} = 150, \quad \bar{T} = 300. \tag{23}$$

Here, $\chi(...)$ is the Heaviside step function, the parameters $\tilde{k}$ and Δk set the central frequency of the pulse $F_1(t)$ and its band $\tilde{k} - \Delta k \leq k \leq \tilde{k} + \Delta k$ ($0.05 \leq k \leq 0.95$), $\tilde{T}$ and $\bar{T}$ are delay and duration of the pulse $\vec{U}_0^i(g,t)$ [32, 34].

Within the frequency range $0.05 \leq k \leq 0.95$, these gratings (we call them gratings of finite thickness) operate in a single mode regime [8, 17], namely there are only principal spatial harmonics propagating without decay (harmonics with $n = 0$) in the reflection and transition zones. For the case $h = 3l$, the band gaps' contours (BGs are frequency bands where $W_{00}^T(k) = 0$) are only indicated (see, for example, Figure 3 in work [37]). But they are finally formed by structures containing 10 or more layers of thickness l each, Figure 8b. Before the left boundary of the first such zone, up to $k \approx 0.28$, a photonic crystal of thickness $h = 10l$ works as a homogenous dielectric plate. It is completely transparent for a normally incident plane wave for k corresponding to half-wave resonances along its thickness.

Let's conduct a computational experiment same as described above, but for the pulse $\vec{U}_0^i(g,t)$ covering the frequency band $0.36 \leq k \leq 0.52$ and for $\Phi = 0.1$, $\Phi = 0.2$, and $\Phi = 0.3$. It appears that with Φ increasing, the forbidden zone I is drifting towards smaller values of k with the width preserved, Figure 9. A larger value Φ corresponds to a larger value of the angle $\alpha_0^i = \arcsin(2\pi\Phi/lk)$ of the wave $\vec{U}_0^i(g,k)$ arrival onto the grating. Summarizing the information given in Figures 8a and 9, we can conclude that the crystal with thickness $10l$ does not transmit H-polarized waves, which arrive at the angles $0 \leq \alpha_0^i \leq 56.4°$

for all $0.395 \leq k \leq 0.455$ (the width $B_k = 2(k_{upp} - k_{low})/(k_{upp} + k_{low}) \times 100\%$ of this band is approximately equal to 14%).

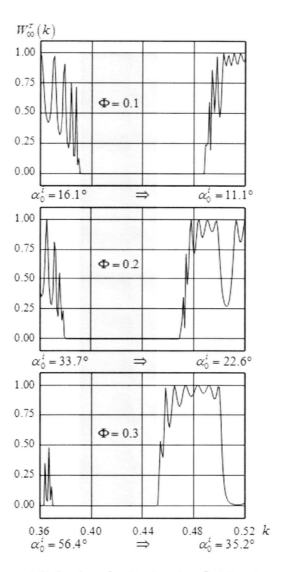

Figure 9. Drifting of the crystal's bandgap I with changing Φ (changing angle of incidence of the primary plane wave α_0^i).

Now excite the crystal with thickness $h = 10l$ with the pulse

$$\vec{U}_1^i(g,t) : H_x^i = v_1(z,t)\varphi_1(y); \quad \Phi = 0, \quad v_1(0,t) = F_1(t),$$
$$\tilde{k} = 0.5, \quad \Delta k = 0.45, \quad \tilde{T} = 150, \quad \overline{T} = 300.$$
(24)

The pulse (24) covers the frequency band $0.05 \leq k \leq 0.95$, and in this case a homogenous plane wave $\vec{U}_1^i(g,k)$ corresponds to it: $\Phi_1 = k/\beta = 1.0$. The field of this wave is associated with the field of the electron flow. At that, the diffraction radiation is represented by harmonics $\vec{U}_0^R(g,k)$ and $\vec{U}_0^T(g,k)$, which go to infinity strictly normally to the periodic structure.

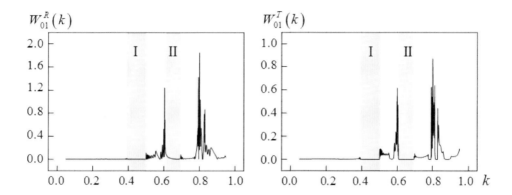

Figure 10. Efficiency of diffraction radiation into the reflection and transition zones of the crystal with finite thickness: $h = 10l$, $\Phi = 0$.

Up to the left boundary of the first forbidden band, the efficiency of wave $\vec{U}_1^i(g,k)$ transformation into propagating spatial harmonics of the crystal structure (diffraction radiation efficiency) is practically zero (Figure 10). The diffraction radiation in the transmission zone is completely absent in the intervals of k variation corresponding to the bandgaps I and II. The value of $W_{01}^R(k)$ in these zones slightly differs from zero. We can confidently say that up to $k \approx 0.5$ (and $\beta \approx 0.5$), the electron flow passing near the crystal limited in thickness generates in its reflection and transmission zones only inhomogeneous waves decaying exponentially with increase of the distance from the periodic boundaries 'vacuum – photon crystal'.

The above-mentioned remains mostly valid and in cases when the parameter Φ is non-zero (Figure 11). The intervals of the frequency parameter k variation for which $W_{01}^T(k) = 0$ move with growth of Φ as well as the bandgaps of the crystal; here the efficiency of diffraction radiation $W_{01}^R(k)$ slightly increases. The values $0.36 \leq k \leq 0.52$ correspond to the electron beam velocities $0.327 \leq \beta \leq 0.473$ for $\Phi = 0.1$, $0.3 \leq \beta \leq 0.433$ for $\Phi = 0.2$, and $0.277 \leq k \leq 0.4$ for $\Phi = 0.3$. The curves $W_{01}^R(k)$ and $W_{01}^T(k)$ presented in Figure 11 had been calculated for these parameters. For such values of k and β, the wave $\vec{U}_0^R(g,k)$ exit the crystal in its reflection zone at the angle $-16.13° \leq \alpha_0 \leq -11.09°$ for $\Phi = 0.1$, $-33.75° \leq \alpha_0 \leq -22.62°$ for $\Phi = 0.2$, and $-56.44° \leq \alpha_0 \leq -35.23°$ for $\Phi = 0.3$.

The frequencies $0.93 \leq k \leq 0.99$ (see Figure 12) correspond to the relativistic velocities of electrons: $0.93 \leq \beta \leq 0.99$ for $\Phi = 0$, $0.929 \leq \beta \leq 0.989$ for $\Phi = 0.001$, and

$0.921 \leq \beta \leq 0.98$ for $\Phi = 0.01$. With these values of k and β, $-0.62°$ is the maximum value of the angle α_0 between the normal to the periodic interface between the media and the direction of the wave $\vec{U}_0^R(g,k)$ when it exits the crystal into the reflection zone. With the k growth, the number M of waves propagating in the crystal and connecting its reflection and transmission zones also increases. It is known [8, 17], that a pair of values $\{N, M\}$, where N is the number of spatial harmonics $\vec{U}_n^R(g,k)$ and $\vec{U}_n^T(g,k)$ propagating in the zones $z > 0$ and $z < -h$ without attenuation, is one of the most general characteristics of the processes of monochromatic waves scattering by periodic structures. The greater the difference $M - N$, the greater the number of resonances of various types involved in the formation of the grating response to any external excitation. As a rule, the energy characteristics $W_{01}^R(k)$ and $W_{01}^T(k)$ rush to their local or global extremes when one or another resonance occurs (this is due to the excitation of free oscillations of the field responsible for this resonance in the grating).

Free oscillations of the field $\vec{U}(g, \bar{k}_m) = \{\vec{E}(g, \bar{k}_m), \vec{H}(g, \bar{k}_m)\}$ (sometimes they are called eigen oscillations) in those periodic structures that are considered in this section, for any fixed real value Φ can exist for no more than a countable set of natural complex frequencies $\bar{k} \in K$ without finite points of accumulation. Here, K is the Riemann surface, to which the solution of the stationary problem, corresponding to the initial boundary value problem (16), is analytically continued from real values of the frequency parameter k [8, 40]. The Q-factor of the oscillation, corresponding to $\bar{k} = \text{Re}\,\bar{k} + i\,\text{Im}\,\bar{k}$, located in the first (physical) sheet of the surface K (here $\text{Im}\,\bar{k} \leq 0$ for all $\bar{k}$), is defined as $Q = \text{Re}\,\bar{k} / 2|\text{Im}\,\bar{k}|$.

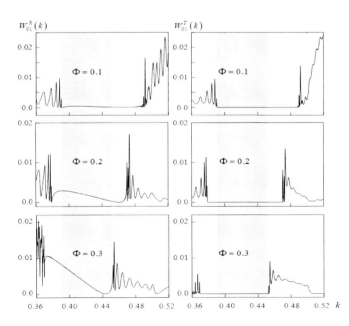

Figure 11. Efficiency of diffraction radiation into the reflection and transition zones of the crystal with thickness $h = 10l$ for $\Phi = 0.1$, $\Phi = 0.2$, and $\Phi = 0.3$.

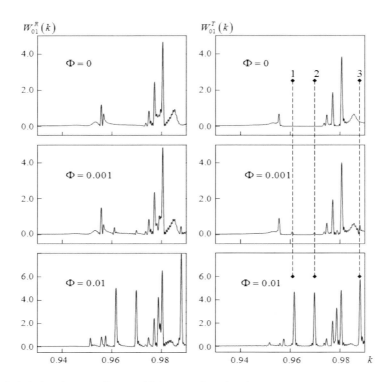

Figure 12. Relativistic electron flow. Efficiency of diffraction radiation near the normal to the flat interface 'vacuum – crystal': $h = 10l$.

The excitation of oscillations with sufficiently high Q (frequency of the exciting signal is $k \approx \operatorname{Re}\bar{k}$) leads to sharp (resonant) changes in the energy characteristics of the periodic structure. The result of these changes in the situation presented in Figure 12 are high enough values of $W_{01}^R(k)$ and $W_{01}^T(k)$ characterizing the efficiency of diffraction radiation.

Consider the occurrence and development of the resonances, indicated in Figure 12 by the numbers 1, 2 and 3. The structure under consideration is symmetrical with respect to the planes $y = ml/2$, $m = 0, \pm 1, \pm 2, \ldots$. Therefore, when it is excited by a wave $\vec{U}_1^i(g,k)$, in the case of $\Phi = 0$, in the corresponding part of space, the wave formations of two symmetry classes occur; but $\vec{U}_0^R(g,k)$ and $\vec{U}_0^T(g,k)$ belong only to one of them. In the wave formations belonging to this symmetry class, there are no free oscillations that could cause a resonance rise of the values $W_{01}^R(k)$ and $W_{01}^T(k)$ in the frequency interval $0.96 \leq k \leq 0.973$; here equality $W_{01}^T(k) \equiv 0$ holds. When passing to the values $\Phi \neq 0$, after corresponding changes in configuration of their field, the wave formations of two different classes of symmetry are combined into one general class of symmetry. In this class, there are already free oscillations, contributing to the resonant growth of $W_{01}^R(k)$ and $W_{01}^T(k)$ at the frequencies close to the real components $\operatorname{Re}\bar{k}$ of their eigen frequencies $\bar{k}$. In the case of resonance 1 (it occurs on the frequency $k = 0.96165$ for $\Phi = 0.01$), this is the oscillation under study, when the crystal is excited with the pulse

$$\ddot{U}_1^i(g,t): H_x^i = v_1(z,t)\varphi_1(y); \quad \Phi = 0.01, \quad v_1(0,t) = \exp\left[-(t-\tilde{T})^2/4\tilde{\alpha}^2\right] \times$$
$$\times \cos\left[\tilde{k}(t-\tilde{T})\right]\chi(\overline{T}-t) = F_2(t), \tag{25}$$
$$\tilde{k} = 0.96165, \quad \tilde{\alpha} = 550, \quad \tilde{T} = 2500, \quad \overline{T} = 5000, \quad T = 15000.$$

This pulse covers a very narrow frequency band $0.95756 \le k \le 0.96574$ (the bandwidth is 0.85%), which does not include real values of eigen frequencies of other free field oscillations. Therefore, after turning off the source (25), the field of free oscillation dominates in the total field of the structure (Figure 13a), which causes the resonance under study. The behavior of the function $\operatorname{Re} u_{01}^+(t)\chi(t-\overline{T})$, $t > \overline{T}$ (see Figure 13b) and the enveloping function $A \exp\left[\operatorname{Im}\overline{k}(t-\overline{T})\right]$ define unambiguously the frequencies $\operatorname{Re}\overline{k} \approx 0.9617$, $\operatorname{Im}\overline{k} \approx -0.00105$ and $Q = \operatorname{Re}\overline{k}/2|\operatorname{Im}\overline{k}| \approx 460$ [34, 44–46].

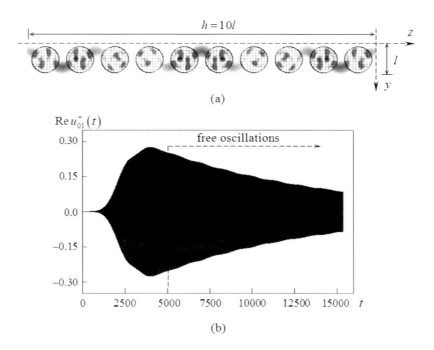

Figure 13. Excitation of the crystal structure by the pulse (25). (a) Pattern of $H_x(g,t)$, $g \in \Omega$, $t = 14000$ corresponding to eigen oscillation. (b) Behavior of $\operatorname{Re} u_{01}^+(t)$, $0 \le t \le 15000$.

Now we shut the access to the crystal's transition zone with a metal substrate; the crystal thickness is $h = 10l$ (see Figure 2b), and it is excited by the pulse covering the band $0.1 \le k \le 1.1$. In the interval $0.9 \le k \le 1.1$ of this band, the wave $\ddot{U}_1^i(g,k)$, whose field is identified with the field of electrons flying over a periodic structure with speed β, generates plane waves $\ddot{U}_0^R(g,k)$ and $\ddot{U}_{-1}^R(g,k)$, propagating in the direction of increasing z at the angles $\alpha_0(k)$ and $\alpha_{-1}(k)$. All principal characteristics of diffraction radiation, manifesting in this frequency range, are presented in Figure 14. On the frequency $k = 0.9948$ (in Figure 14, the resonance 1 corresponds to this frequency), $W_{01}^R(k) \approx 0.07$ and $W_{-11}^R(k) \approx 5.0$, which means that almost all energy generated in the system 'crystal – electrons flow' is carried away into free space by the wave $\vec{U}_{-1}^R(g,k)$ ($\alpha_{-1} \approx 64.78°$ – reverse radiation). On the frequency $k = 1.0668$ (in Figure 14, the resonance 2 corresponds to it), $W_{01}^R(k) \approx 4.12$ and $W_{-11}^R(k) \approx 7.0$, which means that the generated energy is divided approximately in equal parts between the waves $\vec{U}_0^R(g,k)$ and $\vec{U}_{-1}^R(g,k)$, propagating from the periodic interface at the angles $\alpha_0 \approx -5.38°$ and $\alpha_{-1} \approx 57.53°$. On the frequency $k = 1.0728$ (in Figure 14, the resonance 3 corresponds to it), $W_{01}^R(k) \approx 11.6$ and $W_{-11}^R(k) \approx 2.1$, which means that the forward radiation ($\alpha_0 \approx -5.35°$) is significantly more powerful than the reverse one ($\alpha_{-1} \approx 57.03°$).

$$\vec{U}_1^i(g,t): H_x^i = v_1(z,t)\varphi_1(y); \quad \Phi = 0.1, \quad v_1(0,t) = F_1(t),$$
$$\tilde{k} = 0.6, \quad \Delta k = 0.5, \quad \tilde{T} = 150, \quad \bar{T} = 300, \tag{26}$$

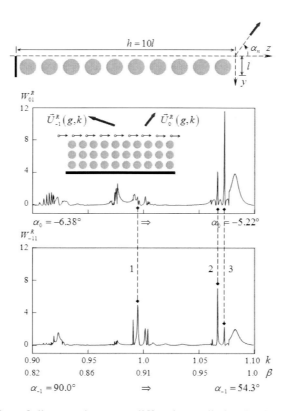

Figure 14. Characteristics of direct and reverse diffraction radiation in the system 'electron flow – crystal on metal substrate': $\Phi = 0.1$.

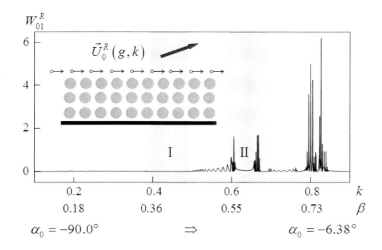

Figure 15. Regime of single-mode conversion of the inhomogeneous plane wave $\vec{U}_1^i(g,k)$ into the homogeneous wave $\vec{U}_0^R(g,k)$, propagating without attenuation in the reflection zone of the crystal on metal substrate: $\Phi = 0.1$.

In the part of the frequency band $0.1 \leq k \leq 0.9$ covered by the pulse (26), the wave $\vec{U}_1^i(g,k)$ generates only one plane wave $\vec{U}_0^R(g,k)$, propagating at the angle $\alpha_0(k)$ in the direction of growing z. All principal characteristics of the effects of diffraction radiation are presented in Figure 15. Up to the first bandgap of the crystal, the magnitude $W_{01}^R(k)$ varies within the limits $0.0 \div 0.01$. For frequencies k from the first crystal's bandgap, the relation $W_{01}^R(k) \equiv 0$ holds. The frequencies of the second bandgap, in contrast, are characterized by sharp bursts of the $W_{01}^R(k)$ magnitude (up to $W_{01}^R(k) \approx 1.5$) at the left end of the band and at one of its inner sections, which are then replaced by intervals with a smooth change of $W_{01}^R(k)$ within $0.05 \div 0.15$ and within a section where $W_{01}^R(k) \equiv 0$. The highest value $W_{01}^R(k) = 6.21$ corresponds to the frequency $k = 0.8268$, the electron flow velocity $\beta \approx 0.752$, and the angle $\alpha_0 \approx -6.95$ of the $\vec{U}_0^R(g,k)$ wave's departure.

Photonic-Crystal Structures for Electron-Wave Systems Generating the Smith-Purcell Radiation

One of the main technological problems arising in the creation of electronic devices for terahertz range is associated with the decrease in geometric dimensions of electrodynamic systems' elements and, in particular, with the decrease in periods of various retardation structures frequently used in such devices. New opportunities can be realized using the results of rapidly developing physics and technology of artificial media with unusual properties – photonic crystals. In such crystals, to allow interaction with an electron beam, the voids are arranged, and they have pronounced waveguide properties in bandgaps of an ideal periodic structure [37]. In addition, exploring such local crystals' defects, it is possible to create open resonant structures with rather high quality factors. Thus, a hollow channel in a photonic crystal (voids in crystal structure) can be used to transmit a linear electron flow. On frequencies outside of forbidden zones of a periodic structure, this flow will generate Smith-Purcell or Vavilov-Cherenkov radiation into space surrounding the channel. And on frequencies from forbidden zones, the electron flow will generate slow and fast waves running through the channel. Since at present there are sufficiently developed technologies for producing crystals with a characteristic cell size smaller than micron, it seems very promising to use these periodic structures in resonant and non-resonant terahertz devices. To do so, it is necessary to ensure that slow waves of a certain polarization exist in the waveguide channel of a crystal: if their phase velocity coincides with the electron flow speed (synchronism mode), the efficiency of diffraction radiation or the excitation efficiency of waves traveling along the channel increases manifold.

Calculations of dispersion diagrams for waves in regular and defected photonic crystals were made using the freely distributed MIT Photonic Bands software package, which is based on the plane wave method and is widely used for electrodynamic modeling of various photonic crystal. The simplest way to form a waveguide in an infinite photonic crystal is to create a linear defect, namely changes in physical properties of one or more elements in adjacent layers. Figures 16 and 17 show several versions of hollow waveguides suitable for the implementation of diffraction radiation generated by a flat density-modulated electron beam. Dispersion curves of guided waves in the forbidden zones (darkening and oblique shading) and transparency zones of corresponding regular photonic crystals are also presented in Figures 16 and 17. Here, Φ_y is the longitudinal propagation constants of H-polarized waves ($E_x = H_y = H_z = 0$) in a linear crystal with defects' characteristic size a (a is distance between the axes of hollow circular cylinders of radius $r = 0.45a$ in a medium with permittivity $\varepsilon = 12.0$). Dashed tilted lines are the light lines, they separate domains that correspond to the bulk (fast) and surface (slow) wave regimes of photonic crystal waveguide. In the detailed images of the structures under consideration all the proportions are preserved.

Calculations presented in [47–53] showed that the minimum phase velocity of slow eigen waves in channels with simple periodic boundaries, made by removing a plane-parallel layer of finite thickness from a crystal, is close to the velocity of light. To obtain a larger retardation rate for such waves, we changed configuration of periodic boundaries in several ways. Two of them are presented in Figures 16 and 17. In the lower fragment of Figure 16, hollow cylinders bordering the waveguide channel have radius smaller ($0.25a$) than elements of the regular part of the photonic crystal. As a result, it was possible to reduce the phase velocity $\beta = k/\Phi_y$ of the wave presented by the dispersion curve 1 from $\beta \approx 0.82$ (upper fragment) to $\beta \approx 0.64$ (lower fragment). The points on the plane of the variables k and Φ_y corresponding to this numerical experiment are marked with crosses in Figure 16.

The dispersion characteristics of the channel, which boundaries are lamellar gratings, are presented in Figure 17. For H-polarized waves, in the frequency range under consideration, the corresponding regular photonic crystal has two forbidden bands marked by oblique hatching. Dispersion curves crossing these zones correspond to the modes localized in the region of periodicity defect. Phase velocities of all slow waves are easily determined from the data given in Figure 17. Thus for the cross-marked point on the dispersion curve from the first bandgap of a regular crystal, the phase velocity of the corresponding surface wave is approximately equal to 0.68. Configuration of the wave field at this point ($E_y(y,z)$ pattern) is shown in the bottom of Figure 17 on the fragment 2. The fragment 1 shows configuration of the wave field from the second forbidden zone with propagation constant $\Phi_y a/2\pi = 0.44$. The corresponding dispersion curve is located closer to the low limit of forbidden zone.

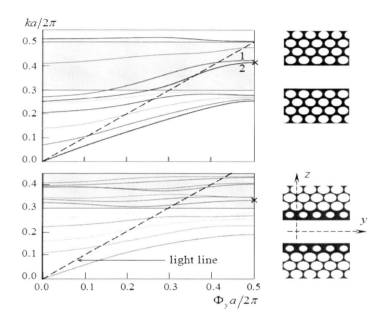

Figure 16. Dispersion characteristics of wave guiding channels in 2-D photonic crystal.

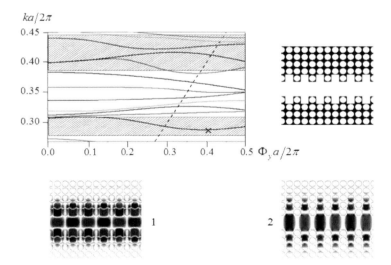

Figure 17. Same as in Figure 16, but for another geometry of waveguiding channel's periodic boundaries.

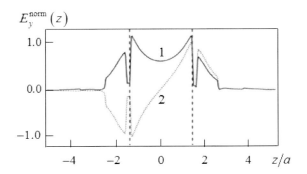

Figure 18. Longitudinal components of the slow waves' electric field represented by the dispersion curves 1 and 2 on the upper fragment of Figure 16 in the cross section of corresponding waveguide channel.

Symmetric (curve 1) and antisymmetric (curve 2) with respect to the plane $x0y$, bulk (fast) and surface (slow)modes are realized in the frequency region $3.0 < ka/2\pi < 4.5$ in the waveguide which dispersion characteristic are presented in the upper fragment of Figure 16. It should be noted that only symmetric modes can effectively interact with linear electron beams due to nonzero field intensity on the waveguide axis. Figure 18 shows spatial transverse distributions of the electric field longitudinal component E_y corresponding to this modes in the points where $\Phi_y a/2\pi = 0.45$. Vertical dashed lines indicate the boundaries of waveguide hollow channel. Field intensity drastically decreases within two periods of the photonic crystal waveguide boundary. Maximum intensity of electric field occurs on the boundaries of waveguide channel. Electric field value for the symmetric mode on the waveguide channel axis is about two times less than maximum value on the boundaries of hollow channel. Therefore, an electron beam occupying one half of the channel width can provide sufficient interaction impedance.

So, 2-D photonic crystals waveguides with different configurations of hollow channel periodic boundaries are investigated. Dispersion characteristics of these waveguides are calculated. The focus was set on slow wave modes in bandgaps of photonic crystals. Beams of charged particles interact with those modes only in devices implementing the excitation of slow and fast electromagnetic waves traveling through channels.

The Fine Structure of Smith-Purcell Radiation

In the final section of this chapter, we present experimental results that reveal several not yet mentioned details important for the design of real-world systems with Smith-Purcell radiation. That is the question of so-called 'fine structure of diffraction radiation', which is usually out of consideration when solving many applied problems, and this may lead to results that are very far from expected ones. This phenomenon appears because an electron flow moving with speed β, which is modulated by a high-frequency field and focused by a magnetic field, generates not only the wave exciting a periodic structure, but a combination of waves which includes space-charge waves (SCWs) and cyclotron waves (CWs). The SCWs and CWs phase velocities, $\beta_{\pm m}^{\text{scw}}$ and $\beta_{\pm m}^{\text{cw}}$, are determined by the relations

$$\beta_{\pm m}^{\text{scw}} = \beta / \left(1 \pm \omega^{\text{plasma}} \omega^{-1}\right) < 1 \text{ and } \beta_{\pm m}^{\text{cw}} = \beta / \left(1 \pm m \omega^{\text{cyclotr}} \omega^{-1}\right) < 1. \tag{27}$$

Here, m is the number of a cyclotron wave (with $m = 0$ that is so-called synchronic electron wave, SEW), ω^{plasma} and ω^{cyclotr} are the plasma and cyclotron frequencies, ω is a frequency corresponding to the wavenumber k. ω^{plasma} is proportional to the square root of the current density divided by β, and ω^{cyclotr} is proportional to the external magnetic field induction. It is clear that each of these waves generate its own radiation field consisting of propagating harmonics $\vec{U}_n^R(g,k,\tilde{\beta})$ (n is such that $\text{Re}\Gamma_n(k,\tilde{\beta}) > 0$) when moving with different phase velocities $\tilde{\beta}$ over a reflective, for example, periodic (with period l) structure. Supposing, same as before, $\Phi_1 = 2\pi/l(1+\Phi) = k/\tilde{\beta}$, we derive $\Phi_n = 2\pi/l(n+\Phi) = \Phi_1 + 2\pi(n-1)/l$ and formulate the condition guaranteeing that a harmonic with the number n will be propagating: $k > |k/\tilde{\beta} + 2\pi(n-1)/l|$. It is obvious that this can only be a harmonic with a non-positive number. The angle $\alpha_n(k,\tilde{\beta})$, at which this harmonic exits the grating, is determined by the relation (see previous sections) $\alpha_n(k,\tilde{\beta}) = -\arcsin(\Phi_n/k) = -\arcsin(1/\tilde{\beta} + 2\pi(n-1))$.

Differences in the spatial orientation of harmonics corresponding to different waves associated with a density-modulated and focused electron beam allow experimental studies of the fine structure of generated field. For millimeter waves, it was first recorded and studied in details in 1982–1987 [54–58]. The obtained results significantly changed the concept of both the spatial-angular distribution and the polarization characteristics of radiation field. For the first time peculiarities in radiation processes associated with the magnitude and direction of magnetic field focusing an electron beam have been also investigated. We note that the question of the influence of space charge waves on characteristics of Smith-Purcell radiation have been raised earlier in [20], but the authors concluded that due to low density of electron current in the optical range, the corresponding effects may be neglected.

Electrons generating cyclotron waves are twisted into a spiral, which moves forward with the speed β and rotates in a fixed cross-section with the frequency ω^{cyclotr} [59]. This affects polarization characteristics of the radiation field corresponding to such waves.

An experimental study of an electron beam radiation in the millimeter wavelength range have been carried out on a quasi-optical complex described in [56] (see also Figure 19). The complex was placed in the gap of a focusing electromagnet $70mm$ in length. The measurements have been carried out according to the following scheme. The tape-shaped electron beam 1 with the cross section $3.8 \times 0.1 mm^2$ and the current density less than $50 A/cm^2$ was formed by the electron gun 2 with the accelerating voltage $1.0 \div 4.0 kV$, and after passing near the reflective grating 3 (lamellar grating with period $l = 0.4mm$ and length $40mm$) was deposited on the collector 4. The beam was focused by magnetic field, which induction could be changed within the interval $0.08 \div 0.5T$ in the central part of the gap.

There was also the possibility of reversing the direction of the magnetic induction vector. The electron flow was modulated by the short-focus, compact diffraction radiation generator with a spherical mirror 5 (radius of the mirror is $15mm$) and a segment of the diffraction grating $18mm$ in length. The modulator was tuned by changing the distance between the grating and the spherical mirror.

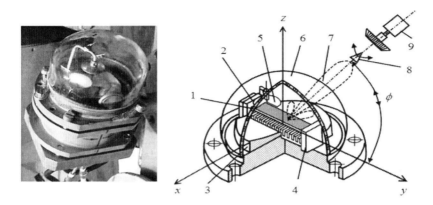

Figure 19. Electron-vacuum unit of the experimental setup and its schematic representation.

The radiation investigated in the experiment was generated by the diffraction of the density-modulated electron beam field by a part of the reflective grating $22mm$ in length, not related to the modulator. Structurally, the modulator, the grating, the electron gun and the collector are in vacuum inside the thin-walled hemisphere glass cylinder 6, which center is aligned with the grating's center. The measuring unit of the complex enabled the determination of the following principal characteristics of the diffraction radiation 7: the angle ϕ between the direction of the electron beam velocity vector and the direction of the radiation wave propagation in the plane $y0z$, the orientation and width of corresponding lobe of radiation pattern, the polarization state of the radiated field, and the radiation power. The interval of measured values ϕ was limited by the electromagnet coils and was $50 \div 130°$. The error of angular measurements did not exceed $0.5°$. The angle of rotation of the polarization plane of the radiated field was set with $\pm 5'$ accuracy. The velocity β of the electron flow was regulated by the anode voltage of the gun and the frequency of modulation k ($f = 48 \div 80 GHz$), via mechanical adjustment of the mirror 5 position. The polarization analyzer consisted of the conical horn 8, a waveguide junction connected to it with the rotating section 9 loaded onto a semiconductor amplitude modulator, and a crystal detector with an indicator. The measurement error of the orientation angle of the polarization ellipse was about $1°$. The maximum intensity in certain lobes of the diagram peaked $1.0 \div 3.0 mW/cm^2$. To eliminate the reflections of the wave emitted from the device, metal surfaces of the installation have been shielded with an absorbing material.

In the general case, a multilobe structure was observed, due to the excitation of electron waves in the electron beam in the measured patterns of diffraction radiation. At currents of the order $30 A/cm^2$ and the focusing magnetic fields of $0.25 \div 0.5T$, the difference between the angles of radiation excited by fast (minus sign) and slow (plus sign) spatial charge waves propagating in the stream was equal to $15 \div 25°$, and for SCWs and synchronic electron waves was equal to $8 \div 12°$. Mechanical adjustments of the modulator provided the excitation of radiation by one or several electron waves at fixed electron beam velocity β.

In one of the experiments with the accelerating voltage $1100V$ and the frequency $f = 52.6 GHz$ in the working range of angles ϕ, the radiation was excited by two electron waves, namely fast SCW and SEW. For the voltage $2255V$ and $f = 67.5 GHz$, the radiation was excited only by slow SCW. For the voltage $2545V$ and $f = 75.0 GHz$, there were three lobes corresponding to the excitation of radiation of slow SCW ($\phi \approx 71°$), fast SCW ($\phi \approx 94°$), and SEW ($\phi \approx 83°$) in the radiation pattern. The measured radiation angles were somewhat different from the theoretical ones, which is obviously caused by the reduction of plasma frequency due to finite dimensions of the electron beam, and the measurement error.

We have studied experimentally the change in intensity of slow SCW radiation and the magnitude of the angle γ between the major axis of polarization ellipse of the radiated field and the electron velocity vector when the accelerating voltage changes within the modulator generation zone $2180 \div 2350V$. In this experiment, the ellipticity coefficient varied within the interval $0.0 \div 0.15$ for the entire range of accelerating voltage, i.e., the polarization remained almost linear. Significant changes of γ occurred in the regime of soft generation. The range of γ variation decreased when the amplitude of the field in the modulator increased. The increasing accelerating voltage sets the value of γ at its minimum $\gamma = 12 \div 15°$. With the decrease in the electron current density, the range of γ over the generation zone of the modulator decreased while maintaining the general form of the corresponding dependence. The small deviation in the radiation field polarization from H-polarization is apparently due to the contribution of cyclotron (transverse) electron waves to this radiation. This assumption is confirmed by experimental studies of the influence of focusing magnetic field on the polarization state of diffraction radiation. It has been established that with increasing magnetic field $\vec{H}$, the angle γ for radiation generated by fast and slow SCWs increases and depends on the direction of $\vec{H}$. A characteristic feature of the results obtained is the presence of a jump in the orientation of the polarization plane of radiation when the direction of the focusing magnetic field is reversed.

A distinctive feature of synchronic electron waves is the fact that they propagate with the phase velocity β, and it is difficult to detect or identify them using diffraction radiation patterns. However, a theoretical analysis showed [55] that under the influence of a space charge, SEWs can be split into left- and right-polarized waves with separated values of phase velocities (fast and slow SEWs). The implementation of this possibility has been confirmed experimentally. It is shown that with well-defined relations between the current density and the magnitude of the focusing magnetic field in the electron stream, one can detect the presence of two SEWs differing in the amplitude and polarization structure. The orientation γ of the polarization ellipses for these signals was almost the same in magnitude and opposite in sign. When studying the behavior of γ and the ellipticity coefficient in parameters region corresponding to the splitting, it was found that the high sensitivity of ellipticity coefficient to changes in the beam parameters is a feature distinguishing the radiation field generated by SEWs from the field generated by spatial charge waves.

The statement is confirmed by the results presented in Figure 20. The ellipticity coefficient varied from 0.01 to 1.0, and the radiation field polarization changed from linear to circular. As the accelerating voltage rises, slow SEW ($\gamma = 20 \div 40°$) was excited first, and then, when going through the dip ($U \approx 2520V$) associated with the splitting, fast SEW ($\gamma = -20 \div -40°$) is excited too. The arrangement of polarization planes in main lobes of the diffraction radiation pattern, which is symmetric with respect to the motion path of unperturbed electrons, suggests that in this experiment we observed the fine structure of diffraction radiation caused by the excitation of synchronous electron waves, SEWs. The orientation angle γ of the polarization ellipse of their radiation field depends on magnitude and direction of the focusing magnetic field.

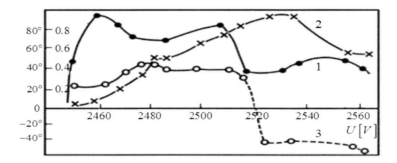

Figure 20. Normalized power (curve 1), the ellipticity coefficient (curve 2), and the rotation angle γ of polarization ellipse (curve 3) of the SWEs' radiation field when the accelerating voltage U is changing.

Conclusion

An electron flow moving near any periodic structure is a common thing in many instruments and devices of vacuum electronics and, in general, is already rather well-studied [2, 20, 21, 47]. Here, in this chapter, we study peculiar features of this problem, which are specified by properties of materials used to construct periodic structures interacting with the flow. These are artificial materials that exhibit electromagnetic properties which are not observed in natural, traditional materials. And the main point of this work is to demonstrate the possibility to simulate accurately the interaction of moving charged particle beams with structures made of such materials, and the possibility to obtain reliable information about features of spatial-temporal and spatial-frequency transformations of electromagnetic waves associated with this interaction. A density-modulated electron beam moving near a periodic boundary dividing natural and artificial media generates Smith-Purcell or Vavilov-Cherenkov radiation. A number of physical systems where such radiation occurs are detailed in the chapter, including plasma-like medium and photonic crystals bordered with vacuum. A significant part of the chapter is devoted to the fine structure of Smith-Purcell radiation, it is an important, but often overlooked topic. Problems of numerical modeling and analysis of Smith-Purcell and

Vavilov-Cherenkov radiation are considered in this chapter within the framework of the given current approximation. To avoid disadvantages of this approximation, the authors have developed and tested methods of frequency and time domains, which are based on the same principles as the methods that have been briefly described above.

References

Agranovich, Zalman, Marchenko, Volodymyr, and Viktor Shestopalov. 1962. "Diffraction of Electromagnetic Waves on Plane Metal Gratings." *Zhurnal Tekhnicheskoi Fiziki* 32:381–94.

Budanov, Valentin, et al. 1977. *Characteristics of Diffraction Radiation of Different Reflecting Gratings*. Kharkiv: IRE, Academy of Sciences of Ukraine, Preprint no.83.

Cherenkov, Pavel. 1937. "Visible Radiation Produced by Electrons Moving in a Medium with Velocities Exceeding that of Light." *Physical Review* 52:378–79.

Frank, Il'ya. 1988. *Vavilov-Cherenkov Radiation*. Moscow: Nauka.

Granet, Gerard, et al. 2015. "Resonances in Reverse Vavilov-Cherenkov Radiation Produced by Electron Beam Passage over Periodic Interface." *International Journal of Antennas and Propagation* 2015:ID 784204.

Jelley, John. 1958. *Cherenkov Radiation and its Applications*. London: Pergamon Press.

Kravchenko, Viktor, Sirenko, Kostyantyn, and Yuriy Sirenko. 2011. *Electromagnetic Wave Transformation and Radiation by the Open Resonant Structures. Modeling and Analysis of Transient and Steady-State Processes*. Moscow: Fizmathlit.

Lopukhin, Vladimir, and Anatoliy Roshal. 1968. *Electron-Beam Parametric Amplifiers*. Moscow: Sovetskoye Radio.

Masalov, Sergey. 1980. "On a Possibility of Using an Echelette in the Diffraction Radiation Generators." *Ukrainskiy Fizicheskiy Zhurnal* 25(4):570–74.

Mazur, Volodymyr, et al. 2018. "Diffraction Antennas. Linear Structures on the Basis of a Modified Goubau Line." *Telecommunications and Radio Engineering* 77(16):1397–408.

Melezhik, Petro, et al. 2006. "Radiation from Surface with Periodic Boundary of Metamaterials Excited by a Current." *Progress in Electromagnetics Research* 65:1–14.

Melezhik, Petro, et al. 2007. "Periodic Boundary of Metamaterial: Eigen Regimes and Resonant Radiation." *Journal of Optics A*: *Pure and Applied Optics* 9:403–09.

Melezhik, Petro, et al. 2010. "Analytic Regularization Method." In *Modern Theory of Gratings. Resonant Scattering: Analysis Techniques and Phenomena*, edited by Yuriy Sirenko and Staffan Strom, 43–172. New York: Springer.

Melezhik, Petro, et al. 2018. "Cherenkov Radiation Based Antenna with theFunnel-Shaped Directional Pattern." *Electromagnetics* 38(1):34–44.

Ney, Michel, et al. 2017. "2-D Photonic Crystals: Electromagnetic Models of the Method of Exact Absorbing Conditions." *Telecommunications andRadio Engineering* 76(3):185–207.

Odarenko, Evgeniy, and Alexandr Shmat'ko. 2011. "Slow-Wave PBG Structures for Terahertz Electronics." Paper presented at the *IEEE Crimean International Conference on Microwave and Telecommunication Technology,Sevastopol*, Ukraine, September 11–16.

Odarenko, Evgeniy, and Alexandr Shmat'ko. 2012. "Photonic Crystal Waveguides in O-Type Electron Devices." Paper presented at the *IEEE Crimean International Conference on Microwave and Telecommunication Technology,* Sevastopol, Ukraine, September 10–14.

Odarenko, Evgeniy, and Alexandr Shmat'ko. 2013. "Enhancement of the Interaction Efficiency in O-Type Electeronical Devices with Photonic Crysral Slow-Wave Systems." Paper presented at the I*EEE Crimean International Conference on Microwave and Telecommunication Technology*, Sevastopol, Ukraine, September 9–13.

Odarenko, Evgeniy, and Alexandr Shmat'ko. 2016. "Novel THz Sources with Profiled Focusing Field and Photonic Crystal Electrodynamic Systems." Paper presented at the *IEEE International Conference on Modern Problems of Radio Engineering, Telecommucations, and Computer Science*, Lviv-Slavsko, Ukraine, February 23–26.

Odarenko, Evgeniy, and Alexandr Shmat'ko. 2016. "Photonic Crystal and Bragg Waveguides for THz Electron Devices." Paper presented at the *IEEE International Conference on Laser and Fiber-Optical Networks Modeling*, Odessa, Ukraine, September 12–15.

Pazynin, Vadim, et al. 2017. "The Exact Absorbing Conditions in Initial Boundary Value Problems of Computational Electrodynamics. Survey." *Fizicheskie Osnovy Priborostroenija* 6(4):4–35.

Perov, Andrey, Sirenko, Yuriy, and Nataliya Yashina. 1999. "Explicit Conditions for Virtual Boundaries in Initial Boundary Value Problems in the Theory of Wave Scattering." *Journal of Electromagnetic Waves and Applications* 13(10):1343–71.

Sashkova, Yulia, et al. 2018. "Analysis of Slow Wave Modes in Modified Photonic Crystal Waveguides Using the MPB Package." Paper presented at the *IEEE International Conference on Mathematical Methods in Electromagnetic Theory*, Kyiv, Ukraine, July 2–5.

Sautbekov, Seil, et al. 2015. "Diffraction Radiation Effects: A Theoretical and Experimental Study." *Antennas and Propagation Magazine, IEEE* 57(5):73–93.

Sautbekov, Seil, et al. 2016. "Diffraction Radiation Phenomena: Physical Analysis and Application," In *Electromagnetic waves in Complex Systems. Selected Theoretical and Applied Problems*, edited by Yuriy Sirenko, and Lyudmyla Velychko, 387–442. New York: Springer.

Sautbekov, Seil, et al. 2018. "Diffraction Antennas. Synthesis of Radiating Elements." *Telecommunications andRadio Engineering* 77(11):925–43.

Shestopalov, Viktor. 1971. *The Method of the Riemann-Hilbert Problem in the Theory of Electromagnetic Wave Diffraction and Propagation*. Kharkiv: Kharkiv State University Press.

Shestopalov, Viktor.1976. *Diffraction Electronics*. Kharkiv: Kharkiv State University Press.

Shestopalov, Viktor. 1998. *The Smith-Purcell Effect*. New York: Nova Science Publishes Inc.

Shestopalov, Viktor, and Yuriy Sirenko. 1989. *Dynamic Theory of Gratings*. Kyiv: Naukova Dumka.

Shestopalov, Viktor, et al. 1973. *Wave Diffraction by Gratings*. Kharkiv: Kharkiv State University Press.

Shestopalov, Viktor, et al. 1986. *Resonance Wave Scattering. Diffraction Gratings*. Kyiv: Naukova Dumka.

Shestopalov, Viktor, et al. 1991. *Diffraction Radiation Generators*. Kyiv: Naukova Dumka.

Shestopalov, Viktor, et al. 1997. *New Solution Methods for Direct and Inverse Problems of the Diffraction Theory. Analytical Regularization of the Boundary Value Problems in Electromagnetic Theory*. Kharkiv: Osnova.

Shestopalov, Viktor, Kirilenko, Anatoliy, and Sergey Masalov. 1984. *Matrix Convolution-Type Equations in the Diffraction Theory*. Kyiv: Naukova Dumka.

Shmat'ko, Alexandr. 2008. *Electron-Wave Systems of the Millimeter Range. Vol.1*. Kharkiv: Kharkiv National University Press.

Sirenko, Kostyantyn, and Yuriy Sirenko. 2016. "The Exact Absorbing Conditions Method in the Analysis of Open Electrodynamic Structures." In *Electromagnetic waves in Complex Systems. Selected Theoretical and Applied Problems*, edited by Yuriy Sirenko, and Lyudmyla Velychko, 225–326. New York: Springer.

Sirenko, Kostyantyn, et al. 2018. "Comparison of Exact and Approximate Absorbing Conditions for Initial Boundary Value Problems of the Electromagnetic Theory of Gratings." *Telecommunications and Radio Engineering* 77(18):1581–95.

Sirenko, Kostyantyn, Sirenko, Yuriy, and Anatoliy Yevdokymov. 2018. "Diffraction Antennas. A Ridged Dielectric Waveguide." *Telecommunications and Radio Engineering* 77(10):839–52.

Sirenko, Kostyantyn, Sirenko, Yuriy, and Nataliya Yashina. 2010. "Modeling and Analysis of Transients in Periodic Gratings. I. Fully Absorbing Boundaries for 2-D Open Problems." *Journal of the Optical Society of America A* 27:532–43.

Sirenko, Konstyantyn, and Yuriy Sirenko. 2005. "Exact 'Absorbing' Conditions in the Initial Boundary Value Problems of the Theory of Open Waveguide Resonators." *Computational Mathematics and Mathematical Physics* 45:490–506.

Sirenko, Yuriy. 2010. "Basic Statements." In *Modern Theory of Gratings. Resonant Scattering: Analysis Techniques and Phenomena*, edited by Yuriy Sirenko and Staffan Strom, 1–42. New York: Springer.

Sirenko, Yuriy, and Anatoliy Yevdokymov. 2018. "Diffraction Antennas. Linear Structures on the Basis of a Ridged Waveguide." *Telecommunications and Radio Engineering* 77(14):1203–29.

Sirenko, Yuriy, and Lyudmyla Velychko. 2001. "The Features of Resonant Scattering of Plane Inhomogeneous Waves by Gratings: Model Problem for Relativistic Diffraction Electronics." *Telecommunications and Radio Engineering* 55(3):33–39.

Sirenko, Yuriy, and Nayaliya Yashina. 2003. "Time Domain Theory of Open Waveguide Resonators: Canonical Problems and a Generalized Matrix Technique." *Radio Science* 38:VIC 26-1–VIC 26-12.

Sirenko, Yuriy, Strom, Staffan, and Nataliya Yashina. 2007. *Modeling and Analysis of Transient Processes in Open Resonant Structures. New Methods and Techniques*. New York: Springer.

Sirenko, Yuriy, Strom, Staffan, and Nataliya Yashina. 2010. "Modeling and Analysis of Transients in Periodic Structures: Fully Absorbing Boundaries for 2-D Open Problems." In *Modern Theory of Gratings. Resonant Scattering: Analysis Techniques and Phenomena*, edited by Yuriy Sirenko and Staffan Strom, 211–334. New York: Springer.

Sirenko, Yuriy, Velychko, Lyudmyla, and Fatih Erden. 2004. "Time-Domain and Frequency-Domain Methods Combined in the Study of Open Resonance Structures of Complex Geometry." *Progress In Electromagnetics Research* 44:57–79.

Smith, Steve, and Edward Purcell. 1953. "Visible Light from Localized Surface Charges Moving Across a Grating." *Physical Review* 92(4):1069–73.

Tretyakov, Oleg, Tretyakova, Svitlana, and Viktor Shestopalov. 1965. "Electromagnetic Wave Radiation by Electron beam Mowing over Diffraction Grating." *Radiotehnika I Elektronika* 10(7):1233–43.

Velychko, Lyudmyla, and Yuriy Sirenko. 2009. "Controlled Changes in Spectra of Open Quasi-Optical Resonators." *Progress In Electromagnetics Research B* 16:85–105.

Velychko, Lyudmyla, Sirenko, Yuriy, and Olena Velychko. 2006. "Time-Domain Analysis of Open Resonators. Analytical Grounds." *Progress In Electromagnetics Research* 61:1–26.

Vertiy, Alexey. 1985. *"Study and Applications of Resonant Quasi-Optical Systems for Millimeter Wave Physics."* Doctor of Sciences diss., Kharkiv State University.

Vertiy, Alexey, and Viktor Shestopalov. 1982. "Polarization Effects in Diffraction Radiation Generators – Free Electron Lasers." *Soviet Physics Doklady* 262(5):1124–27.

Vertiy, Alexey, et al. 1985. "The Fine Structure of Diffraction Emission in Millimeter Wave Region." *Radiophysics and Quantum Electronics* 28(10):888–94.

Vertiy, Alexey, et al. 1987. "Study of Polarization Characteristics of Radiation from a Diffraction Radiation Generator."*Radiophysics and Quantum Electronics* 30(1):85–92.

Vertiy, Alexey, Tsvyk, Alexey, and Viktor Shestopalov. 1985. "Experimental Observation of the Diffraction Radiation Effect in the Millimeter-Wave Range." *Soviet Physics Doklady* 280(2):343–47.

Yevdokymov, Anatoliy. 2013. "Diffraction Radiation Antennas." *Fizicheskie Osnovy Priborostroenija* 2(1):108–25.

Yevdokymov, Anatoliy, et al. 2018. "Antennas of Diffraction Radiation on the Basis of a Groove Transmission Line." *Fizicheskie Osnovy Priborostroenija* 7(1):24–36.

Reviewed by Prof. Panayiotis Frangos, School of Electrical and Computer Engineering, National Technical University of Athens, Athens, Greece. E-mail: pfrangos@central.ntua.gr. Tel.: +302107723694. Academic profile: https://www.ece.ntua.gr/en/staff/48.

In: An Essential Guide to Electrodynamics
Editor: Norma Brewer

ISBN: 978-1-53615-705-5
© 2019 Nova Science Publishers, Inc.

Chapter 6

THE CORNELL POTENTIAL IN LEE-WICK INSPIRED ELECTRODYNAMICS

Anais Smailagic[1] *and Euro Spallucci*[2,*]
[1]INFN, Sezione di Trieste,
Trieste, Italy
[2]Dipartimento di Fisica, Gruppo Teorico,
Università di Trieste, and INFN, Sezione di Trieste,
Trieste, Italy

Abstract

In the seventies, Lee and Wick proposed an interesting modification of classical electrodynamics that renders it finite at the quantum level. At the classical level, this modified theory leads to a regular linear potential at short distances while also reproducing the Coulomb potential at large distances. It is shown that a suitable modification of the Lee-Wick idea can also lead to a linear potential at large distances. For this purpose, we study an Abelian model that "simulates" the QCD confining phase while maintaining the Coulomb behavior at short distances.

This chapter is organized in three parts. In the first part, we present a pedagogical derivation of the static potential in the Lee-Wick model between two heavy test charges using the Hamiltonian formulation. In the second part, we describe a modification of the Lee-Wick idea leading to the standard Cornell potential. In the third part, we consider the effect of replacing a point-like charge with a smeared Gaussian-type source, that renders the electrostatic potential finite as $r \to 0$.

1. Introduction

Classical Electrodynamics is an example of a very satisfactory theoretical description of electromagnetic phenomena which agrees with experimental data. Nevertheless, there are still some underlying difficulties related to the point-like description of elementary charges. Even at the quantum level, this classical problem remains in the form of divergent Feynman integrals. This requires an "*ad hoc*" procedure of "sweeping under the rug" the divergent

---

[*]Corresponding Author's E-mail: euro@ts.infn.it.

quantities. It is fair to notice that the predictions of Quantum Electrodynamics are still in excellent agreement with the experimental results, as long as, a perturbative approach remains valid.

To overcome the problem of divergences, the Maxwell Lagrangian needs suitable modifications in the strong field regime, while preserving the standard form in the weak-field regime. One possible modification was proposed by Lee and Wick [1, 2]. However, this modification has introduced new problems into the theory. In particular, it led to a Yukawa-like correction to the potential between charges. In detail, the combination of Coulomb and Yukawa terms introduces a linearly rising potential at short distances while leaving the $1/r$ behavior at large distances.

This is an interesting result because it can give hints to the different, still unsolved, problem of searching for a linearly confining potential in a theory of strong interactions.

Although the theory of strong interactions has its widely accepted description in terms of non-Abelian gauge fields, so far, no one has been able to extract a confining potential in this formulation.

In a phenomenological framework, it is customary to use the sum of an attractive Coulomb part and a linearly rising long-distance part to describe the mass spectrum of heavy quark-anti quark bound states. This is known as the Cornell potential [3].

It is interesting to note that the Lee-Wick model has the same sum of the two potentials, but with short/large distances regimes exchanged.

On the basis of this observation, one could consider a modification of the Lee-Wick model leading to the Cornell potential between charges.

A first, modest, attempt to implement this idea will be limited to classical electrodynamics, where the technical intricacies of non-Abelian gauge theories are absent. Any positive outcome in this direction will be helpful eventually to tackle the more complicated non-Abelian situation.

However, exchanging the two regimes, will reintroduce the problem of the short distances behavior of the Coulomb potential. In order to finally solve this problem, we propose a smearing of the source compatible with the finite size of any physical charge particle. This will remove the singularity at the origin, while not effecting the linear long-distances behavior.

The chapter is organized as follows: in Sect. (2) we review the Hamiltonian formulation of the Lee-Wick model. That formulation is a particularly useful way to factor out the static potential from the dynamical degrees of freedom. We than recover the Yukawa correction to the Coulomb potential and obtain the linear behavior at short distances. We also propose an alternative interpretation of the Lee-Wick potential as due to the polarization of the vacuum through the distances dependent (or "*running*") dielectric constant.

In Sect. (3) we modify the Lee-Wick model in a way to exchange short/long distances regimes. This leads to the desired long distances "confinement" of charges. The price to pay, is the reintroduction of a divergent behavior at short distances. In the final part of this Section we cure the aforementioned singular behavior by attributing a finite size to the charge source.

Finally, in the Appendix, we give a list of formulae used to obtain the results in this chapter.

2. Lee-Wick Model

Higher order derivative theories have been often proposed as a cure for the ultraviolet divergences in quantum field theories. A remarkable example is given by higher order derivative gravity where quadratic curvature terms are instrumental in getting a renormalizable quantum theory. However, the presence of higher order derivatives usually violates the unitarity of the theory. In the seventies, Lee and Wick [1, 2] proposed a way to reconcile UV finiteness and unitarity by claiming that the extra poles in the propagator describe unstable (heavy) particles. These particles eventually decay restoring the unitarity of the theory.

Lee-Wick electrodynamics is described by the higher derivative Lagrangian density [1]

$$L_{LW} = -\frac{1}{4} F_{\mu\nu} \left(1 - \frac{\partial^2}{m^2} \right) F^{\mu\nu} - eJ^\mu A_\mu \qquad (1)$$

where m is a new constant with a dimension of mass in natural units. In the literature, it is often referred to as the heavy photon mass. We consider this terminology misleading since the theory is explicitly gauge invariant. A true photon mass stands in front a quadratic term in A_μ which is absent in the Lagrangian (1). A proper interpretation of m is through the introduction of a characteristic length $l_0 \equiv 1/m$ that indicates the distances range $r < 1/m$, where the Lee-Wick correction dominates over the Maxwell term.

A well known problem in the covariant formulation of gauge theories is the presence of propagating non-physical degrees of freedom. The usual way to deal with these states, while maintaining covariance, is to introduce Fadeev-Popov ghosts. Alternatively, one may give up explicit covariance in favor of exposing only the physical degrees of freedom of the gauge field using the Hamiltonian formulation of the theory. This is the approach we shall follow in this chapter in order to extract the static, classical interaction energy between two test charges.

There are different ways to achieve this goal in the quantum theory [4, 5, 6, 7, 8, 9, 10]. The Hamiltonian formalism will explicitly factorize the static interaction from the gauge field's dynamical degrees of freedom.

We start by writing down the Lagrangian for the Lee-Wick model (1) as:

$$\begin{aligned} L_{LW} &= -\frac{1}{2} \left(\partial^0 A^i - \partial^i A^0 \right) \left(1 - \frac{\partial^2}{m^2} \right) \left(\partial_0 A_i - \partial_i A_0 \right) + \\ &\quad -\frac{1}{2} \left(\partial^k A^i - \partial^i A^k \right) \left(1 - \frac{\partial^2}{m^2} \right) \left(\partial_k A_i - \partial_i A_k \right) - e\rho A_0 - ej^k A_k \end{aligned} \qquad (2)$$

Introducing the electric and magnetic fields

$$E^i \equiv \partial^0 A^i - \partial^i A^0 = -\dot{A}^i + \partial^i A_0 , \qquad (3)$$

$$B_i \equiv \frac{1}{2} \epsilon_{ijk} \left(\partial^j A^k - \partial^k A^j \right) \qquad (4)$$

the Lee-Wick Lagrangian (1) reads

---
[1] We use the metric signature convention $- + + +$.

$$L_{LW} = -\frac{1}{2}E^i\left(1 - \frac{\partial^2}{m^2}\right)E_i - \frac{1}{2}B^i\left(1 - \frac{\partial^2}{m^2}\right)B_i - e\rho A_0 - ej^k A_k \qquad (5)$$

To write the corresponding Hamiltonian we define the canonically conjugate momenta. One can see immediately that there is no kinetic term for A_0. Thus, only A_i has a conjugate momentum

$$\Pi^i \equiv \frac{\delta L_{LW}}{\delta \partial_0 A_i} = -\left(1 - \frac{\partial^2}{m^2}\right)E^i \qquad (6)$$

By Legendre transforming L_{LW} one finds the Hamiltonian

$$\begin{aligned}H_{LW} &= \left(E^i + \partial^i A_0\right)\left(1 - \frac{\partial^2}{m^2}\right)E_i - L_{LW}\\ &= \frac{1}{2}E^i\left(1 - \frac{\partial^2}{m^2}\right)E_i - \frac{1}{2}B^i\left(1 - \frac{\partial^2}{m^2}\right)B_i - A_0\left[e\rho - \left(1 - \frac{\partial^2}{m^2}\right)\partial_i E^i\right] - ej^k A_k\end{aligned} \qquad (7)$$

In (7) A_0 is simply a Lagrange multiplier enforcing the modified Gauss law:

$$\frac{\delta H_{LW}}{\delta A_0} = 0 \longrightarrow \left(1 - \frac{\partial^2}{m^2}\right)\partial_i E^i = -e\rho \qquad (8)$$

Equation (8) can be rewritten as Gauss law with a running charge

$$\partial_i E^i = -e\left(\frac{m^2}{-\partial^2 + m^2}\right)\rho \qquad (9)$$

This is the customary interpretation given in quantum field theory where the interaction strength becomes scale dependent due to the vacuum polarization effects. Taking this for granted, it seems more consistent to attribute the strength variation to the dielectric properties of the medium, i.e., the vacuum, rather than to the charge itself. For this purpose, it is worth recalling that one usually works in a system of units where $1/4\pi\varepsilon_0 = 1$. Restoring the dielectric constant, equation (9) is written as

$$\partial_i E^i = -\frac{e}{\varepsilon_0}\left(\frac{m^2}{-\partial^2 + m^2}\right)\rho \equiv \frac{e}{\varepsilon\left[\partial^2/m^2\right]} \qquad (10)$$

In equation (10) e is a point-like charge, commonly described by the distributional density $\rho(\vec{r}) = e\delta(\vec{r})$, while ε is a modified effective dielectric constant accounting for the polarization effects through the substitution

$$\frac{1}{\varepsilon_0} \longrightarrow \frac{1}{\varepsilon\left[\partial^2/m^2\right]} \qquad (11)$$

We now proceed to calculate the Coulomb-like potential $\phi(r)$ from (10)

$$\vec{E} = -\nabla\phi, \qquad (12)$$

$$\phi(r) = -\frac{e}{\varepsilon_0} \int \frac{d^3k}{(2\pi)^3} \frac{1}{\vec{k}^2} \frac{m^2}{\vec{k}^2 + m^2} e^{i\vec{k}\cdot\vec{r}},$$

$$= -\frac{e}{\varepsilon_0} \int \frac{d^3k}{(2\pi)^3} \left[\frac{1}{\vec{k}^2} - \frac{1}{\vec{k}^2 + m^2} \right] e^{i\vec{k}\cdot\vec{r}} \quad (13)$$

To evaluate the second term one can use the Schwinger parametrization:

$$\frac{1}{\vec{k}^2 + m^2} = \int_0^\infty ds\, e^{-s(\vec{k}^2 + m^2)} \quad (14)$$

and then perform a Gaussian integration over $\vec{k}$. An additional rescaling $m^2 s \equiv \tau$ gives the integral

$$\longrightarrow \frac{1}{r} \int_0^\infty \frac{d\tau}{\tau^{1/2}} e^{-\tau} e^{-m^2 r^2/4\tau} \equiv \frac{1}{r} I(m^2) \quad (15)$$

The first integral in (13) is simply given by $I(m=0)$. The complete potential is

$$\phi(r) = -\frac{e}{4\pi^{3/2}\varepsilon_0} \frac{1}{r} \int_0^\infty \frac{d\tau}{\tau^{1/2}} e^{-\tau} \left(1 - e^{-m^2 r^2/4\tau}\right) \quad (16)$$

Let us look at the asymptotic behavior of (16). In the small distances limit $mr \ll 1$ one finds a linear term:

$$\phi(r) = -\frac{e}{4\pi^{3/2}\varepsilon_0} \frac{1}{r} \int_0^\infty \frac{d\tau}{\tau^{1/2}} e^{-\tau} \left(1 - 1 + \frac{m^2 r^2}{4\tau} + \ldots\right)$$

$$= \frac{em^2}{8\pi\varepsilon_0} r + O(r^3) \quad (17)$$

This result shows that the Lee-Wick modification regularizes the short-distances behavior of the, otherwise divergent, electrostatic potential.

A relatively simple way to evaluate (16) is to notice that the exponential function is quickly vanishing both at the lower and upper integration limits. This leaves only a non-zero contribution in a narrow strip around its maximal value τ_0. Therefore, the integral can be calculated by expanding the exponent around $\tau_0 = mr/2$ as

$$\int_0^\infty \frac{d\tau}{\tau^{1/2}} e^{-\tau - r^2 m^2/4\tau} \equiv \int_0^\infty \frac{d\tau}{\tau^{1/2}} e^{-\omega(\tau)}, \quad \omega(\tau) \equiv \tau + \frac{m^2 r^2}{4\tau} \quad (18)$$

$$\omega'(\tau) = 0 \longrightarrow \tau_0 = \frac{mr}{2} \quad (19)$$

With the above definitions, one finds

$$\int_0^\infty \frac{d\tau}{\tau^{1/2}} e^{-\tau - r^2 m^2/4\tau} = \frac{1}{\tau_0^{1/2}} e^{-\omega(\tau_0)} \int_{-\infty}^\infty d\tau\, e^{-\omega''(\tau_0)(\tau-\tau_0)^2/2}$$

$$= \frac{1}{\tau_0^{1/2}} e^{-\omega(\tau_0)} \left(\frac{2\pi}{\omega''(\tau_0)}\right)^{1/2} \quad (20)$$

The potential (17) is found to be

$$\phi(r) = -\frac{e}{4\pi\varepsilon_0}\frac{1}{r}\left(1 - e^{-mr}\right) \tag{21}$$

The result (21) shows that the emerging potential is composed of a Coulomb part and a Yukawa-like correction.

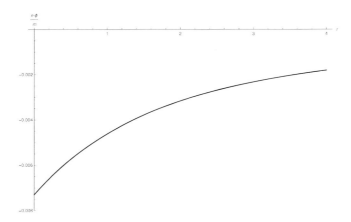

Figure 1. Plot of the Coulomb-like potential (16).

Figure (1) shows that the potential is linear and finite near $r = 0$, while approaches the Coulomb $1/r$ behavior at large distances. Accordingly, we see that $\varepsilon(r)$ diverges as $r \to 0$ and tends to the usual value of the vacuum dielectric constant ε_0 for $r \gg 1/m$.

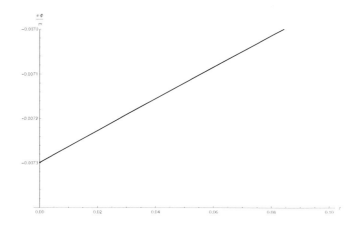

Figure 2. Zoom of the Figure(1) near $r = 0$, showing the linear behavior of the potential.

Equation (21) can be rewritten in a Coulomb-like form with a position-dependent dielectric constant

$$\phi(r) = \frac{e}{4\pi\varepsilon(r)\,r}, \qquad \varepsilon(r) = \frac{\varepsilon_0}{(1 - e^{-mr})} \tag{22}$$

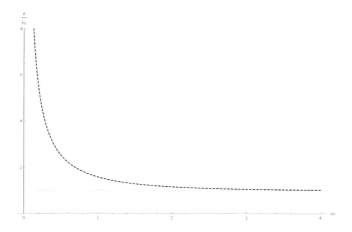

Figure 3. Plot of the position-dependent relative dielectric constant (22).

This modified dielectric constant describes the "vacuum" polarization induced by the test charge due to virtual pair creation in its proximity. Figure (4) below indicates an unbounded increase in $\varepsilon(r)$ as the distance decreases. (This unphysical behavior of the dielectric constant is due to the choice of an idealized point-like charges. In the next subsection we shall improve this behavior by assuming a smeared (non-point like) source.)

3. Cornell Potential

In the previous Section we showed that it is possible to obtain a linear potential via a suitable modification of the current-gauge field interaction term. While preserving gauge invariance, this modified interaction accounts for the vacuum dielectric properties at different length scales. This is a re-interpretation of the original Lee-Wick model, leading to a linear behavior at short distances.

On the other hand, to reproduce confinement, one needs a linear potential at *large* distances. At the phenomenological level, the Cornell potential, reproducing a linear behavior, is currently adopted to fit the quark/anti-quark bound state spectrum [3].

$$g_s V_C(r) = -\frac{4}{3}\frac{\alpha_s}{r} + \sigma r, \tag{23}$$

The first term, which dominates the short distances behavior, is a Coulomb potential with the fine structure constant being replaced by the *strong coupling* constant $\alpha_s = g_s^2/4\pi$. On the other hand, the second term provides a constant confining force. The constant σ represents the *tension* of the color flux tube connecting the quark/anti-quark pair.

In QCD, it is not at all evident that confinement is actually an Abelian *"phenomenon"* as shown in [11, 12, 13, 14]. Nevertheless, isolating the commuting degrees of freedom in a Yang-Mills gauge theory is a complicated task and recovering a linear potential is technically even more difficult. This is a good motivation to look for an equivalent model in a simpler framework.

Following the results of the previous Section, we introduce a possible redefinition of the Lee-Wick action which exchanges the short and long distance behavior. The new Lagrangian is

$$L = -\frac{1}{4} F_{\mu\nu} \frac{-\partial^2}{-\partial^2 + m^2} F^{\mu\nu} - e J^\mu A_\mu \tag{24}$$

It is known that confinement kicks in at an energy determined by the dynamically generated QCD scale $\Lambda_{QCD} \approx 200 MeV$. To simulate confinement in the Abelian model (24) we take $m \sim \Lambda_{QCD}$.

In order to calculate the new potential we start as in (13)

$$\phi(r) = -\frac{e}{\varepsilon_0} \int \frac{d^3k}{(2\pi)^3} \frac{1}{\vec{k}^2} \frac{\vec{k}^2 + m^2}{\vec{k}^2} e^{i\vec{k}\cdot\vec{r}},$$

$$= -\frac{e}{\varepsilon_0} \int \frac{d^3k}{(2\pi)^3} \frac{1}{\vec{k}^2} \left(1 + \frac{m^2}{\vec{k}^2} \right) e^{i\vec{k}\cdot\vec{r}} \tag{25}$$

Again, the first term in (25) gives the Coulomb part of the potential

$$\int \frac{d^3k}{(2\pi)^3} \frac{1}{\vec{k}^2} e^{i\vec{k}\cdot\vec{r}} = \frac{1}{4\pi r} \tag{26}$$

This part dominates the short-distances behavior.
The second term leads to a linearly rising potential

$$m^2 \int \frac{d^3k}{(2\pi)^3} \frac{1}{(\vec{k}^2)^2} e^{i\vec{k}\cdot\vec{r}} = \frac{m^2}{(2\pi)^3} \int_0^\infty ds s \int d^3k e^{-s\vec{k}^2} e^{i\vec{k}\cdot\vec{r}},$$

$$= \frac{m^2}{8\pi^{3/2}} \int_0^\infty ds s^{-1/2} e^{-r^2/4s},$$

$$= \frac{m^2}{8\pi^{3/2}} \frac{r}{2} \Gamma(-1/2) = -\frac{m^2}{8\pi} r \tag{27}$$

The complete result for ϕ is encompassed in

$$e\phi(r) = -\frac{e^2}{4\pi\varepsilon_0} \frac{1}{r} \left(1 - \frac{m^2}{2} r^2 \right) \tag{28}$$

The correspondence with the Cornell potential (23) is established through the following identifications

$$e^2 \iff \frac{4}{3} \alpha_s, \tag{29}$$

$$m^2 e^2 \iff \frac{4}{3}\alpha_s \Lambda_{QCD}^2 , \tag{30}$$

$$\sigma = \frac{1}{4\pi\varepsilon_0}\frac{4}{3}g_s^2 \Lambda_{QCD}^2 \tag{31}$$

It is interesting to notice the singular behavior of the "color" dielectric constant $\varepsilon(r)$ as r approaches the "*critical*" distance $r^* = \sqrt{2}/m$

$$\varepsilon(r) = \frac{\varepsilon_0}{1 - m^2 r^2/2} \tag{32}$$

The Coulomb region corresponds to $0 \leq r < r^*$, where the running dielectric constant $\varepsilon(r)$ is positive and diverges at the critical distance r^*. Beyond this point begins the "large distances" region $r > r^*$, where $\varepsilon(r)$ is negative and vanishes as $r \to \infty$. This behavior can be surprising at first, but its possible physical explanation could be that it is impossible to separate two charges beyond a distances at which the creation of a new pair is energetically favorable. Thus, the infinite discontinuity in $r = r^*$ marks a " phase transition " between the Coulomb and the confining phase.

3.1. Gaussian Source

As shown in the previous Section, a suitable modification of the Maxwell action leads to the Cornell confining potential at large distances. However, the Coulomb potential still emerges at short distances with the well-known singular behavior at the origin. A similar pattern is encountered both in the Newtonian gravitational potential and in its relativistic extension. In General Relativity this problem acquires a particular relevance since it leads to the, so-called, *curvature singularity* issue. This invalidates basic assumptions of the space-time being a smooth, differentiable, manifold. On a physical ground, even in the case when the singularity is "*hidden*" by an event horizon, the very existence of the singularity prevents a consistent description of the final stage of black-hole evaporation.

Recently, this problem has been successfully solved by the natural assumption that the matter source of the gravitational field cannot be point-like. It has been replaced by modeling a "particle" as a Gaussian matter/energy distribution. This choice was motivated by the known property of Gaussian states in Quantum Mechanics, as being the best approximation of a classical, point-like object [15, 16, 17, 18, 19, 20, 21].

Based on this experience, we would like to apply the same idea to obtain a *regular* Cornell potential. The physical motivation is that a confining potential is instrumental to describe the properties of hadrons, which are, surely, not point-like objects.

Thus, we start by replacing the Dirac delta-function by a Gaussian of a finite width l_0

$$\delta(\vec{r}) \longrightarrow \frac{1}{(2\pi l_0^2)^{3/2}} e^{-r^2/4l_0^2} \tag{33}$$

The form of the charge distribution turns out to have a particularly simple form in the momentum space

$$\rho(\vec{k}) = e^{-\vec{k}^2 l_0^2} , \tag{34}$$

At this point, it is important to mention that we have two length scales in the theory: l_0 and $1/m$. The former is needed to regularize the theory at short-distances and is assumed to be $l_0 \ll 1/m$, while the latter gives a dominant contribution at large distances. Now, we shall repeat the steps from the previous Sections using (34) in (25) to obtain

$$\begin{aligned}\phi(r,l_0) &= -\frac{e}{\varepsilon_0}\int\frac{d^3k}{(2\pi)^3}\frac{1}{\vec{k}^2}\frac{\vec{k}^2+m^2}{\vec{k}^2}e^{i\vec{k}\cdot\vec{r}}e^{-l_0^2\vec{k}^2}, \\ &= -\frac{e}{\varepsilon_0}\int\frac{d^3k}{(2\pi)^3}\frac{1}{\vec{k}^2}\left(1+\frac{m^2}{\vec{k}^2}\right)e^{i\vec{k}\cdot\vec{r}}e^{-l_0^2\vec{k}^2}, \\ &\equiv \phi(r,l_0;m=0)+\phi(r,l_0;m)\end{aligned} \quad (35)$$

Let us begin by evaluating the m-independent part of the integral (35):

$$\begin{aligned}\phi(r,l_0;m=0) &= -\frac{e}{\varepsilon_0}\int\frac{d^3k}{(2\pi)^3}\frac{1}{\vec{k}^2}e^{i\vec{k}\cdot\vec{r}}e^{-l_0^2\vec{k}^2}, \\ &= -\frac{e}{\varepsilon_0}\int_0^\infty ds\int\frac{d^3k}{(2\pi)^3}e^{i\vec{k}\cdot\vec{r}}e^{-(l_0^2+s)\vec{k}^2}, \\ &= -\frac{e}{\varepsilon_0}\frac{1}{8\pi^{3/2}}\int_{l_0^2}^\infty d\tau\,\tau^{-3/2}e^{-r^2/4\tau},\ \tau=s+l_0^2, \quad (36)\\ &= -\frac{e}{\varepsilon_0}\frac{1}{4\pi\sqrt{\pi}}\frac{1}{r}\gamma\left(\frac{1}{2};\frac{r^2}{4l_0^2}\right), \quad (37)\end{aligned}$$

The asymptotic behavior of $\phi(r;m=0)$ is (see Appendix):

$$\phi(r,l_0;m=0) \sim -\frac{e}{4\pi\varepsilon_0}\frac{1}{r},\qquad r\longrightarrow\infty, \quad (38)$$

$$\sim -\frac{e}{\varepsilon_0}\frac{1}{4\pi^{3/2}l_0}\left(1-\frac{r^2}{4l_0^2}+\dots\right),\quad r\ll l_0 \quad (39)$$

$$\phi(r=0,l_0;m=0) = -\frac{e}{4\pi^{3/2}\varepsilon_0 l_0} \quad (40)$$

Equations (39) and (40) show that the m-independent part of the total potential corresponds to the regular Coulomb part. The width l_0 of the Gaussian source turns out to be a natural cut-off eliminating the singularity in $r=0$.
On the other hand, the m-dependent part of the potential is crucial to obtain a linear behavior at large distances.

$$\begin{aligned}\phi(r,l_0;m) &= -\frac{e}{\varepsilon_0}\int\frac{d^3k}{(2\pi)^3}\frac{m^2}{(\vec{k}^2)^2}e^{i\vec{k}\cdot\vec{r}}e^{-l_0^2\vec{k}^2}, \\ &= -\frac{e}{\varepsilon_0}\frac{m^2}{8\pi^{3/2}}\int_0^\infty ds\frac{s}{(s+l_0^2)^{3/2}}e^{-r^2/4(s+l_0^2)},\end{aligned}$$

$$= -\frac{e}{\varepsilon_0} \frac{m^2}{8\pi^{3/2}} \int_{l_0^2}^{\infty} \frac{d\tau}{\tau^{1/2}} \left(1 - \frac{l_0^2}{\tau}\right) e^{-r^2/4\tau}, \tau = s + l_0^2 \quad (41)$$

$$\equiv \phi(r, l_0 = 0; m) + \tilde{\phi}(r, l_0; m) \quad (42)$$

As before, we compute the two contributions in (42) separately starting with $\tilde{\phi}(r, l_0; m)$:

$$\tilde{\phi}(r, l_0; m) = -\frac{e}{\varepsilon_0} \frac{m^2 l_0^2}{8\pi^{3/2}} \int_{l_0^2}^{\infty} \frac{d\tau}{\tau^{3/2}} e^{-r^2/4\tau} = \frac{m^2 l_0^2}{4\pi^{3/2}} \frac{1}{r} \gamma\left(\frac{1}{2}; \frac{r^2}{4l_0^2}\right) \quad (43)$$

$$\sim -\frac{e}{\varepsilon_0} \frac{m^2 l_0}{4\pi^{3/2}} \left(1 - \frac{r^2}{6l_0^2} + \ldots\right), \quad r \ll l_0 \quad (44)$$

Then, we calculate the remaining l_0-independent part

$$\phi(r, l_0 = 0; m) = -\frac{e}{\varepsilon_0} \frac{m^2}{8\pi^{3/2}} \frac{r}{2} \int_0^{r^2/4l_0^2} \frac{du}{u^{3/2}} e^{-u}, \quad u = r^2/4\tau,$$

$$= -\frac{e}{\varepsilon_0} \frac{m^2}{8\pi^{3/2}} \frac{r}{2} \gamma\left(-1/2; r^2/4l_0^2\right), \quad (45)$$

$$\sim \frac{em^2}{8\pi \varepsilon_0} r, \quad r \gg l_0 \quad (46)$$

Equation (46) exhibits the expected linear behavior at large distances. Therefore, the complete potential is given by

$$\phi(r, l_0; m) =$$
$$-\frac{e}{4\pi^{3/2} \varepsilon_0} \frac{1}{r} \left[\left(1 - m^2 l_0^2\right) \gamma\left(1/2; r^2/4l_0^2\right) + \frac{m^2}{4} r^2 \gamma\left(-1/2; r^2/4l_0^2\right) \right] \quad (47)$$

As it was anticipated, the spread of the source modifies only the short-distances behavior of the potential rendering the potential finite at the origin:

$$\phi(r = 0) = -\frac{e}{4\pi^{3/2} \varepsilon_0 l_0} \left(1 - m^2 l_0^2\right) \approx -\frac{e}{4\pi^{3/2} \varepsilon_0 l_0} \quad (48)$$

Our previous assumption $m l_0 \ll 1$ assures the attractive character of the interaction near the origin.

4. Conclusion

At the beginning of this work we have briefly reviewed the Lee-Wick modification of ordinary electrodynamics leading to a linear electrostatic potential at short distances, while preserving the Coulomb form at large distances. We have also shown that the higher derivative correction to the kinetic term can be alternatively shifted into the interaction term. In this way, the kinetic term retains its canonical form and all the problems, at the quantum level, related to the presence of higher order derivatives are avoided. The new interaction

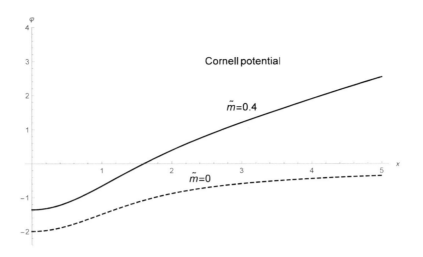

Figure 4. Plot of the potential in eq.(48) for $\widetilde{m} \equiv ml_0 = 0$ (regular Coulomb potential at short distances), and $\widetilde{m} \equiv ml_0 = 0.4$, i.e., regular Cornell potential.

term can be seen either as a non-point-like source of the field or as an effective dielectric constant depending on the distances from the source charge. In the latter interpretation the source remains point-like while the electric properties of the surrounding vacuum change with distances.

From this vantage point, one can further proceed to define a suitably modified interaction term. This leads to the exchange between the short and long distances behavior of the original Lee-Wick model. In this way, one obtains the Cornell potential between electric charges. Within this framework, one simulates the behavior expected to take place for color charges in Quantum Chromodynamics. Without pretending to construct a phenomenologically relevant model we, nevertheless, hope this work gives useful insights into the more realistic, but so far unsolved, case of Yang-Mills gauge theories.

In the last part of the work, we transferred the results, obtained in General Relativity, in order to remove the singular behavior of the Cornell potential in $r = 0$. For this purpose, we have replaced point-like charges by distributed ones. The final result is a modified Cornell potential which remains linear at large distances but turns out to be finite in $r = 0$.

5. Useful Formulas

The Fourier integrals are computed using the Schwinger formula:

$$\frac{1}{\left(\vec{k}^{\,2}\right)^{\alpha}} = \frac{1}{\Gamma(\alpha)} \int_0^\infty ds\, s^{\alpha-1} e^{-s\vec{k}^{\,2}} \qquad (49)$$

Gauss Integral:

$$\int d^n x\, e^{-\frac{1}{2}x A x} e^{B \cdot x} = e^{\frac{1}{4} B A^{-1} B} \left(\frac{\pi}{\det(A)} \right)^{n/2} \qquad (50)$$

Lower incomplete Gamma Function definition

$$\gamma(\alpha; z) \equiv \int_0^z dt\, t^{\alpha-1} e^{-t}, \qquad \text{Re}\,\alpha > 0 \qquad (51)$$

The asymptotic behavior of $\gamma(\alpha; z)$ at small z is

$$\gamma(\alpha; z) \sim \frac{1}{\alpha} z^\alpha \qquad (52)$$

Upper incomplete Gamma Function definition

$$\Gamma(\alpha; z) \equiv \int_z^\infty dt\, t^{\alpha-1} e^{-t} \qquad (53)$$

$$\gamma(\alpha; z) + \Gamma(\alpha; z) = \Gamma(\alpha), \qquad \alpha \Gamma(\alpha) = \Gamma(\alpha+1) \qquad (54)$$

where $\Gamma(\alpha)$ is the Euler Gamma Function.

$$\Gamma\left(\frac{1}{2}\right) = \sqrt{\pi} \qquad (55)$$

References

[1] Lee T. D. and Wick G. C., "Negative Metric and the Unitarity of the S Matrix," *Nucl. Phys. B* 9, 209 (1969).

[2] Lee T. D. and Wick G. C., "Finite Theory of Quantum Electrodynamics," *Phys. Rev. D* 2, 1033 (1970).

[3] Eichten E., Gottfried K., Kinoshita T., Lane K. D. and Yan T. M, "Charmonium: The Model," *Phys. Rev. D* 17, 3090 (1978), (Erratum: *Phys. Rev. D* 21, 313 (1980)).

[4] Accioly A., Gaete P., Helayel-Neto J., Scatena E. and Turcati R., "Exploring Lee-Wick finite electrodynamics," arXiv:1012.1045 [hep-th].

[5] Accioly A., Gaete P., Helayel-Neto J., Scatena E. and Turcati R., "Investigations in the Lee-Wick electrodynamics," *Mod. Phys. Lett. A* 26, 1985 (2011).

[6] Gaete P., "Remarks on gauge invariant variables and interaction energy in QED," *Phys. Rev. D* 59, 127702 (1999).

[7] Gaete P. and Schmidt I., "From screening to confinement in a gauge invariant formalism," *Phys. Rev. D* 61, 125002 (2000).

[8] Gaete P. and Spallucci E., "Confinement from gluodynamics in curved space-time," *Phys. Rev. D* 77, 027702 (2008).

[9] Gaete P., Helayel-Neto J., "On scale symmetry breaking and confinement in D=3 models," *J. Phys. A* 41, 425401 (2008).

[10] Gaete P. and Spallucci E., "From screening to confinement in a Higgs-like model," *Phys. Lett. B* 675, 145 (2009).

[11] Luscher M., "The Secret Long Range Force in Quantum Field Theories With Instantons," *Phys. Lett.* 78B, 465 (1978).

[12] Kondo K. I., "Abelian projected effective gauge theory of QCD with asymptotic freedom and quark confinement," *Phys. Rev. D* 57, 7467 (1998).

[13] Kondo K. I., "Abelian projected effective gauge theory of QCD with asymptotic freedom and quark confinement," *Prog. Theor. Phys. Suppl.* 131, 243 (1998).

[14] Kondo K. I., Kato S., Shibata A. and Shinohara T., "Quark confinement: Dual superconductor picture based on a non-Abelian Stokes theorem and reformulations of YangMills theory," *Phys. Rept.* 579, 1 (2015).

[15] P. Nicolini, A. Smailagic and E. Spallucci, *Phys. Lett. B* **632**, 547 (2006).

[16] S. Ansoldi, P. Nicolini, A. Smailagic and E. Spallucci, *Phys. Lett. B* **645**, 261 (2007).

[17] S. Ansoldi, "Spherical black holes with regular center: A Review of existing models including a recent realization with Gaussian sources," arXiv:0802.0330 [gr-qc].

[18] E. Spallucci, A. Smailagic and P. Nicolini, *Phys. Lett. B* **670**, 449 (2009).

[19] P. Nicolini, *Int. J. Mod. Phys. A* **24**, 1229 (2009).

[20] P. Nicolini and E. Spallucci, *Class. Quant. Grav.* **27**, 015010 (2010).

[21] E. Spallucci and A. Smailagic, *Int. J. Mod. Phys. D* **26**, no. 07, 1730013 (2017).

In matrix form:

$$\begin{bmatrix} E'_x & 0 & 0 & 0 \\ * & \frac{E'_y}{c} & B'_z & * \\ B'_y & * & * & \frac{E'_z}{c} \\ 0 & 0 & 0 & B'_x \end{bmatrix} = \begin{bmatrix} 1 & 0 & 0 & 0 \\ 0 & \cosh\frac{g\tau}{c} & -\sinh\frac{g\tau}{c} & 0 \\ 0 & \sinh\frac{g\tau}{c} & \cosh\frac{g\tau}{c} & 0 \\ 0 & 0 & 0 & 1 \end{bmatrix} \begin{bmatrix} E_x & 0 & 0 & 0 \\ \frac{E_z}{c} & \frac{E_y}{c} & B_z & B_y \\ B_y & B_z & \frac{E_y}{c} & \frac{E_z}{c} \\ 0 & 0 & 0 & B_x \end{bmatrix}$$ (2.3)

The elements marked with asterisks represent entities without any meaning. From (2.2) we obtain immediately:

$$\mathbf{E}'^2 - c^2\mathbf{B}'^2 = \mathbf{E}^2 - c^2\mathbf{B}^2 \qquad (2.4)$$

1.3. Planar Wave Transformation and Speed of Light in a Uniformly Accelerated Frame

In this section we apply the formalism derived in the previous paragraph in order to obtain the transform of a planar wave from an inertial frame S into an accelerated frame S'. Assume that a planar wave is propagating along the y axis in the inertial frame S. The wave has the electric component $\mathbf{E}_x$ and the magnetic component $\mathbf{B}_z$ along the x and z axes, respectively. The components equations are (see Figure 1.1):

$$\begin{aligned} \mathbf{E}_x &= E_{0x}\cos(\omega t - k_y y + \varphi)\mathbf{e}_x \\ \mathbf{B}_z &= B_{0z}\cos(\omega t - k_y y + \varphi)\mathbf{e}_z \\ \mathbf{E}_y &= \mathbf{E}_z = 0 \\ \mathbf{B}_x &= \mathbf{B}_y = 0 \end{aligned} \qquad (3.1)$$

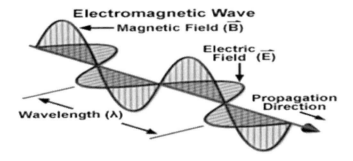

Figure 1.1. The planar electromagnetic wave.

According to reference [1] the transformation for the particular case of accelerated motion along the x-axis from $S'(\tau)$ into S is:

$$\begin{pmatrix} x \\ y \\ z \\ t \end{pmatrix} = Phy\_rectilinear \begin{pmatrix} x' \\ y' \\ z' \\ t' \end{pmatrix} + \begin{bmatrix} \frac{c^2}{g}(\cosh\frac{g\tau}{c}-1) \\ 0 \\ 0 \\ \frac{c}{g}\sinh\frac{g\tau}{c} \end{bmatrix} \quad (3.2)$$

$$Phy\_rectilinear = \begin{bmatrix} \cosh\frac{g\tau}{c} & 0 & 0 & c\sinh\frac{g\tau}{c} \\ 0 & 1 & 0 & 0 \\ 0 & 0 & 1 & 0 \\ \frac{1}{c}\sinh\frac{g\tau}{c} & 0 & 0 & \cosh\frac{g\tau}{c} \end{bmatrix} \quad (3.3)$$

so:

$$y = y'$$
$$t = \frac{x'}{c}\sinh\frac{g\tau}{c} + t'\cosh\frac{g\tau}{c} + \frac{c}{g}\sinh\frac{g\tau}{c} \quad (3.4)$$

Substituting (3.4) into (3.1) we obtain:

$$E_x = E_{0x}\cos[(\omega\cosh\frac{g\tau}{c})t' + (\frac{\omega}{c}\sinh\frac{g\tau}{c})x' - k_y y' + \varphi + \frac{\omega c}{g}\sinh\frac{g\tau}{c})] \quad (3.5)$$

On the other hand, in frame S', the wave equation is:

$$\mathbf{E}'_x = E'_{0x}\cos(\omega' t' - k'_x x' - k'_y y' - k'_z z' + \varphi')\mathbf{e}_x \quad (3.6)$$

Since $\mathbf{E}_x = \mathbf{E}'_x$ it follows that:

$$\begin{aligned} E'_{0x} &= E_{0x} \\ \omega' &= \omega\cosh\frac{g\tau}{c} \\ k'_x &= -\frac{\omega}{c}\sinh\frac{g\tau}{c} \\ k'_y &= k_y \\ k'_z &= 0 \\ \varphi' &= \varphi + \frac{\omega c}{g}\sinh\frac{g\tau}{c} \end{aligned} \quad (3.7)$$

In the accelerated frame, the wave vector **k** "gains a (time-varying) component in the x direction as a consequence of the acceleration. Additionally:

$$E'_y = -B_z c \sinh \frac{g\tau}{c}$$
$$E'_z = 0$$
$$B'_y = 0 \qquad (3.8)$$
$$B'_z = B_z \cosh \frac{g\tau}{c}$$

We can see from (3.8) that in the accelerated frame S' the wave exhibits components along the axes y' and z'. These components are a direct effect of the acceleration. From (3.7) we obtain:

$$k'^2 = k_x'^2 + k_y'^2 + k_z'^2 = \frac{\omega^2}{c^2}\sinh^2\frac{g\tau}{c} + \frac{\omega^2}{c^2} = \frac{\omega^2}{c^2}\cosh^2\frac{g\tau}{c} \qquad (3.9)$$

We can now calculate the phase light speed measured in the accelerated frame:

$$v'_p = \frac{\omega'}{k'} = c \qquad (3.10)$$

Finally, we can calculate the aberration in frame S' induced by the acceleration:

$$\cos\theta' = \frac{k'_x}{k'} = \tanh\frac{g\tau}{c} \qquad (3.11)$$

The Poynting vector in the accelerated frame is:

$$\mathbf{S'} = \begin{vmatrix} \mathbf{e}_x & \mathbf{e}_y & \mathbf{e}_z \\ E_x & -B_z c \sinh\frac{g\tau}{c} & 0 \\ B_x & 0 & B_z \cosh\frac{g\tau}{c} \end{vmatrix} = -B_z^2 c \sinh\frac{g\tau}{c}\cosh\frac{g\tau}{c}\mathbf{e}_x - E_x B_z \cosh\frac{g\tau}{c}\mathbf{e}_y + B_x B_z c \sinh\frac{g\tau}{c}\mathbf{e}_z \qquad (3.12)$$

1.4. General Case of Uniform Acceleration in an Arbitrary Direction

In a prior paper we have shown [12] that the particular transformation (2.1) can be generalized for the case of arbitrary direction constant acceleration $\mathbf{g} = (g_x, g_y, g_z)$ to:

$$\begin{pmatrix} x \\ y \\ z \\ t \end{pmatrix} = (Tr^{-1} * Phy\_rectilinear * Tr) \begin{pmatrix} x' \\ y' \\ z' \\ t' \end{pmatrix} + N \qquad (4.1)$$

where:

$$Tr = Rot(\mathbf{e}_z)_{-90^0} * Rot(\mathbf{e}_y)_{90^0-\varphi} * Rot_y$$

$$Phy\_rectilinear = \begin{bmatrix} 1 & 0 & 0 & 0 \\ 0 & \cosh\frac{g\tau}{c} & -\sinh\frac{g\tau}{c} & 0 \\ 0 & \sinh\frac{g\tau}{c} & \cosh\frac{g\tau}{c} & 0 \\ 0 & 0 & 0 & 1 \end{bmatrix} \qquad (4.2)$$

$$N = \begin{bmatrix} \frac{c^2}{g}(\cosh\frac{g\tau}{c} - 1) \\ 0 \\ 0 \\ \frac{c}{g}\sinh\frac{g\tau}{c} \end{bmatrix}$$

Introducing the triplet $(a,b,c) = (-\frac{g_z}{g}, 0, \frac{g_x}{g})$ the following expressions hold:

$$Rot_y = \begin{bmatrix} c & 0 & -a & 0 \\ 0 & 1 & 0 & 0 \\ a & 0 & c & 0 \\ 0 & 0 & 0 & 1 \end{bmatrix} \qquad (4.3)$$

$$Rot(\mathbf{e}_y)_{90^0-\varphi} = \begin{bmatrix} \cos(90^0-\varphi) & 0 & -\sin(90^0-\varphi) & 0 \\ 0 & 1 & 0 & 0 \\ \sin(90^0-\varphi) & 0 & \cos(90^0-\varphi) & 0 \\ 0 & 0 & 0 & 1 \end{bmatrix} = \begin{bmatrix} \sin\varphi & 0 & -\cos\varphi & 0 \\ 0 & 1 & 0 & 0 \\ \cos\varphi & 0 & \sin\varphi & 0 \\ 0 & 0 & 0 & 1 \end{bmatrix} \qquad (4.4)$$

$Rot(\mathbf{e}_y)_{90^0-\varphi} * Rot_y$ aligns $\mathbf{g}$ with $\mathbf{e}_y$. The second step is comprised by another rotation around the z-axis by -90^0 that aligns $\mathbf{g}$ with $\mathbf{e}_x$ (Figure 1.2):

$$Rot(\mathbf{e}_z)_{-90^0} = \begin{bmatrix} 0 & 1 & 0 & 0 \\ -1 & 0 & 0 & 0 \\ 0 & 0 & 1 & 0 \\ 0 & 0 & 0 & 1 \end{bmatrix} \quad (4.5)$$

Tr^{-1} reverses all the effects of Tr. Expression (4.1) gives the solution for the general case, of arbitrary acceleration direction. Then, from (2.3) and (4.1) the matrix for the general transform in the case of frame with arbitrary direction acceleration S' into the inertial frame S is simply:

$$\begin{bmatrix} E'_x & 0 & 0 & 0 \\ * & \dfrac{E'_y}{c} & B'_z & * \\ B'_y & * & * & \dfrac{E'_z}{c} \\ 0 & 0 & 0 & B'_x \end{bmatrix} = Tr * \begin{bmatrix} 1 & 0 & 0 & 0 \\ 0 & \cosh\dfrac{g\tau}{c} & -\sinh\dfrac{g\tau}{c} & 0 \\ 0 & \sinh\dfrac{g\tau}{c} & \cosh\dfrac{g\tau}{c} & 0 \\ 0 & 0 & 0 & 1 \end{bmatrix} * Tr^{-1} * \begin{bmatrix} E_x & 0 & 0 & 0 \\ \dfrac{E_z}{c} & \dfrac{E_y}{c} & B_z & B_y \\ B_y & B_z & \dfrac{E_y}{c} & \dfrac{E_z}{c} \\ 0 & 0 & 0 & B_x \end{bmatrix} \quad (4.6)$$

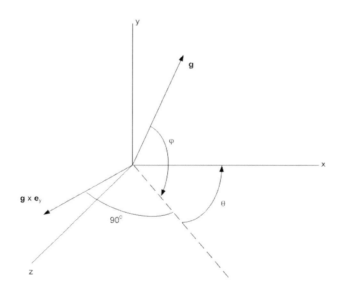

Figure 1.2. General acceleration.

1.5. Application I: The General Expressions for Aberration and for the Doppler Effect

We have seen in section 3 that the relativistic Doppler effect can be derived from the frame invariance of the expression:

$$\Psi = \omega t - (k_x x + k_y y + k_z z) + \varphi \quad (5.1)$$

We have seen in section 4 that the general coordinate transformation from an accelerated frame S' into an inertial frame S is:

$$\begin{pmatrix} x \\ y \\ z \\ t \end{pmatrix} = \begin{bmatrix} a_{11} & a_{12} & a_{13} & a_{14} \\ a_{21} & a_{22} & a_{23} & a_{24} \\ a_{31} & a_{32} & a_{33} & a_{34} \\ a_{41} & a_{42} & a_{43} & a_{44} \end{bmatrix}_g \begin{pmatrix} x' \\ y' \\ z' \\ t' \end{pmatrix} + \begin{bmatrix} \frac{c^2}{g}(\cosh\frac{g\tau}{c}-1) \\ 0 \\ 0 \\ \frac{c}{g}\sinh\frac{g\tau}{c} \end{bmatrix} \quad (5.2)$$

The subscript **g** represents the dependence of the matrix elements $a_{ij} = a_{ij}(\mathbf{g})$ of the acceleration $\mathbf{g} = (g_x, g_y, g_z)$ between frames S and S'. Substituting (5.2) into (5.1) we obtain:

$$\begin{aligned}\Psi &= \omega(\mathbf{a_4}.\mathbf{x'} + \frac{c}{g}\sinh\frac{g\tau}{c}) - k_x(\mathbf{a_1}.\mathbf{x'} + \frac{c^2}{g}(\cosh\frac{g\tau}{c}-1)) - k_y(\mathbf{a_2}.\mathbf{x'}) - k_z(\mathbf{a_3}.\mathbf{x'}) + \varphi = \\ &= (\omega a_{44} - k_x a_{14} - k_y a_{24} - k_z a_{34})t' - (k_x a_{11} + k_y a_{21} + k_z a_{31} - \omega a_{41})x' - \\ &\quad -(k_x a_{12} + k_y a_{22} + k_z a_{32} - \omega a_{42})y' - (k_x a_{13} + k_y a_{23} + k_z a_{33} - \omega a_{43})z' + \\ &\quad +\varphi + \omega\frac{c}{g}\sinh\frac{g\tau}{c} - k_x\frac{c^2}{g}(\cosh\frac{g\tau}{c}-1)\end{aligned} \quad (5.3)$$

On the other hand, in frame S':

$$\Psi' = \omega't' - (k'_x x' + k'_y y' + k'_z z') + \varphi' = \Psi \quad (5.4)$$

Comparing (5.3) and (5.4) we obtain the general expressions of the relativistic Doppler effect between the inertial frame S and the accelerated frame S':

$$\begin{aligned} \omega' &= \omega a_{44} - k_x a_{14} - k_y a_{24} - k_z a_{34} \\ k'_x &= k_x a_{11} + k_y a_{21} + k_z a_{31} - \omega a_{41} \\ k'_y &= k_x a_{12} + k_y a_{22} + k_z a_{32} - \omega a_{42} \\ k'_z &= k_x a_{13} + k_y a_{23} + k_z a_{33} - \omega a_{43} \\ \varphi' &= \varphi + \omega\frac{c}{g}\sinh\frac{g\tau}{c} - k_x\frac{c^2}{g}(\cosh\frac{g\tau}{c}-1) \end{aligned} \quad (5.5)$$

In matrix form:

$$\begin{bmatrix} \omega' \\ k'_x \\ k'_y \\ k'_z \end{bmatrix} = \begin{bmatrix} a_{44} & -a_{14} & -a_{24} & -a_{34} \\ -a_{41} & a_{11} & a_{21} & a_{31} \\ -a_{42} & a_{12} & a_{22} & a_{32} \\ -a_{43} & a_{13} & a_{23} & a_{33} \end{bmatrix}_g \begin{bmatrix} \omega \\ k_x \\ k_y \\ k_z \end{bmatrix} \quad (5.6)$$

For the reverse transformations we start with:

$$\begin{pmatrix} x' \\ y' \\ z' \\ t' \end{pmatrix} = \begin{bmatrix} a_{11} & a_{12} & a_{13} & a_{14} \\ a_{21} & a_{22} & a_{23} & a_{24} \\ a_{31} & a_{32} & a_{33} & a_{34} \\ a_{41} & a_{42} & a_{43} & a_{44} \end{bmatrix}^{-1} \begin{pmatrix} x - \dfrac{c^2}{g}(\cosh\dfrac{g\tau}{c} - 1) \\ y \\ z \\ t - \dfrac{c}{g}\sinh\dfrac{g\tau}{c} \end{pmatrix} \quad (5.7)$$

For simplicity, we re-write (5.7) as:

$$\begin{pmatrix} x' \\ y' \\ z' \\ t' \end{pmatrix} = \begin{bmatrix} b_{11} & b_{12} & b_{13} & b_{14} \\ b_{21} & b_{22} & b_{23} & b_{24} \\ b_{31} & b_{32} & b_{33} & b_{34} \\ b_{41} & b_{42} & b_{43} & b_{44} \end{bmatrix}_g \begin{pmatrix} x - \dfrac{c^2}{g}(\cosh\dfrac{g\tau}{c} - 1) \\ y \\ z \\ t - \dfrac{c}{g}\sinh\dfrac{g\tau}{c} \end{pmatrix} \quad (5.8)$$

The subscript **g** represents the dependence of the matrix elements $a_{ij} = a_{ij}(\mathbf{g})$ of the acceleration $\mathbf{g} = (g_x, g_y, g_z)$ between frames S and S'. In the accelerated frame S':

$$\Psi' = \omega' t' - (k'_x x' + k'_y y' + k'_z z') + \varphi' \quad (5.9)$$

Substituting (5.8) into (5.9):

$$\begin{aligned}\Psi' &= \omega'(\mathbf{b}_4 \cdot \mathbf{x} - \dfrac{c}{g}\sinh\dfrac{g\tau}{c}) - k'_x(\mathbf{b}_1 \cdot \mathbf{x} - \dfrac{c^2}{g}(\cosh\dfrac{g\tau}{c} - 1)) - k'_y(\mathbf{b}_2 \cdot \mathbf{x}) - k'_z(\mathbf{b}_3 \cdot \mathbf{x}) + \varphi' = \\ &= (\omega' b_{44} - k'_x b_{14} - k'_y b_{24} - k'_z b_{34})t - (k'_x b_{11} + k'_y b_{21} + k'_z b_{31} - \omega' b_{41})x - \\ &\quad -(k'_x b_{12} + k'_y b_{22} + k'_z b_{32} - \omega' b_{42})y - (k'_x b_{13} + k'_y b_{23} + k'_z b_{33} - \omega' b_{43})z + \\ &\quad + \varphi' - \omega'\dfrac{c}{g}\sinh\dfrac{g\tau}{c} + k'_x\dfrac{c^2}{g}(\cosh\dfrac{g\tau}{c} - 1)\end{aligned} \quad (5.10)$$

On the other hand, in frame S:

$$\Psi = \omega t - (k_x x + k_y y + k_z z) + \varphi = \Psi' \tag{5.11}$$

Comparing (5.10) and (5.11) we obtain the general expressions of the relativistic Doppler effect between the inertial frame S' and the accelerated frame S:

$$\begin{aligned}
\omega &= \omega' b_{44} - k'_x b_{14} - k'_y b_{24} - k'_z b_{34} \\
k_x &= k'_x b_{11} + k'_y b_{21} + k'_z b_{31} - \omega' b_{41} \\
k_y &= k'_x b_{12} + k'_y b_{22} + k'_z b_{32} - \omega' b_{42} \\
k_z &= k'_x b_{13} + k'_y b_{23} + k'_z b_{33} - \omega' b_{43}
\end{aligned} \tag{5.12}$$

$$\varphi = \varphi' - \omega' \frac{c}{g} \sinh \frac{g\tau}{c} + k'_x \frac{c^2}{g} \left(\cosh \frac{g\tau}{c} - 1 \right)$$

In matrix form:

$$\begin{bmatrix} \omega \\ k_x \\ k_y \\ k_z \end{bmatrix} = \begin{bmatrix} b_{44} & -b_{14} & -b_{24} & -b_{34} \\ -b_{41} & b_{11} & b_{21} & b_{31} \\ -b_{42} & b_{12} & b_{22} & b_{32} \\ -b_{43} & b_{13} & b_{23} & b_{33} \end{bmatrix}_{\mathbf{g}} \begin{bmatrix} \omega' \\ k'_x \\ k'_y \\ k'_z \end{bmatrix} \tag{5.13}$$

1.6. Application II: The Doppler Effect for Emitter and Receiver Accelerated at Different Rates

In this section we treat the general case, the emitter moves with arbitrary acceleration **g1** and the receiver moves with arbitrary acceleration **g2**, both with respect to the inertial frame S.

According to (5.13):

$$\begin{bmatrix} \omega \\ k_x \\ k_y \\ k_z \end{bmatrix} = \begin{bmatrix} b_{44} & -b_{14} & -b_{24} & -b_{34} \\ -b_{41} & b_{11} & b_{21} & b_{31} \\ -b_{42} & b_{12} & b_{22} & b_{32} \\ -b_{43} & b_{13} & b_{23} & b_{33} \end{bmatrix}_{\mathbf{g1}} \begin{bmatrix} \omega' \\ k'_x \\ k'_y \\ k'_z \end{bmatrix}_{emitter} \tag{6.1}$$

According to (5.6):

$$\begin{bmatrix} \omega' \\ k'_x \\ k'_y \\ k'_z \end{bmatrix}_{receiver} = \begin{bmatrix} a_{44} & -a_{14} & -a_{24} & -a_{34} \\ -a_{41} & a_{11} & a_{21} & a_{31} \\ -a_{42} & a_{12} & a_{22} & a_{32} \\ -a_{43} & a_{13} & a_{23} & a_{33} \end{bmatrix}_{g2} \begin{bmatrix} \omega \\ k_x \\ k_y \\ k_z \end{bmatrix} \quad (6.2)$$

From (6.1) and (6.2) we obtain the general form of Doppler effect and aberration for the case of the emitter and the receiver moving with arbitrary accelerations **g1** and **g2** with respect to the same inertial reference frame:

$$\begin{bmatrix} \omega' \\ k'_x \\ k'_y \\ k'_z \end{bmatrix}_{receiver} = \begin{bmatrix} a_{44} & -a_{14} & -a_{24} & -a_{34} \\ -a_{41} & a_{11} & a_{21} & a_{31} \\ -a_{42} & a_{12} & a_{22} & a_{32} \\ -a_{43} & a_{13} & a_{23} & a_{33} \end{bmatrix}_{g2} \begin{bmatrix} b_{44} & -b_{14} & -b_{24} & -b_{34} \\ -b_{41} & b_{11} & b_{21} & b_{31} \\ -b_{42} & b_{12} & b_{22} & b_{32} \\ -b_{43} & b_{13} & b_{23} & b_{33} \end{bmatrix}_{g1} \begin{bmatrix} \omega' \\ k'_x \\ k'_y \\ k'_z \end{bmatrix}_{emitter} \quad (6.3)$$

2. Electrodynamics in Uniformly Accelerated Frames as Viewed from the Accelerated Frame

Real life applications include accelerating and rotating frames more often than the idealized case of inertial frames. Our daily experiments happen in the laboratories attached to the rotating, continuously accelerating Earth. Many books and papers have been dedicated to transformations between particular cases of rectilinear acceleration and/or rotation [14] and to the applications of such formulas [15-25]. In a recent pair of papers, [26-27], we have presented the equations of electrodynamics in an accelerated /rotating frame as viewed from the point of view of the inertial frame of the laboratory. In the current chapter, we are presenting the equations of electrodynamics in an accelerated frame as viewed from the accelerated frame. There is also great interest in producing a general solution that deals with arbitrary orientation of acceleration in the case of rectilinear motion., so we produced the equations for the general case as well The main idea of this chapter is to generate a standard blueprint for a general solution that gives equivalent of the Lorentz transforms for the case of the transforms between an inertial frame and an accelerated frame.

2.1. Transforms between the Accelerated Frame and the Inertial Frame

In this section we will derive the transforms between the accelerated frame and the inertial frame for the electromagnetic tensor. Let S represent an inertial system of coordinates and $S'(\tau)$ an accelerated one. According to reference [14] the transformation for the particular case of accelerated motion along the x-axis from $S'(\tau)$ into S is:

$$\begin{pmatrix} x \\ y \\ z \\ t \end{pmatrix} = Phy\_rectilinear \begin{pmatrix} x' \\ y' \\ z' \\ t' \end{pmatrix} + \begin{bmatrix} \dfrac{c^2}{g}(\cosh\dfrac{g\tau}{c} - 1) \\ 0 \\ 0 \\ \dfrac{c}{g}\sinh\dfrac{g\tau}{c} \end{bmatrix} \quad (2.1)$$

where:

"c" is the speed of light in vacuum, "g" is the proper acceleration, τ is the proper time and

$$Phy\_rectilinear = \begin{bmatrix} \cosh\dfrac{g\tau}{c} & 0 & 0 & c\sinh\dfrac{g\tau}{c} \\ 0 & 1 & 0 & 0 \\ 0 & 0 & 1 & 0 \\ \dfrac{1}{c}\sinh\dfrac{g\tau}{c} & 0 & 0 & \cosh\dfrac{g\tau}{c} \end{bmatrix} \quad (2.2)$$

The electromagnetic potential, by virtue of being a 4-vector transforms the same way:

$$\begin{pmatrix} A_x \\ A_y \\ A_z \\ c\varphi \end{pmatrix} = Phy\_rectilinear \begin{pmatrix} A'_x \\ A'_y \\ A'_z \\ c\varphi' \end{pmatrix} \quad (2.3)$$

In the inertial frame, the differential Maxwell equations in vacuum, in the absence of electric charge, are [1]:

$$-E_x = \frac{\partial A_x}{\partial t} + c^2 \frac{\partial \varphi}{\partial x} \quad (2.4)$$

$$-E_y = \frac{\partial A_y}{\partial t} + c^2 \frac{\partial \varphi}{\partial y} \quad (2.5)$$

$$-E_z = \frac{\partial A_z}{\partial t} + c^2 \frac{\partial \varphi}{\partial z} \quad (2.6)$$

$$\mathbf{B} = curl\mathbf{A} \quad (2.7)$$

Let's start with (2.4):

$$\frac{\partial A_x}{\partial t} = \frac{\partial A_x'}{\partial t}\cosh\frac{g\tau}{c} + \frac{\partial \varphi'}{\partial t} c\sinh\frac{g\tau}{c}$$

$$\frac{\partial A_x'}{\partial t} = \frac{\partial A_x'}{\partial x'}\frac{\partial x'}{\partial t} + \frac{\partial A_x'}{\partial t'}\frac{\partial t'}{\partial t} = \frac{\partial A_x'}{\partial x'}(-c\sinh\frac{g\tau}{c}) + \frac{\partial A_x'}{\partial t'}\cosh\frac{g\tau}{c} \quad (2.8)$$

$$\frac{\partial \varphi'}{\partial t} = \frac{\partial \varphi'}{\partial x'}(-c\sinh\frac{g\tau}{c}) + \frac{\partial \varphi'}{\partial t'}\cosh\frac{g\tau}{c}$$

$$\frac{\partial A_x}{\partial t} = \frac{\partial A_x'}{\partial x'}(-c\sinh\frac{g\tau}{c}\cosh\frac{g\tau}{c}) + \frac{\partial A_x'}{\partial t'}\cosh^2\frac{g\tau}{c} + \frac{\partial \varphi'}{\partial x'}(-c^2\sinh^2\frac{g\tau}{c}) + \frac{\partial \varphi'}{\partial t'} c\sinh\frac{g\tau}{c}\cosh\frac{g\tau}{c}$$

In a similar manner:

$$\frac{\partial \varphi}{\partial x} = \frac{\partial}{\partial x}(\frac{A_x'}{c}\sin\frac{g\tau}{c} + \varphi'\cos\frac{g\tau}{c})$$

$$\frac{\partial A_x'}{\partial x} = \frac{\partial A_x'}{\partial x'}\frac{\partial x'}{\partial x} + \frac{\partial A_x'}{\partial t'}\frac{\partial t'}{\partial x} = \frac{\partial A_x'}{\partial x'}\cos\frac{g\tau}{c} + \frac{\partial A_x'}{\partial t'}\frac{-\sin\frac{g\tau}{c}}{c} \quad (2.9)$$

$$\frac{\partial \varphi'}{\partial x} = \frac{\partial \varphi'}{\partial x'}\cos\frac{g\tau}{c} + \frac{\partial \varphi'}{\partial t'}\frac{-\sin\frac{g\tau}{c}}{c}$$

$$\frac{\partial \varphi}{\partial x} = \frac{\partial A_x'}{\partial x'}\frac{\sin\frac{g\tau}{c}\cos\frac{g\tau}{c}}{c} - \frac{\partial A_x'}{\partial t'}\frac{\sin^2\frac{g\tau}{c}}{c^2} + \frac{\partial \varphi'}{\partial x'}\cos^2\frac{g\tau}{c} - \frac{\partial \varphi'}{\partial t'}\frac{\sin\frac{g\tau}{c}\cos\frac{g\tau}{c}}{c}$$

Substitute (2.8), (2.9) into (2.4):

$$-E_x = \frac{\partial A_x'}{\partial t'} + c^2\frac{\partial \varphi'}{\partial x'}$$
$$E_x = E_x' \quad (2.10)$$

Moving on to (2.5)

$$\frac{\partial A_y}{\partial t} = \frac{\partial A_y'}{\partial t} = \frac{\partial A_y'}{\partial x'}\frac{\partial x'}{\partial t} + \frac{\partial A_x'}{\partial t'}\frac{\partial t'}{\partial t} = \frac{\partial A_y'}{\partial x'}(-c\sinh\frac{g\tau}{c}) + \frac{\partial A_y'}{\partial t'}\cosh\frac{g\tau}{c} \quad (2.11)$$

$$\frac{\partial \varphi}{\partial y} = \frac{\partial}{\partial y'}(\frac{A_x'}{c}\sinh\frac{g\tau}{c} + \varphi'\cosh\frac{g\tau}{c}) = \frac{\partial A_x'}{\partial y'}\frac{\sinh\frac{g\tau}{c}}{c} + \frac{\partial \varphi'}{\partial y'}\cosh\frac{g\tau}{c}$$

Substitute (2.11) into (2.5):

$$E_y = -(\frac{\partial A_y'}{\partial t'} + c^2 \frac{\partial \varphi'}{\partial y'})\cosh\frac{g\tau}{c} + \left(\frac{\partial A_y'}{\partial x'} - \frac{\partial A_x'}{\partial y'}\right)c\sinh\frac{g\tau}{c}$$

$$E_y' = -(\frac{\partial A_y'}{\partial t'} + c^2 \frac{\partial \varphi'}{\partial y'}) \qquad (2.12)$$

$$E_y = E_y'\cosh\frac{g\tau}{c} + \left(\frac{\partial A_y'}{\partial x'} - \frac{\partial A_x'}{\partial y'}\right)c\sinh\frac{g\tau}{c}$$

In a similar manner we obtain from (2.6):

$$E_z = E_z'\cosh\frac{g\tau}{c} + \left(\frac{\partial A_z'}{\partial x'} - \frac{\partial A_x'}{\partial z'}\right)c\sinh\frac{g\tau}{c} \qquad (2.13)$$

From (2.7) we obtain:

$$B_x = \frac{\partial A_z}{\partial y} - \frac{\partial A_y}{\partial z} = \frac{\partial A_z'}{\partial y'} - \frac{\partial A_y'}{\partial z'} = B_x' \qquad (2.14)$$

$$\begin{aligned}B_z &= \frac{\partial A_y}{\partial x} - \frac{\partial A_x}{\partial y} = \frac{\partial A_y'}{\partial x} - \frac{\partial A_x}{\partial y'} = (\frac{\partial A_y'}{\partial x'} - \frac{\partial A_x'}{\partial y'})\cosh\frac{g\tau}{c} - \frac{\partial A_y'}{\partial t'}\frac{\sinh\frac{g\tau}{c}}{c} - \\ &- \frac{\partial \varphi'}{\partial y'}c\sinh\frac{g\tau}{c} = (\frac{\partial A_y'}{\partial x'} - \frac{\partial A_x'}{\partial y'})\cosh\frac{g\tau}{c} - \frac{\sinh\frac{g\tau}{c}}{c}(\frac{\partial A_y'}{\partial t'} + c^2\frac{\partial \varphi'}{\partial y'}) = \\ &= B_z'\cosh\frac{g\tau}{c} + E_y'\frac{\sinh\frac{g\tau}{c}}{c}\end{aligned} \qquad (2.15)$$

$$B_z' = \frac{\partial A_y'}{\partial x'} - \frac{\partial A_x'}{\partial y'} \qquad (2.16)$$

Therefore:

$$E_y = E_y'\cosh\frac{g\tau}{c} + B_z'c\sinh\frac{g\tau}{c} \qquad (2.17)$$

In a similar manner we obtain:

$$B_y = B_y'\cosh\frac{g\tau}{c} - E_z'\frac{\sinh\frac{g\tau}{c}}{c} \qquad (2.18)$$

$$B'_y = \frac{\partial A'_x}{\partial z'} - \frac{\partial A'_z}{\partial x'} \tag{2.19}$$

$$E_z = E'_z \cosh\frac{g\tau}{c} - B'_y c \sinh\frac{g\tau}{c} \tag{2.20}$$

Putting everything together:

$$\begin{aligned}
E_x &= E'_x \\
E_y &= E'_y \cosh\frac{g\tau}{c} + B'_z c \sinh\frac{g\tau}{c} \\
E_z &= E'_z \cosh\frac{g\tau}{c} - B'_y c \sinh\frac{g\tau}{c} \\
B_x &= B'_x \\
B_y &= B'_y \cosh\frac{g\tau}{c} - E'_z \frac{\sinh\frac{g\tau}{c}}{c} \\
B_z &= B'_z \cosh\frac{g\tau}{c} + E'_y \frac{\sinh\frac{g\tau}{c}}{c}
\end{aligned} \tag{2.21}$$

Notice the resemblance with the standard Lorentz transforms in [1], for example:

$$\begin{aligned}
E_x &= E'_x \\
E_y &= \gamma(E'_y + V\frac{H'_z}{c}) \\
E_z &= \gamma(E'_z - V\frac{H'_y}{c}) \\
H_x &= H'_x \\
H_y &= \gamma(H'_y - V\frac{E'_z}{c}) \\
H_z &= \gamma(H'_z + V\frac{E'_y}{c})
\end{aligned} \tag{2.22}$$

2.2. Consequences

In this section we present a few consequences of the transforms between the accelerated frame and the inertial frame.

2.2.1. Maxwell Laws in Accelerated Frame

$$\mathbf{B}' = \text{curl}\mathbf{A}'$$
$$\mathbf{E}' = -\frac{\partial \mathbf{A}'}{\partial t'} - \nabla \varphi' \qquad (2.23)$$

The above follows immediately from (2.14), (2.15) and (2.19).

2.2.2. The Gauge Invariance Condition in a Uniformly Accelerated Frame

$$\text{div}\mathbf{A}' + \frac{\partial \varphi'}{\partial t'} = \text{div}\mathbf{A} + \frac{\partial \varphi}{\partial t} = 0 \qquad (2.24)$$

$$\text{div}\mathbf{A}' = \frac{\partial A'_x}{\partial x'} + \frac{\partial A'_y}{\partial y'} + \frac{\partial A'_z}{\partial z'} = \frac{\partial A'_x}{\partial x'} + \frac{\partial A_y}{\partial y} + \frac{\partial A_z}{\partial z} \qquad (2.25)$$

$$\frac{\partial A'_x}{\partial x'} = \frac{\partial A_x}{\partial x}\cosh^2\frac{g\tau}{c} + \frac{\partial A_x}{\partial t}\frac{\sinh\frac{g\tau}{c}\cosh\frac{g\tau}{c}}{c} - \frac{\partial \varphi}{\partial x}c\sinh\frac{g\tau}{c}\cosh\frac{g\tau}{c} - \frac{\partial \varphi}{\partial t}\sinh^2\frac{g\tau}{c} \qquad (2.26)$$

$$\frac{\partial \varphi'}{\partial t'} = -\frac{\partial A_x}{\partial x}\sinh^2\frac{g\tau}{c} - \frac{\partial A_x}{\partial t}\frac{\sinh\frac{g\tau}{c}\cosh\frac{g\tau}{c}}{c} + \frac{\partial \varphi}{\partial x}c\sinh\frac{g\tau}{c}\cosh\frac{g\tau}{c} + \frac{\partial \varphi}{\partial t}\cosh^2\frac{g\tau}{c} \qquad (2.27)$$

$$\frac{\partial A'_x}{\partial x'} + \frac{\partial \varphi'}{\partial t'} = \frac{\partial A_x}{\partial x} + \frac{\partial \varphi}{\partial t} \qquad (2.28)$$

Equality (2.28) results into:

$$\text{div}\mathbf{A}' + \frac{\partial \varphi'}{\partial t'} = \text{div}\mathbf{A} + \frac{\partial \varphi}{\partial t} = 0 \qquad (2.29)$$

Equalities (2.23) result into Maxwell's wave equations having the same exact form in the accelerated frame as the equations in the inertial frames with the immediate consequence that light speed in vacuum in a uniformly accelerated frame is "c". Indeed, (2.23) results into:

$$\frac{1}{c^2}\frac{\partial^2 \mathbf{E}'}{\partial t'^2} - \nabla^2 \mathbf{E}' = 0$$
$$\frac{1}{c^2}\frac{\partial^2 \mathbf{B}'}{\partial t'^2} - \nabla^2 \mathbf{B}' = 0 \qquad (2.30)$$

2.2.3. The Lorentz Force in an Accelerated Frame

In the inertial frame, the Lorentz force has the expression:

$$\mathbf{F} = q(\mathbf{E} + \mathbf{v} \times \mathbf{B}) \tag{2.31}$$

$$\mathbf{v} = v_x \mathbf{e}_x$$

$$\mathbf{F} = q(E_x \mathbf{e}_x + E_y \mathbf{e}_y + E_z \mathbf{e}_z + \begin{bmatrix} \mathbf{e}_x & \mathbf{e}_y & \mathbf{e}_z \\ v_x & 0 & 0 \\ B_x & B_y & B_z \end{bmatrix}) = q(E_x \mathbf{e}_x + (E_y - v_x B_z)\mathbf{e}_y + (E_z + v_x B_y)\mathbf{e}_z)$$

We know that:

$$dx = dx' \cosh\frac{g\tau}{c} + cdt' \sinh\frac{g\tau}{c}$$

$$dt = \frac{1}{c}\sinh\frac{g\tau}{c}dx' + dt'\cosh\frac{g\tau}{c} \tag{2.32}$$

$$v_x = \frac{dx}{dt} = \frac{\frac{dx'}{dt'}\cosh\frac{g\tau}{c} + c\sinh\frac{g\tau}{c}}{\frac{1}{c}\sinh\frac{g\tau}{c}\frac{dx'}{dt'} + \cosh\frac{g\tau}{c}} = \frac{v'_x \cosh\frac{g\tau}{c} + c\sinh\frac{g\tau}{c}}{\frac{v'_x}{c}\sinh\frac{g\tau}{c} + \cosh\frac{g\tau}{c}}$$

The formula:

$$v_x = \frac{v'_x \cosh\frac{g\tau}{c} + c\sinh\frac{g\tau}{c}}{\frac{v'_x}{c}\sinh\frac{g\tau}{c} + \cosh\frac{g\tau}{c}} \tag{2.33}$$

Ties the speed v'_x of the particle in the accelerated frame to its measured speed in the inertial frame v_x. Using the above, we obtain:

$$E_y - v_x B_z = E'_y \cosh\frac{g\tau}{c} + B'_z c \sinh\frac{g\tau}{c} - (\frac{E'_y}{c}\sinh\frac{g\tau}{c} + B'_z \cosh\frac{g\tau}{c})\frac{v'_x \cosh\frac{g\tau}{c} + c\sinh\frac{g\tau}{c}}{\frac{v'_x}{c}\sinh\frac{g\tau}{c} + \cosh\frac{g\tau}{c}} =$$

$$= \frac{E'_y - v'_x B'_z}{\frac{v'_x}{c}\sinh\frac{g\tau}{c} + \cosh\frac{g\tau}{c}} \tag{2.34}$$

Similarly:

$$E_z + v_x B_y = \frac{E'_z + v'_x B'_y}{\frac{v'_x}{c} \sinh \frac{g\tau}{c} + \cosh \frac{g\tau}{c}} \qquad (2.35)$$

So, we can write:

$$\mathbf{F}'_\| = \mathbf{F}_\| = qE_x \mathbf{e}_x$$
$$\mathbf{F}'_\perp = \frac{\mathbf{F}_\perp}{\frac{v'_x}{c} \sinh \frac{g\tau}{c} + \cosh \frac{g\tau}{c}} = q \frac{(E'_y - v'_x B'_z)\mathbf{e}_y + (E'_z + v'_x B'_y)\mathbf{e}_z}{\frac{v'_x}{c} \sinh \frac{g\tau}{c} + \cosh \frac{g\tau}{c}} \qquad (2.36)$$

Expressions (2.36) represent the transformation of the Lorentz force between the inertial and the accelerated frame.

2.2.4. Bremsstrahlung

Bremsstrahlung is the *electromagnetic radiation* produced by the *deceleration* of a charged particle. The moving particle loses *kinetic energy,* which is converted into a *photon,* it is the process of producing the energy radiation. Bremsstrahlung has a *continuous spectrum,* which becomes more intense and whose peak intensity shifts toward higher frequencies as the change of the energy of the decelerated particles increases. The term is frequently used in the more narrow sense of radiation from electrons (from whatever source) slowing in matter. In astrophysics, Bremsstrahlung refers to radiation emitted from zones of the universe characterized by a high concentration of plasma. The radiation in this case is created by charged particles that are free; i.e., not part of an ion, atom or molecule, both before and after the deflection (*acceleration*) that caused the emission. In any case, the total radiated power is given by [28]

$$P = \frac{q^2 \gamma^6}{6\pi\varepsilon_0 c}(\dot{\beta}^2 - (\vec{\beta} \times \dot{\vec{\beta}})^2)$$
$$\gamma = \frac{1}{\sqrt{1-\beta^2}} \qquad (2.37)$$
$$\vec{\beta} = \frac{\vec{v}}{c}$$
$$\dot{\vec{\beta}} = \frac{\vec{g}}{c}$$

In the case where velocity is parallel to acceleration (for example, linear motion), the formula simplifies to [28]:

$$P = \frac{q^2 g^2 \gamma^6}{6\pi\varepsilon_0 c^3} \qquad (2.38)$$

For the case of acceleration perpendicular to the velocity (as in the case of synchrotrons), the formula simplifies to:

$$P = \frac{q^2 g^2 \gamma^4}{6\pi\varepsilon_0 c^3} \qquad (2.39)$$

In either case, we do not observe any significant X-rays radiated from the free electrons in the Earth atmosphere due to several factors:

- the speed of the electrons is low (γ is small, very close to unity),
- the deceleration "g" is very small due to the absence of fields or matter that could affect the free electrons
- the electron density per unit of volume is small
- the presence of c^{-3}

By contrast, this is not the case in astrophysics. where we have observed *significant* emission from certain galaxies' intra-cluster medium due to thermal bremsstrahlung. This radiation is in the energy range of X-rays and can be easily observed with space-based telescopes such as Chandra X-ray Observatory, XMM-Newton, ROSAT, ASCA, EXOSAT, Suzaku, RHESSI and future missions like IXO[29] and Astro-H [30]. The reasons for observing such effects are:

- much higher charge density
- much larger speeds and accelerations (due to the presence of strong magnetic fields)

In addition to the changes in frequency (energy), we also observe light polarization effects, due to the presence of the magnetic fields mentioned above.

2.3. Planar Wave Transformation and Speed of Light in a Uniformly Accelerated Frame

In this section we apply the formalism derived in the previous paragraph in order to obtain the transform of a planar wave. Assume that a planar wave is propagating along the y' axis in the accelerated frame $S'(\tau)$. The wave has the electric component $\mathbf{E}'_x$ and the magnetic component $\mathbf{B}'_z$ along the x' and z' axes, respectively. The components equations are:

$$\begin{aligned}\mathbf{E}'_x &= E'_{0x}\cos(\omega' t' - k'_y y' + \varphi')\mathbf{e}_x \\ \mathbf{B}'_z &= B'_{0z}\cos(\omega' t' - k'_y y' + \varphi')\mathbf{e}_z\end{aligned} \qquad (3.1)$$

Inverting transforms (2.1) we obtain:

$$y' = y$$
$$t' = -\frac{x}{c}\sinh\frac{g\tau}{c} + t\cosh\frac{g\tau}{c} - \frac{a}{c}\sinh\frac{g\tau}{c} + b\cosh\frac{g\tau}{c}$$
$$a = \frac{c^2}{g}(\cosh\frac{g\tau}{c} - 1) \tag{3.2}$$
$$b = \frac{c}{g}\sinh\frac{g\tau}{c}$$

Substituting (3.2) into (3.1) we obtain:

$$E'_x = E'_{0x}\cos[(\omega'\cosh\frac{g\tau}{c})t - (\frac{\omega'}{c}\sinh\frac{g\tau}{c})x - k'_y y + \varphi' + \omega'(b\cosh\frac{g\tau}{c} - \frac{a}{c}\sinh\frac{g\tau}{c})] \tag{3.3}$$

On the other hand, in frame S, the wave equation is:

$$\mathbf{E}_x = E_{0x}\cos(\omega t - k_x x - k_y y - k_z z + \varphi)\mathbf{e}_x \tag{3.4}$$

Since $\mathbf{E}_x = \mathbf{E}'_x$ it follows that:

$$E'_{0x} = E_{0x}$$
$$\omega = \omega'\cosh\frac{g\tau}{c}$$
$$k_x = \frac{\omega'}{c}\sinh\frac{g\tau}{c} = \frac{\omega}{c}\tanh\frac{g\tau}{c} \tag{3.5}$$
$$k_y = k'_y$$
$$k_z = 0$$
$$\varphi = \varphi' + \omega'(b\cosh\frac{g\tau}{c} - \frac{a}{c}\sinh\frac{g\tau}{c})$$

The formula $\omega = \omega'\cosh\frac{g\tau}{c}$ represents the Doppler effect due to acceleration. We see that the pulsation decreases in time by the factor $\cosh\frac{g\tau}{c}$ in a frame that is uniformly accelerated in the same direction of the propagation of the electromagnetic wave.

From (3.8) we obtain:

$$\frac{\omega^2}{c^2} = k^2 = k_x^2 + k_y^2 + k_z^2 = \frac{\omega^2}{c^2}\tanh^2\frac{g\tau}{c} + k_y'^2 \tag{3.6}$$

Therefore, in frame $S'(\tau)$ the wave vector is:

$$k' = k_y' = \frac{\omega}{c\cosh\frac{g\tau}{c}} \tag{3.7}$$

We can now calculate the phase light speed in the accelerated frame:

$$v_p' = \frac{\omega'}{k'} = \frac{\omega}{k} = c \tag{3.8}$$

So, the light speed in the accelerated frame equals the light speed in the inertial frame, c.
We can now proceed to calculating the amplitude and the phase transformation between the inertial and the accelerated frame:

$$\varphi = \varphi' + \omega'(b\cosh\frac{g\tau}{c} - \frac{a}{c}\sinh\frac{g\tau}{c}) = \varphi' + \frac{\omega c}{g}\tanh\frac{g\tau}{c}$$
$$\varphi' = \varphi - \frac{\omega c}{g}\tanh\frac{g\tau}{c} \tag{3.9}$$

The magnetic field component transforms as:

$$B_{0z} = B_{0z}'\cosh\frac{g\tau}{c} \tag{3.10}$$

So, the absolute value of the Poynting vector transforms as:

$$S = E_{0x}B_{0z} = S'\cosh\frac{g\tau}{c} \tag{3.11}$$

So the electromagnetic flux in the accelerated frame decreases by with respect to the flux in the inertial frame. Finally, we can calculate the aberration in frame S induced by the acceleration:

$$\cos\theta = \frac{k_x}{\sqrt{k_x^2 + k_y^2}} = \tanh\frac{g\tau}{c} \tag{3.12}$$

2.4. General Case of Uniform Acceleration in an Arbitrary Direction

In a prior paper we have shown [25] that the particular transformation (2.1) can be generalized for the case of arbitrary direction constant acceleration $\mathbf{g} = (g_x, g_y, g_z)$ to:

$$\begin{pmatrix} x \\ y \\ z \\ t \end{pmatrix} = (Tr^{-1} * Phy\_rectilinear * Tr) \begin{pmatrix} x' \\ y' \\ z' \\ t' \end{pmatrix} + N \tag{4.1}$$

where:

$$Tr = Rot(\mathbf{e}_z)_{-90^0} * Rot(\mathbf{e}_y)_{90^0-\varphi} * Rot_y \tag{4.2}$$

Introducing the triplet $(a, b, c) = (-\frac{g_z}{g}, 0, \frac{g_x}{g})$ the following expressions hold:

$$Rot_y = \begin{bmatrix} c & 0 & -a & 0 \\ 0 & 1 & 0 & 0 \\ a & 0 & c & 0 \\ 0 & 0 & 0 & 1 \end{bmatrix} \tag{4.3}$$

$$Rot(\mathbf{e}_y)_{90^0-\varphi} = \begin{bmatrix} \cos(90^0-\varphi) & 0 & -\sin(90^0-\varphi) & 0 \\ 0 & 1 & 0 & 0 \\ \sin(90^0-\varphi) & 0 & \cos(90^0-\varphi) & 0 \\ 0 & 0 & 0 & 1 \end{bmatrix} = \begin{bmatrix} \sin\varphi & 0 & -\cos\varphi & 0 \\ 0 & 1 & 0 & 0 \\ \cos\varphi & 0 & \sin\varphi & 0 \\ 0 & 0 & 0 & 1 \end{bmatrix} \tag{4.4}$$

$Rot(\mathbf{e}_y)_{90^0-\varphi} * Rot_y$ aligns $\mathbf{g}$ with $\mathbf{e}_y$. The second step is comprised by another rotation around the z-axis by -90^0 that aligns $\mathbf{g}$ with $\mathbf{e}_x$ (Figure 2.1):

$$Rot(\mathbf{e}_z)_{-90^0} = \begin{bmatrix} 0 & 1 & 0 & 0 \\ -1 & 0 & 0 & 0 \\ 0 & 0 & 1 & 0 \\ 0 & 0 & 0 & 1 \end{bmatrix} \tag{4.5}$$

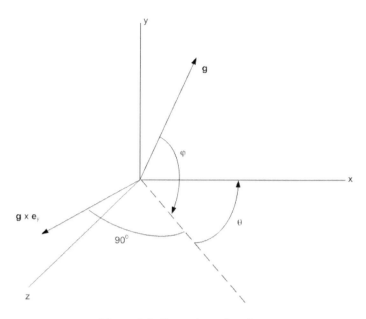

Figure 2.1. General acceleration.

Tr^{-1} reverses all the effects of Tr. Expression (4.1) gives the solution for the general case, of arbitrary acceleration direction. The net effect is that the derivative operators become more complicated:

$$\frac{\partial}{\partial t'} = \frac{\partial}{\partial x}\frac{\partial x}{\partial t'} + \frac{\partial}{\partial t}\frac{\partial t}{\partial t'} + \frac{\partial}{\partial y}\frac{\partial y}{\partial t'} + \frac{\partial}{\partial z}\frac{\partial z}{\partial t'} \qquad (4.6)$$

$$\frac{\partial}{\partial x'} = \frac{\partial}{\partial x}\frac{\partial x}{\partial x'} + \frac{\partial}{\partial t}\frac{\partial t}{\partial x'} + \frac{\partial}{\partial y}\frac{\partial y}{\partial x'} + \frac{\partial}{\partial z}\frac{\partial z}{\partial x'} \qquad (4.7)$$

$$\begin{bmatrix} \frac{\partial x}{\partial x'} & \frac{\partial x}{\partial y'} & \frac{\partial x}{\partial z'} & \frac{\partial x}{\partial t'} \\ \frac{\partial y}{\partial x'} & \frac{\partial y}{\partial y'} & \frac{\partial y}{\partial z'} & \frac{\partial y}{\partial t'} \\ \frac{\partial z}{\partial x'} & \frac{\partial z}{\partial y'} & \frac{\partial z}{\partial z'} & \frac{\partial z}{\partial t'} \\ \frac{\partial t}{\partial x'} & \frac{\partial t}{\partial y'} & \frac{\partial t}{\partial z'} & \frac{\partial t}{\partial t'} \end{bmatrix} = Tr^{-1} * Phy\_rectilinear * Tr \qquad (4.8)$$

Using (4.8) in (4.6), (4.7) gives the general forms of the transforms (2.21) for the electromagnetic field tensor. Next, we will show a very nice way of getting the general transforms. We start by writing (2.21) in the form:

$$\begin{bmatrix} E_x & 0 & 0 & 0 \\ * & \dfrac{E_y}{c} & B_z & * \\ B_y & * & * & \dfrac{E_z}{c} \\ 0 & 0 & 0 & B_x \end{bmatrix} = \begin{bmatrix} 1 & 0 & 0 & 0 \\ 0 & \cosh\dfrac{g\tau}{c} & \sinh\dfrac{g\tau}{c} & 0 \\ 0 & -\sinh\dfrac{g\tau}{c} & \cosh\dfrac{g\tau}{c} & 0 \\ 0 & 0 & 0 & 1 \end{bmatrix} \begin{bmatrix} E'_x & 0 & 0 & 0 \\ \dfrac{E'_z}{c} & \dfrac{E'_y}{c} & B'_z & B'_y \\ B'_y & B'_z & \dfrac{E'_y}{c} & \dfrac{E'_z}{c} \\ 0 & 0 & 0 & B'_x \end{bmatrix} \quad (4.9)$$

The elements marked with asterisks represent entities without any meaning. We do not care about them. Then, the matrix for the general transform is simply:

$$Tr^{-1} * \begin{bmatrix} 1 & 0 & 0 & 0 \\ 0 & \cosh\dfrac{g\tau}{c} & \sinh\dfrac{g\tau}{c} & 0 \\ 0 & -\sinh\dfrac{g\tau}{c} & \cosh\dfrac{g\tau}{c} & 0 \\ 0 & 0 & 0 & 1 \end{bmatrix} * Tr \quad (4.10)$$

3. Electrodynamics in Uniformly Rotating Frames

3.1. Introduction

Real life applications include accelerating and rotating frames more often than the idealized case of inertial frames. Our daily experiments happen in the laboratories attached to the rotating, continuously accelerating Earth. Many books and papers have been dedicated to transformations between particular cases of rectilinear acceleration and/or rotation [31] and to the applications of such formulas [32-42]. The main idea of this chapter is to generate a standard blueprint for a general solution that gives equivalent of the Lorentz transforms for the case of the transforms between an inertial frame and a uniformly rotating frame.

3.2. Uniformly Rotating Motion – the Transforms of the Electromagnetic Field

In this section we discuss the case of the particle moving in an arbitrary plane, with the normal given by the constant angular velocity $\boldsymbol{\omega}(a,b,c)$. According to Moller [31], the simpler case when $\boldsymbol{\omega}$ is aligned with the z-axis produces the transformation between the rotating frame $S'(\tau)$ attached to the particle and an inertial, non-rotating frame S attached to the center of rotation:

Electrodynamics in Uniformly Accelerated/Rotating Frames

$$\begin{pmatrix} x \\ y \\ z \\ t \end{pmatrix} = Phy\_rotation \begin{pmatrix} x' \\ y' \\ z' \\ t' \end{pmatrix} \quad (2.1)$$

where:

$$Phy\_rotation = \begin{bmatrix} \cos\alpha\cos\beta + \gamma\sin\alpha\sin\beta & \sin\alpha\cos\beta - \gamma\cos\alpha\sin\beta & 0 & u\gamma\sin\beta \\ \cos\alpha\sin\beta - \gamma\sin\alpha\cos\beta & \sin\alpha\sin\beta + \gamma\cos\alpha\cos\beta & 0 & -u\gamma\cos\beta \\ 0 & 0 & 1 & 0 \\ \dfrac{u\gamma\sin\alpha}{c^2} & -\dfrac{u\gamma\cos\alpha}{c^2} & 0 & \gamma \end{bmatrix}$$

$$\gamma = \frac{1}{\sqrt{1 - \dfrac{u^2}{c^2}}}$$

$$u = r\omega$$
$$\alpha = \omega\gamma\tau$$
$$\beta = \omega\gamma^2\tau$$

(2.2)

We also need the following:

$$\frac{\partial}{\partial x'} = \frac{\partial}{\partial x}\frac{\partial x}{\partial x'} + \frac{\partial}{\partial y}\frac{\partial y}{\partial x'} + \frac{\partial}{\partial z}\frac{\partial z}{\partial x'} + \frac{\partial}{\partial t}\frac{\partial t}{\partial x'}$$

$$\frac{\partial}{\partial y'} = \frac{\partial}{\partial x}\frac{\partial x}{\partial y'} + \frac{\partial}{\partial y}\frac{\partial y}{\partial y'} + \frac{\partial}{\partial z}\frac{\partial z}{\partial y'} + \frac{\partial}{\partial t}\frac{\partial t}{\partial y'}$$

$$\frac{\partial}{\partial z'} = \frac{\partial}{\partial x}\frac{\partial x}{\partial z'} + \frac{\partial}{\partial y}\frac{\partial y}{\partial z'} + \frac{\partial}{\partial z}\frac{\partial z}{\partial z'} + \frac{\partial}{\partial t}\frac{\partial t}{\partial z'}$$

$$\frac{\partial}{\partial t'} = \frac{\partial}{\partial t}\frac{\partial t}{\partial t'} + \frac{\partial}{\partial x}\frac{\partial x}{\partial t'} + \frac{\partial}{\partial y}\frac{\partial y}{\partial t'} + \frac{\partial}{\partial z}\frac{\partial z}{\partial t'}$$

$$\begin{bmatrix} \dfrac{\partial x}{\partial x'} & \dfrac{\partial x}{\partial y'} & \dfrac{\partial x}{\partial z'} & \dfrac{\partial x}{\partial t'} \\ \dfrac{\partial y}{\partial x'} & \dfrac{\partial y}{\partial y'} & \dfrac{\partial y}{\partial z'} & \dfrac{\partial y}{\partial t'} \\ \dfrac{\partial z}{\partial x'} & \dfrac{\partial z}{\partial y'} & \dfrac{\partial z}{\partial z'} & \dfrac{\partial z}{\partial t'} \\ \dfrac{\partial t}{\partial x'} & \dfrac{\partial t}{\partial y'} & \dfrac{\partial t}{\partial z'} & \dfrac{\partial t}{\partial t'} \end{bmatrix} = \begin{bmatrix} \cos\alpha\cos\beta + \gamma\sin\alpha\sin\beta & \sin\alpha\cos\beta - \gamma\cos\alpha\sin\beta & 0 & u\gamma\sin\beta \\ \cos\alpha\sin\beta - \gamma\sin\alpha\cos\beta & \sin\alpha\sin\beta + \gamma\cos\alpha\cos\beta & 0 & -u\gamma\cos\beta \\ 0 & 0 & 1 & 0 \\ \dfrac{u\gamma\sin\alpha}{c^2} & -\dfrac{u\gamma\cos\alpha}{c^2} & 0 & \gamma \end{bmatrix}$$

(2.3)

We also need the inverse transformation, from frame $S'(\tau)$ into frame S because it gives us:

$$\begin{bmatrix} \frac{\partial x'}{\partial x} & \frac{\partial x'}{\partial y} & \frac{\partial x'}{\partial z} & \frac{\partial x'}{\partial t} \\ \frac{\partial y'}{\partial x} & \frac{\partial y'}{\partial y} & \frac{\partial y'}{\partial z} & \frac{\partial y'}{\partial t} \\ \frac{\partial z'}{\partial x} & \frac{\partial z'}{\partial y} & \frac{\partial z'}{\partial z} & \frac{\partial z'}{\partial t} \\ \frac{\partial t'}{\partial x} & \frac{\partial t'}{\partial y} & \frac{\partial t'}{\partial z} & \frac{\partial t'}{\partial t} \end{bmatrix} = \begin{bmatrix} \cos\alpha\cos\beta+\gamma\sin\alpha\sin\beta & \cos\alpha\sin\beta-\gamma\sin\alpha\cos\beta & 0 & -u\gamma\sin\alpha \\ \sin\alpha\cos\beta-\gamma\cos\alpha\sin\beta & \sin\alpha\sin\beta+\gamma\cos\alpha\cos\beta & 0 & u\gamma\cos\alpha \\ 0 & 0 & 1 & 0 \\ -\frac{u\gamma\sin\beta}{c^2} & \frac{u\gamma\cos\beta}{c^2} & 0 & \gamma \end{bmatrix} \quad (2.4)$$

In order to simplify the calculations, we will use the notation:

$$\begin{bmatrix} \cos\alpha\cos\beta+\gamma\sin\alpha\sin\beta & \sin\alpha\cos\beta-\gamma\cos\alpha\sin\beta & 0 & u\gamma\sin\beta \\ \cos\alpha\sin\beta-\gamma\sin\alpha\cos\beta & \sin\alpha\sin\beta+\gamma\cos\alpha\cos\beta & 0 & -u\gamma\cos\beta \\ 0 & 0 & 1 & 0 \\ \frac{u\gamma\sin\alpha}{c^2} & -\frac{u\gamma\cos\alpha}{c^2} & 0 & \gamma \end{bmatrix} = \begin{bmatrix} a_{11} & a_{12} & 0 & a_{14} \\ a_{21} & a_{22} & 0 & a_{24} \\ 0 & 0 & 1 & 0 \\ a_{41} & a_{42} & 0 & a_{44} \end{bmatrix}$$

$$\begin{bmatrix} \cos\alpha\cos\beta+\gamma\sin\alpha\sin\beta & \cos\alpha\sin\beta-\gamma\sin\alpha\cos\beta & 0 & -u\gamma\sin\alpha \\ \sin\alpha\cos\beta-\gamma\cos\alpha\sin\beta & \sin\alpha\sin\beta+\gamma\cos\alpha\cos\beta & 0 & u\gamma\cos\alpha \\ 0 & 0 & 1 & 0 \\ -\frac{u\gamma\sin\beta}{c^2} & \frac{u\gamma\cos\beta}{c^2} & 0 & \gamma \end{bmatrix} = \begin{bmatrix} b_{11} & b_{12} & 0 & b_{14} \\ b_{21} & b_{22} & 0 & b_{24} \\ 0 & 0 & 1 & 0 \\ b_{41} & b_{42} & 0 & b_{44} \end{bmatrix}$$

The electromagnetic potential, by virtue of being a 4-vector transforms the same way as described by (2.1):

$$\begin{pmatrix} A_x \\ A_y \\ A_z \\ \varphi \end{pmatrix} = Phy\_rotation \begin{pmatrix} A'_x \\ A'_y \\ A'_z \\ \varphi' \end{pmatrix} \quad (2.5)$$

In the inertial frame, the differential Maxwell equations in vacuum, in the absence of electric charge, are [31]:

$$-E_x = \frac{\partial A_x}{\partial t} + c^2 \frac{\partial \varphi}{\partial x} \quad (2.6)$$

$$-E_y = \frac{\partial A_y}{\partial t} + c^2 \frac{\partial \varphi}{\partial y} \quad (2.7)$$

$$-E_z = \frac{\partial A_z}{\partial t} + c^2 \frac{\partial \varphi}{\partial z} \quad (2.8)$$

$$\mathbf{B} = curl\mathbf{A} \tag{2.9}$$

Let's start with (2.6):

$$\frac{\partial A_x}{\partial t} = a_{11}b_{14}\frac{\partial A_x'}{\partial x'} + a_{11}b_{24}\frac{\partial A_x'}{\partial y'} + a_{11}b_{44}\frac{\partial A_x'}{\partial t'} + a_{12}b_{14}\frac{\partial A_y'}{\partial x'} + a_{12}b_{24}\frac{\partial A_y'}{\partial y'} + a_{12}b_{44}\frac{\partial A_y'}{\partial t'} +$$
$$+ a_{14}b_{14}\frac{\partial \varphi'}{\partial x'} + a_{14}b_{24}\frac{\partial \varphi'}{\partial y'} + a_{14}b_{44}\frac{\partial \varphi'}{\partial t'} \tag{2.10}$$

$$\frac{\partial \varphi}{\partial x} = a_{41}b_{11}\frac{\partial A_x'}{\partial x'} + a_{41}b_{21}\frac{\partial A_x'}{\partial y'} + a_{41}b_{41}\frac{\partial A_x'}{\partial t'} + a_{42}b_{11}\frac{\partial A_y'}{\partial x'} + a_{42}b_{21}\frac{\partial A_y'}{\partial y'} + a_{42}b_{41}\frac{\partial A_y'}{\partial t'} +$$
$$+ a_{44}b_{11}\frac{\partial \varphi'}{\partial x'} + a_{44}b_{21}\frac{\partial \varphi'}{\partial y'} + a_{44}b_{41}\frac{\partial \varphi'}{\partial t'} \tag{2.11}$$

$$\frac{\partial A_x}{\partial t} + c^2\frac{\partial \varphi}{\partial x} = (a_{11}b_{44} + c^2 a_{41}b_{41})\frac{\partial A_x'}{\partial t'} + (a_{14}b_{14} + c^2 a_{44}b_{11})\frac{\partial \varphi'}{\partial x'} + (a_{12}b_{44} + c^2 a_{42}b_{41})\frac{\partial A_y'}{\partial t'} +$$
$$+ (a_{14}b_{24} + c^2 a_{44}b_{21})\frac{\partial \varphi'}{\partial y'} + (a_{11}b_{24} + c^2 a_{41}b_{21})\frac{\partial A_x'}{\partial y'} + (a_{12}b_{14} + c^2 a_{42}b_{11})\frac{\partial A_y'}{\partial x'} =$$
$$= (\gamma\cos\alpha\cos\beta + \sin\alpha\sin\beta)\{\frac{\partial A_x'}{\partial t'} + c^2\frac{\partial \varphi'}{\partial x'}\} + (\gamma\sin\alpha\cos\beta - \cos\alpha\sin\beta)\{\frac{\partial A_y'}{\partial t'} + c^2\frac{\partial \varphi'}{\partial y'}\} +$$
$$+ (a_{11}b_{24} + c^2 a_{41}b_{21})\frac{\partial A_x'}{\partial y'} + (a_{12}b_{14} + c^2 a_{42}b_{11})\frac{\partial A_y'}{\partial x'} =$$
$$= (\gamma\cos\alpha\cos\beta + \sin\alpha\sin\beta)E_x' + (\gamma\sin\alpha\cos\beta - \cos\alpha\sin\beta)E_y' - u\gamma\cos\beta B_z' \tag{2.12}$$

So:

$$E_x = (\gamma\cos\alpha\cos\beta + \sin\alpha\sin\beta)E_x' + (\sin\alpha\cos\beta - \cos\alpha\sin\beta)E_y' - (u\gamma\cos\beta)B_z' \tag{2.13}$$

Moving on to the second Maxwell equation, (2.7):

$$\frac{\partial A_y}{\partial t} = a_{21}b_{14}\frac{\partial A_x'}{\partial x'} + a_{21}b_{24}\frac{\partial A_x'}{\partial y'} + a_{21}b_{44}\frac{\partial A_x'}{\partial t'} + a_{22}b_{14}\frac{\partial A_y'}{\partial x'} + a_{22}b_{24}\frac{\partial A_y'}{\partial y'} + a_{22}b_{44}\frac{\partial A_y'}{\partial t'} +$$
$$+ a_{24}b_{14}\frac{\partial \varphi'}{\partial x'} + a_{24}b_{24}\frac{\partial \varphi'}{\partial y'} + a_{24}b_{44}\frac{\partial \varphi'}{\partial t'} \tag{2.14}$$

$$\frac{\partial \varphi}{\partial y} = a_{41}b_{12}\frac{\partial A_x'}{\partial x'} + a_{41}b_{22}\frac{\partial A_x'}{\partial y'} + a_{41}b_{42}\frac{\partial A_x'}{\partial t'} + a_{42}b_{12}\frac{\partial A_y'}{\partial x'} + a_{42}b_{22}\frac{\partial A_y'}{\partial y'} + a_{42}b_{42}\frac{\partial A_y'}{\partial t'} +$$
$$+ a_{44}b_{12}\frac{\partial \varphi'}{\partial x'} + a_{44}b_{22}\frac{\partial \varphi'}{\partial y'} + a_{44}b_{42}\frac{\partial \varphi'}{\partial t'} \tag{2.15}$$

$$\frac{\partial A_y}{\partial t} + c^2 \frac{\partial \varphi}{\partial y} = (a_{21}b_{44} + c^2 a_{41}b_{42})\frac{\partial A'_x}{\partial t'} + (a_{14}b_{14} + c^2 a_{44}b_{12})\frac{\partial \varphi'}{\partial x'} +$$
$$+ (a_{22}b_{44} + c^2 a_{42}b_{42})\frac{\partial A'_y}{\partial t'} + (a_{14}b_{24} + c^2 a_{44}b_{22})\frac{\partial \varphi'}{\partial y'} +$$
$$+ (a_{21}b_{24} + c^2 a_{41}b_{22})\frac{\partial A'_x}{\partial y'} + (a_{22}b_{24} + c^2 a_{42}b_{12})\frac{\partial A'_y}{\partial x'} =$$
$$= (\gamma \cos\alpha \sin\beta - \sin\alpha \cos\beta)\{\frac{\partial A'_x}{\partial t'} + c^2 \frac{\partial \varphi'}{\partial x'}\} +$$
$$+ (\gamma \sin\alpha \sin\beta + \cos\alpha \cos\beta)\{\frac{\partial A'_y}{\partial t'} + c^2 \frac{\partial \varphi'}{\partial y'}\} +$$
$$+ (a_{21}b_{24} + c^2 a_{41}b_{21})\frac{\partial A'_x}{\partial y'} + (a_{22}b_{24} + c^2 a_{42}b_{11})\frac{\partial A'_y}{\partial x'} =$$
$$= (\gamma \cos\alpha \sin\beta - \sin\alpha \cos\beta)E'_x + (\gamma \sin\alpha \sin\beta + \cos\alpha \cos\beta)E'_y - (u\gamma \sin\beta)B'_z$$
(2.16)

$$E_y = (\gamma \cos\alpha \sin\beta - \sin\alpha \cos\beta)E'_x + (\gamma \sin\alpha \sin\beta + \cos\alpha \cos\beta)E'_y - (u\gamma \sin\beta)B'_z \quad (2.17)$$

Moving on to (2.8):

$$\frac{\partial A_z}{\partial t} = \frac{\partial A'_z}{\partial t} = \frac{\partial A'_z}{\partial x'}\frac{\partial x'}{\partial t} + \frac{\partial A'_z}{\partial y'}\frac{\partial y'}{\partial t} + \frac{\partial A'_z}{\partial z'}\frac{\partial z'}{\partial t} + \frac{\partial A'_z}{\partial t'}\frac{\partial t'}{\partial t} = \frac{\partial A'_z}{\partial x'}b_{14} + \frac{\partial A'_z}{\partial y'}b_{24} + \frac{\partial A'_z}{\partial t'}b_{44} \quad (2.18)$$

$$\frac{\partial \varphi}{\partial z} = \frac{\partial}{\partial z}(A'_x a_{41} + A'_y a_{42} + \varphi' a_{44})$$
$$\frac{\partial A'_x}{\partial z} = \frac{\partial A'_x}{\partial z'}$$
$$\frac{\partial A'_y}{\partial z} = \frac{\partial A'_y}{\partial z'} \quad (2.19)$$
$$\frac{\partial \varphi'}{\partial z} = \frac{\partial \varphi'}{\partial z'}$$
$$\frac{\partial \varphi}{\partial z} = \frac{\partial A'_x}{\partial z'}a_{41} + \frac{\partial A'_y}{\partial z'}a_{42} + \frac{\partial \varphi'}{\partial z'}a_{44}$$

$$\frac{\partial A_z}{\partial t} + c^2 \frac{\partial \varphi}{\partial z} = \frac{\partial A'_z}{\partial x'}b_{14} + \frac{\partial A'_z}{\partial y'}b_{24} + \frac{\partial A'_z}{\partial t'}b_{44} + c^2(\frac{\partial A'_x}{\partial z'}a_{41} + \frac{\partial A'_y}{\partial z'}a_{42} + \frac{\partial \varphi'}{\partial z'}a_{44}) =$$
$$= \frac{\partial A'_z}{\partial t'}b_{44} + c^2 a_{44}\frac{\partial \varphi'}{\partial z'} + (\frac{\partial A'_z}{\partial x'}b_{14} + c^2 \frac{\partial A'_x}{\partial z'}a_{41}) + (\frac{\partial A'_z}{\partial y'}b_{24} + c^2 \frac{\partial A'_y}{\partial z'}a_{42}) = \quad (2.20)$$
$$= \gamma(\frac{\partial A'_z}{\partial t'} + c^2 \frac{\partial \varphi'}{\partial z'}) + u\gamma \sin\alpha(\frac{\partial A'_x}{\partial z'} - \frac{\partial A'_z}{\partial x'}) + u\gamma \cos\alpha(\frac{\partial A'_z}{\partial y'} - \frac{\partial A'_y}{\partial z'}) =$$
$$= \gamma(E'_z - u\cos\alpha B'_x - u\sin\alpha B'_y)$$

Therefore:

$$E_z = \gamma(E'_z - uB'_x \cos\alpha - uB'_y \sin\alpha) \qquad (2.21)$$

$$B_z = \frac{\partial A_y}{\partial x} - \frac{\partial A_x}{\partial y} \qquad (2.22)$$

$$\frac{\partial A_y}{\partial x} = \frac{\partial}{\partial x}(A'_x a_{21} + A'_y a_{22} + \varphi' a_{24}) \qquad (2.23)$$

$$\frac{\partial A'_x}{\partial x} = \frac{\partial A'_x}{\partial x'} b_{11} + \frac{\partial A'_x}{\partial y'} b_{21} + \frac{\partial A'_x}{\partial t'} b_{41} \qquad (2.24)$$

$$\frac{\partial A'_y}{\partial x} = \frac{\partial A'_y}{\partial x'} b_{11} + \frac{\partial A'_y}{\partial y'} b_{21} + \frac{\partial A'_y}{\partial t'} b_{41} \qquad (2.25)$$

$$\frac{\partial \varphi'}{\partial x} = \frac{\partial \varphi'}{\partial x'} b_{11} + \frac{\partial \varphi'}{\partial y'} b_{21} + \frac{\partial \varphi'}{\partial t'} b_{41} \qquad (2.26)$$

$$\frac{\partial A_y}{\partial x} = a_{21}(\frac{\partial A'_x}{\partial x'} b_{11} + \frac{\partial A'_x}{\partial y'} b_{21} + \frac{\partial A'_x}{\partial t'} b_{41}) + a_{22}(\frac{\partial A'_y}{\partial x'} b_{11} + \frac{\partial A'_y}{\partial y'} b_{21} + \frac{\partial A'_y}{\partial t'} b_{41}) +$$
$$+ a_{24}(\frac{\partial \varphi'}{\partial x'} b_{11} + \frac{\partial \varphi'}{\partial y'} b_{21} + \frac{\partial \varphi'}{\partial t'} b_{41}) \qquad (2.27)$$

$$\frac{\partial A_x}{\partial y} = \frac{\partial}{\partial y}(A'_x a_{11} + A'_y a_{12} + \varphi' a_{14}) \qquad (2.28)$$

$$\frac{\partial A'_x}{\partial y} = \frac{\partial A'_x}{\partial x'} b_{12} + \frac{\partial A'_x}{\partial y'} b_{22} + \frac{\partial A'_x}{\partial t'} b_{42} \qquad (2.29)$$

$$\frac{\partial A'_y}{\partial y} = \frac{\partial A'_y}{\partial x'} b_{12} + \frac{\partial A'_y}{\partial y'} b_{22} + \frac{\partial A'_y}{\partial t'} b_{42} \qquad (2.30)$$

$$\frac{\partial \varphi'}{\partial y} = \frac{\partial \varphi'}{\partial x'} b_{12} + \frac{\partial \varphi'}{\partial y'} b_{22} + \frac{\partial \varphi'}{\partial t'} b_{42} \qquad (2.31)$$

$$\frac{\partial A_x}{\partial y} = a_{11}(\frac{\partial A_x^{'}}{\partial x^{'}}b_{12} + \frac{\partial A_x^{'}}{\partial y^{'}}b_{22} + \frac{\partial A_x^{'}}{\partial t^{'}}b_{42}) + a_{12}(\frac{\partial A_y^{'}}{\partial x^{'}}b_{12} + \frac{\partial A_y^{'}}{\partial y^{'}}b_{22} + \frac{\partial A_y^{'}}{\partial t^{'}}b_{42}) + \tag{2.32}$$

$$+ a_{14}(\frac{\partial \varphi^{'}}{\partial x^{'}}b_{12} + \frac{\partial \varphi^{'}}{\partial y^{'}}b_{22} + \frac{\partial \varphi^{'}}{\partial t^{'}}b_{42})$$

$$B_z = \frac{\partial A_y}{\partial x} - \frac{\partial A_x}{\partial y} = (a_{24}b_{41} - a_{14}b_{42})\frac{\partial \varphi^{'}}{\partial t^{'}} + (a_{21}b_{41}\frac{\partial A_x^{'}}{\partial t^{'}} - a_{14}b_{12}\frac{\partial \varphi^{'}}{\partial x^{'}}) + (a_{22}b_{41}\frac{\partial A_y^{'}}{\partial t^{'}} - a_{14}b_{22}\frac{\partial \varphi^{'}}{\partial y^{'}}) -$$

$$- (a_{11}b_{42}\frac{\partial A_x^{'}}{\partial t^{'}} - a_{24}b_{11}\frac{\partial \varphi^{'}}{\partial x^{'}}) - (a_{12}b_{42}\frac{\partial A_y^{'}}{\partial t^{'}} - a_{24}b_{21}\frac{\partial \varphi^{'}}{\partial y^{'}}) + (a_{21}b_{11} - a_{11}b_{12})\frac{\partial A_x^{'}}{\partial x^{'}} +$$

$$+ (a_{22}b_{21} - a_{12}b_{22})\frac{\partial A_y^{'}}{\partial y^{'}} + (a_{21}b_{21}\frac{\partial A_x^{'}}{\partial y^{'}} - a_{12}b_{21}\frac{\partial A_y^{'}}{\partial x^{'}}) - (a_{11}b_{22}\frac{\partial A_x^{'}}{\partial y^{'}} - a_{22}b_{11}\frac{\partial A_y^{'}}{\partial x^{'}}) = \tag{2.33}$$

$$= [-\frac{u\gamma \sin \beta}{c^2}(\cos\alpha\sin\beta - \gamma\sin\alpha\cos\beta) - \frac{u\gamma \cos\beta}{c^2}(\cos\alpha\cos\beta + \gamma\sin\alpha\sin\beta)](\frac{\partial A_x^{'}}{\partial t^{'}} + c^2\frac{\partial \varphi^{'}}{\partial x^{'}}) +$$

$$+ [\frac{u\gamma \sin\beta}{c^2}(\sin\alpha\sin\beta + \gamma\cos\alpha\cos\beta) + \frac{u\gamma\cos\beta}{c^2}(\sin\alpha\cos\beta - \gamma\cos\alpha\sin\beta)](\frac{\partial A_y^{'}}{\partial t^{'}} + c^2\frac{\partial \varphi^{'}}{\partial y^{'}}) +$$

$$+ \gamma(\frac{\partial A_y^{'}}{\partial x^{'}} - \frac{\partial A_x^{'}}{\partial y^{'}}) = \gamma B_z^{'} - \frac{u\gamma\cos\alpha}{c^2}E_x^{'} - \frac{u\gamma\sin\alpha}{c^2}E_y^{'}$$

$$B_z = \gamma B_z^{'} - \frac{u\gamma \cos\alpha}{c^2}E_x^{'} - \frac{u\gamma\sin\alpha}{c^2}E_y^{'} \tag{2.34}$$

$$B_x = \frac{\partial A_z}{\partial y} - \frac{\partial A_y}{\partial z} \tag{2.35}$$

$$\frac{\partial A_z}{\partial y} = \frac{\partial A_z^{'}}{\partial y} = \frac{\partial A_z^{'}}{\partial x^{'}}\frac{\partial x^{'}}{\partial y} + \frac{\partial A_z^{'}}{\partial y^{'}}\frac{\partial y^{'}}{\partial y} + \frac{\partial A_z^{'}}{\partial t^{'}}\frac{\partial t^{'}}{\partial y} = \frac{\partial A_z^{'}}{\partial x^{'}}b_{12} + \frac{\partial A_z^{'}}{\partial y^{'}}b_{22} + \frac{\partial A_z^{'}}{\partial t^{'}}b_{42} \tag{2.36}$$

$$\frac{\partial A_y}{\partial z} = \frac{\partial A_y}{\partial z^{'}} = \frac{\partial}{\partial z^{'}}(A_x^{'}a_{21} + A_z^{'}a_{22} + \varphi^{'}a_{24}) = \frac{\partial A_x^{'}}{\partial z^{'}}a_{21} + \frac{\partial A_y^{'}}{\partial z^{'}}a_{22} + \frac{\partial \varphi^{'}}{\partial z^{'}}a_{24} \tag{2.37}$$

$$B_x = \frac{\partial A_z^{'}}{\partial x^{'}}b_{12} + \frac{\partial A_z^{'}}{\partial y^{'}}b_{22} + \frac{\partial A_z^{'}}{\partial t^{'}}b_{42} - \frac{\partial A_x^{'}}{\partial z^{'}}a_{21} - \frac{\partial A_y^{'}}{\partial z^{'}}a_{22} - \frac{\partial \varphi^{'}}{\partial z^{'}}a_{24} =$$

$$= \frac{\partial A_z^{'}}{\partial t^{'}}b_{42} - \frac{\partial \varphi^{'}}{\partial z^{'}}a_{24} + (\frac{\partial A_z^{'}}{\partial y^{'}}b_{22} - \frac{\partial A_y^{'}}{\partial z^{'}}a_{22}) + (\frac{\partial A_z^{'}}{\partial x^{'}}b_{12} - \frac{\partial A_x^{'}}{\partial z^{'}}a_{21}) = \tag{2.38}$$

$$= -\frac{u\gamma \cos\beta}{c^2}E_z^{'} + (\sin\alpha\sin\beta + \gamma\cos\alpha\cos\beta)B_x^{'} + (-\cos\alpha\sin\beta + \gamma\sin\alpha\cos\beta)B_y^{'}$$

$$B_x = -\frac{u\gamma\cos\beta}{c^2}E_z^{'} + (\sin\alpha\sin\beta + \gamma\cos\alpha\cos\beta)B_x^{'} + (-\cos\alpha\sin\beta + \gamma\sin\alpha\cos\beta)B_y^{'} \tag{2.39}$$

$$B_y = \frac{\partial A_x}{\partial z} - \frac{\partial A_z}{\partial x} \tag{2.40}$$

Electrodynamics in Uniformly Accelerated/Rotating Frames

$$\frac{\partial A_x}{\partial z} = \frac{\partial A_x}{\partial z'} = \frac{\partial}{\partial z'}(A'_x a_{11} + A'_z a_{12} + \varphi' a_{14}) = \frac{\partial A'_x}{\partial z'} a_{11} + \frac{\partial A'_y}{\partial z'} a_{12} + \frac{\partial \varphi'}{\partial z'} a_{14} \quad (2.41)$$

$$\frac{\partial A_z}{\partial x} = \frac{\partial A'_z}{\partial x} = \frac{\partial A'_z}{\partial x'}\frac{\partial x'}{\partial x} + \frac{\partial A'_z}{\partial y'}\frac{\partial y'}{\partial x} + \frac{\partial A'_z}{\partial t'}\frac{\partial t'}{\partial x} = \frac{\partial A'_z}{\partial x'} b_{11} + \frac{\partial A'_z}{\partial y'} b_{21} + \frac{\partial A'_z}{\partial t'} b_{41} \quad (2.42)$$

$$\frac{\partial A_x}{\partial z} - \frac{\partial A_z}{\partial x} = (\frac{\partial A'_x}{\partial z'} a_{11} - \frac{\partial A'_z}{\partial x'} b_{11}) + (\frac{\partial A'_y}{\partial z'} a_{12} - \frac{\partial A'_z}{\partial y'} b_{21}) + \frac{\partial \varphi'}{\partial z'} a_{14} - \frac{\partial A'_z}{\partial t'} b_{41} = \quad (2.43)$$
$$= -\frac{u\gamma \sin\beta}{c^2} E'_z + (-\sin\alpha\cos\beta + \gamma\cos\alpha\sin\beta)B'_x + (\cos\alpha\cos\beta + \gamma\sin\alpha\sin\beta)B'_y$$

$$B_y = -\frac{u\gamma \sin\beta}{c^2} E'_z + (\gamma\cos\alpha\sin\beta - \sin\alpha\cos\beta)B'_x + (\cos\alpha\cos\beta + \gamma\sin\alpha\sin\beta)B'_y \quad (2.44)$$

To summarize:

$$\begin{aligned}
E_x &= (\gamma\cos\alpha\cos\beta + \sin\alpha\sin\beta)E'_x + (\gamma\sin\alpha\cos\beta - \cos\alpha\sin\beta)E'_y - u\gamma\cos\beta B'_z \\
B_x &= (\gamma\cos\alpha\cos\beta + \sin\alpha\sin\beta)B'_x + (\gamma\sin\alpha\cos\beta - \cos\alpha\sin\beta)B'_y - u\gamma\cos\beta \frac{E'_z}{c^2} \\
E_y &= (\gamma\sin\alpha\sin\beta + \cos\alpha\cos\beta)E'_y + (\gamma\cos\alpha\sin\beta - \sin\alpha\cos\beta)E'_x - u\gamma\sin\beta B'_z \\
B_y &= (\gamma\sin\alpha\sin\beta + \cos\alpha\cos\beta)B'_y + (\gamma\cos\alpha\sin\beta - \sin\alpha\cos\beta)B'_x - u\gamma\sin\beta \frac{E'_z}{c^2} \\
E_z &= \gamma E'_z - u\gamma\cos\alpha B'_x - u\gamma\sin\alpha B'_y \\
B_z &= \gamma B'_z - u\gamma\cos\alpha \frac{E'_x}{c^2} - u\gamma\sin\alpha \frac{E'_y}{c^2}
\end{aligned} \quad (2.45)$$

The rotation "mixes" the components of the electromagnetic tensor in a way that is different from the cases of inertial motion or uniformly accelerated motion. Notice the lack of resemblance with the standard Lorentz transforms in [31], for example:

$$\begin{aligned}
E_x &= E'_x \\
E_y &= \gamma(E'_y + V\frac{H'_z}{c}) \\
E_z &= \gamma(E'_z - V\frac{H'_y}{c}) \\
H_x &= H'_x \\
H_y &= \gamma(H'_y - V\frac{E'_z}{c}) \\
H_z &= \gamma(H'_z + V\frac{E'_y}{c})
\end{aligned} \quad (2.46)$$

Notice the lack of resemblance with the transforms for uniformly accelerated motion:

$$E_x = E'_x, B_x = B'_x$$

$$E_y = E'_y \cosh \frac{g\tau}{c} + B'_z c \sinh \frac{g\tau}{c}$$

$$E_z = E'_z \cosh \frac{g\tau}{c} - B'_y c \sinh \frac{g\tau}{c} \qquad (2.47)$$

$$B_y = B'_y \cosh \frac{g\tau}{c} - E'_z \frac{\sinh \frac{g\tau}{c}}{c}$$

$$B_z = B'_z \cosh \frac{g\tau}{c} + E'_y \frac{\sinh \frac{g\tau}{c}}{c}$$

3.2.2. Consequences

3.2.2.1. Maxwell Laws in a Uniformly Rotating Frame

From the calculations in the previous section we deduce immediately that:

$$\mathbf{E}' = -\frac{\partial \mathbf{A}'}{\partial t'} - \nabla \varphi'$$
$$\mathbf{B}' = curl\mathbf{A}' \qquad (2.48)$$

3.2.2.2. The Gauge Invariance Condition in a Uniformly Rotating Frame

$$div\mathbf{A}' + \frac{\partial \varphi'}{\partial t'} = div\mathbf{A} + \frac{\partial \varphi}{\partial t} = 0 \qquad (2.49)$$

$$div\mathbf{A}' = \frac{\partial A'_x}{\partial x'} + \frac{\partial A'_y}{\partial y'} + \frac{\partial A'_z}{\partial z'} = \frac{\partial A'_x}{\partial x'} + \frac{\partial A'_y}{\partial y'} + \frac{\partial A_z}{\partial z} \qquad (2.50)$$

$$\frac{\partial A'_x}{\partial x'} = b_{11}(\frac{\partial A_x}{\partial x}a_{11} + \frac{\partial A_x}{\partial y}a_{21} + \frac{\partial A_x}{\partial t}a_{41}) + b_{12}(\frac{\partial A_y}{\partial x}a_{11} + \frac{\partial A_y}{\partial y}a_{21} + \frac{\partial A_y}{\partial t}a_{41}) +$$
$$+ b_{14}(\frac{\partial \varphi}{\partial x}a_{11} + \frac{\partial \varphi}{\partial y}a_{21} + \frac{\partial \varphi}{\partial t}a_{41}) \qquad (2.51)$$

$$\frac{\partial A'_y}{\partial y'} = b_{21}(\frac{\partial A_x}{\partial x}a_{12} + \frac{\partial A_x}{\partial y}a_{22} + \frac{\partial A_x}{\partial t}a_{42}) + b_{22}(\frac{\partial A_y}{\partial x}a_{12} + \frac{\partial A_y}{\partial y}a_{22} + \frac{\partial A_y}{\partial t}a_{42}) +$$
$$+ b_{24}(\frac{\partial \varphi}{\partial x}a_{12} + \frac{\partial \varphi}{\partial y}a_{22} + \frac{\partial \varphi}{\partial t}a_{42}) \qquad (2.52)$$

$$\frac{\partial \varphi'}{\partial t'} = b_{41}(\frac{\partial A_x}{\partial x}a_{14} + \frac{\partial A_x}{\partial y}a_{24} + \frac{\partial A_x}{\partial t}a_{44}) + b_{42}(\frac{\partial A_y}{\partial x}a_{14} + \frac{\partial A_y}{\partial y}a_{24} + \frac{\partial A_y}{\partial t}a_{44}) +$$
$$+ b_{44}(\frac{\partial \varphi}{\partial x}a_{14} + \frac{\partial \varphi}{\partial y}a_{24} + \frac{\partial \varphi}{\partial t}a_{44})$$
(2.53)

$$div\mathbf{A'} + \frac{\partial \varphi'}{\partial t'} = (b_{11}a_{11} + b_{21}a_{12} + b_{41}a_{14})\frac{\partial A_x}{\partial x} + (b_{11}a_{21} + b_{21}a_{22} + b_{41}a_{24})\frac{\partial A_x}{\partial y} + (b_{11}a_{41} + b_{21}a_{42} + b_{41}a_{44})\frac{\partial A_x}{\partial t} +$$
$$+ (b_{12}a_{11} + b_{22}a_{12} + b_{42}a_{14})\frac{\partial A_y}{\partial x} + (b_{12}a_{21} + b_{22}a_{22} + b_{42}a_{24})\frac{\partial A_y}{\partial y} + (b_{12}a_{41} + b_{22}a_{42} + b_{42}a_{44})\frac{\partial A_y}{\partial t} +$$
$$+ (b_{14}a_{11} + b_{24}a_{12} + b_{44}a_{14})\frac{\partial \varphi}{\partial x} + (b_{14}a_{21} + b_{24}a_{22} + b_{44}a_{24})\frac{\partial \varphi}{\partial y} + (b_{14}a_{41} + b_{24}a_{42} + b_{44}a_{44})\frac{\partial \varphi}{\partial t} + \frac{\partial A_z}{\partial z} =$$
$$= \frac{\partial A_x}{\partial x} + \frac{\partial A_y}{\partial y} + \frac{\partial A_z}{\partial z} + \frac{\partial \varphi}{\partial t}$$
(2.54)

Equality (2.54) results into:

$$div\mathbf{A'} + \frac{\partial \varphi'}{\partial t'} = div\mathbf{A} + \frac{\partial \varphi}{\partial t} = 0 \quad (2.55)$$

Equalities (2.48) and (2.55) result into Maxwell's equations having the same exact form in the uniformly rotating frame as the equations in the inertial frames with the immediate consequence that light speed in vacuum in a uniformly a rotating frame is "c". Indeed, (2.48) results into:

$$\frac{1}{c^2}\frac{\partial^2 \mathbf{E'}}{\partial t'^2} - \nabla^2 \mathbf{E'} = 0$$
$$\frac{1}{c^2}\frac{\partial^2 \mathbf{B'}}{\partial t'^2} - \nabla^2 \mathbf{B'} = 0$$
(2.56)

The above means that electromagnetic waves propagate in vacuum, in uniformly rotating frames, at the same speed as they propagate in inertial frames. In the next section we will derive this plus some other very interesting facts through a different approach.

3.3. Planar Wave Transformation and Speed of Light in a Uniformly Rotating Frame

In this section we apply the formalism derived in the previous paragraph in order to obtain the transform of a planar wave. Assume that a planar wave is propagating along the y' axis in the accelerated frame $S'(\tau)$. The wave has the electric component $\mathbf{E}'_x$ and the magnetic component $\mathbf{B}'_z$ along the x' and z' axes, respectively. The components equations are:

$$\mathbf{E}'_x = E'_{0x} \cos(\Omega't' - k'_y y' + \theta')\mathbf{e}_x$$
$$\mathbf{B}'_z = B'_{0z} \cos(\Omega't' - k'_y y' + \theta')\mathbf{e}_z$$
(3.1)

On the other hand:

$$\begin{pmatrix} x' \\ y' \\ z' \\ t' \end{pmatrix} = \begin{bmatrix} b_{11} & b_{12} & 0 & b_{14} \\ b_{21} & b_{22} & 0 & b_{24} \\ 0 & 0 & 1 & 0 \\ b_{41} & b_{42} & 0 & b_{44} \end{bmatrix} \begin{pmatrix} x \\ y \\ z \\ t \end{pmatrix}$$
(3.2)

In frame S, the wave equation is either:

$$E_x = (\gamma \cos\alpha \cos\beta + \sin\alpha \sin\beta)E'_x - (u\gamma \cos\beta)B'_z\} =$$
$$= [(\gamma \cos\alpha \cos\beta + \sin\alpha \sin\beta)E'_{0x} - (u\gamma \cos\beta)B'_{0z}]*$$
$$*\cos[(\Omega'b_{44} - k'_y b_{24})t - (k'_y b_{22} - \Omega'b_{42})y - (k'_y b_{21} - \Omega'b_{41})x + \theta']$$
$$E_y = [(\gamma \cos\alpha \sin\beta - \sin\alpha \cos\beta)E'_{0x} - (u\gamma \sin\beta)B'_{0z}]*$$
$$*\cos[(\Omega'b_{44} - k'_y b_{24})t - (k'_y b_{22} - \Omega'b_{42})y - (k'_y b_{21} - \Omega'b_{41})x + \theta']$$
$$B_z = \gamma(B'_{0z} - \frac{u\cos\alpha}{c^2}E'_{0x})*$$
$$*\cos[(\Omega'b_{44} - k'_y b_{24})t - (k'_y b_{22} - \Omega'b_{42})y - (k'_y b_{21} - \Omega'b_{41})x + \theta']$$
(3.3)

or:

$$\mathbf{E}_x = E_{0x} \cos(\Omega t - k_x x - k_y y - k_z z + \theta)\mathbf{e}_x$$
$$\mathbf{E}_y = E_{0y} \cos(\Omega t - k_x x - k_y y - k_z z + \theta)\mathbf{e}_y$$
$$\mathbf{B}_z = B_{0z} \cos(\Omega t - k_x x - k_y y - k_z z + \theta)\mathbf{e}_z$$
(3.4)

Comparing (3.4) and (3.3) we obtain:

$$E_{0x} = (\gamma \cos\alpha \cos\beta + \sin\alpha \sin\beta)E'_{0x} - (u\cos\beta)B'_{0z}$$
$$\Omega = \Omega'b_{44} - k'_y b_{24}$$
$$k_x = k'_y b_{21} - \Omega'b_{41}$$
$$k_y = k'_y b_{22} - \Omega'b_{42}$$
$$k_z = 0$$
$$\theta = \theta'$$
(3.5)

$$c = \frac{\Omega}{\sqrt{k_x^2 + k_y^2}} = \frac{\Omega' b_{44} - k_y' b_{24}}{\sqrt{(k_y' b_{21} - \Omega' b_{41})^2 + (k_y' b_{22} - \Omega' b_{42})^2}}$$

$$\frac{\Omega'^2}{k'^2}(c^2 b_{41}^2 + c^2 b_{42}^2 - b_{44}^2) - 2\frac{\Omega'}{k'}(c^2 b_{21} b_{41} + c^2 b_{22} b_{42} - b_{24} b_{44}) + c^2 b_{21}^2 + c^2 b_{22}^2 - b_{24}^2 = 0$$

$$c^2(b_{41}^2 + b_{42}^2) - b_{44}^2 = c^2((\frac{u\gamma \sin\beta}{c^2})^2 + (\frac{u\gamma \cos\beta}{c^2})^2) - \gamma^2 = \frac{u^2\gamma^2}{c^2} - \gamma^2 = -1$$

$$c^2(b_{21} b_{41} + b_{22} b_{42}) - b_{24} b_{44} = -u\gamma \sin\beta(\sin\alpha \cos\beta - \gamma \cos\alpha \sin\beta) +$$
$$+u\gamma \cos\beta(\sin\alpha \sin\beta + \gamma \cos\alpha \cos\beta) - u\gamma^2 \cos\alpha = 0$$
$$c^2(b_{21}^2 + b_{22}^2) - b_{24}^2 = c^2[(\sin\alpha \cos\beta - \gamma \cos\alpha \sin\beta)^2 + (\sin\alpha \sin\beta + \gamma \cos\alpha \cos\beta)^2] -$$
$$-(u\gamma \cos\alpha)^2 = c^2(\sin^2\alpha + \gamma^2 \cos^2\alpha) - u^2\gamma^2 \cos^2\alpha = c^2$$

$$-\frac{\Omega'^2}{k'^2} + c^2 = 0$$

$$\frac{\Omega'}{k'} = \pm c \tag{3.6}$$

We can now calculate the phase light speed in the rotating frame:

$$v_p' = \frac{\Omega'}{k'} = \frac{\Omega}{k} = \pm c \tag{3.7}$$

So, the light speed in the rotating frame equals the light speed in the inertial frame, c.

We can now proceed to calculating the amplitude and the phase transformation between the inertial and the rotating frame:

$$\theta = \theta' \tag{3.8}$$

The general equation of the Doppler effect is:

$$\Omega = \Omega' b_{44} - k_y' b_{24} = \Omega'(b_{44} - \frac{b_{24}}{c}) = \Omega'(\gamma - \frac{u\gamma \cos\alpha}{c}) = \gamma(1 - \frac{u \cos\alpha}{c})\Omega'$$

$$\gamma = \frac{1}{\sqrt{1 - \frac{u^2}{c^2}}} \tag{3.9}$$

$$u = r\omega$$

$$\alpha = \omega\gamma\tau$$

The general equations for aberration are:

$$k_x = k_y' b_{21} - \Omega' b_{41} = k' b_{21} - \Omega' b_{41} = k'(b_{21} - b_{41}c) = k'(\sin\alpha \cos\beta - \gamma \cos\alpha \sin\beta - \frac{u\gamma \sin\beta}{c})$$

$$k_y = k_y' b_{22} - \Omega' b_{42} = k' b_{22} - \Omega' b_{42} = k'(b_{22} - b_{42}c) = k'(\sin\alpha \sin\beta + \gamma \cos\alpha \cos\beta + \frac{u\gamma \cos\beta}{c}) \tag{3.10}$$

$$k_z = 0$$

$$\beta = \omega\gamma^2\tau$$

3.4. General Case of Rotation about an Arbitrary Axis

In a prior paper we have shown [42] that the particular transformation (2.1) can be generalized for the case of arbitrary direction. The general case is treated by transforming the problem into the particular case treated in [31] through a transformation into the "canonical case", followed by an application of the transformation from the rotating frame into the inertial frame, ending with the inverse of the first transformation, as shown below:

$$\begin{pmatrix} x \\ y \\ z \\ t \end{pmatrix} = (Rr^{-1} * Phy\_rotation * Rr) \begin{pmatrix} x' \\ y' \\ z' \\ t' \end{pmatrix} \quad (4.1)$$

$$Rr = Rot(\mathbf{e}_x)_{90^0} * Rot(\mathbf{e}_y)_{90^0 - \varphi} * Rot_y \quad (4.2)$$

$Rot(\mathbf{e}_y)_{90^0 - \varphi} * Rot_y$ aligns $\boldsymbol{\omega}$ with $\mathbf{e}_y$. The second step is comprised by another rotation around the x-axis by 90^0 that aligns $\boldsymbol{\omega}$ with $\mathbf{e}_z$:

$$Rot(\mathbf{e}_x)_{90^0} = \begin{bmatrix} 1 & 0 & 0 & 0 \\ 0 & 0 & -1 & 0 \\ 0 & 1 & 0 & 0 \\ 0 & 0 & 0 & 1 \end{bmatrix} \quad (4.3)$$

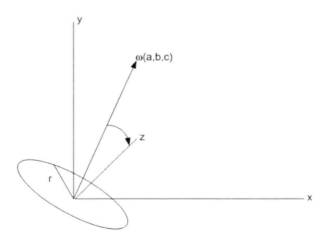

Figure 3.1. Uniform rotation with arbitrary direction of angular velocity.

Expression (4.1) gives the solution for the general case, of arbitrary angular velocity direction. Re-writing (2.45) in tensor form we get:

$$\begin{bmatrix} E_x & B_x & * & * \\ E_y & * & B_y & * \\ * & E_z & * & * \\ * & * & * & B_z \end{bmatrix} = \begin{bmatrix} \gamma\cos\alpha\cos\beta+\sin\alpha\sin\beta & \gamma\sin\alpha\cos\beta-\cos\alpha\sin\beta & -u\gamma\cos\beta & 0 \\ \gamma\cos\alpha\sin\beta-\sin\alpha\cos\beta & \gamma\sin\alpha\sin\beta+\cos\alpha\cos\beta & -u\gamma\sin\beta & 0 \\ u\gamma\cos\alpha & -u\gamma\sin\alpha & 0 & \gamma \\ 0 & -\dfrac{u\gamma\cos\alpha}{c^2} & -\dfrac{u\gamma\sin\alpha}{c^2} & \gamma \end{bmatrix} \begin{bmatrix} E'_x & B'_x & B'_y & 0 \\ E'_y & B'_y & B'_x & E'_x \\ B'_z & \dfrac{E'_z}{c^2} & \dfrac{E'_z}{c^2} & E'_y \\ 0 & E'_z & 0 & B'_z \end{bmatrix} \qquad (4.4)$$

The asterisks represent entries with no particular physical meaning, we do not care about them. Then, the general transform is:

$$Rr^{-1} * \begin{bmatrix} \gamma\cos\alpha\cos\beta+\sin\alpha\sin\beta & \gamma\sin\alpha\cos\beta-\cos\alpha\sin\beta & -u\gamma\cos\beta & 0 \\ \gamma\cos\alpha\sin\beta-\sin\alpha\cos\beta & \gamma\sin\alpha\sin\beta+\cos\alpha\cos\beta & -u\sin\beta & 0 \\ u\gamma\cos\alpha & -u\gamma\sin\alpha & 0 & \gamma \\ 0 & -\dfrac{u\gamma\cos\alpha}{c^2} & -\dfrac{u\gamma\sin\alpha}{c^2} & \gamma \end{bmatrix} * Rr \qquad (4.5)$$

Conclusion

We constructed the general transforms from the uniformly rotating frame into an inertial frame of reference S. The solution is of great interest for real life applications, because our earth-bound laboratories are inertial only in approximation; in real life, the laboratories are accelerated and rotate. We produced a blueprint for generalizing the solutions for the arbitrary cases and we concluded with an application that explains the general case of planar electromagnetic waves. A very interesting consequence is the fact that light speed in vacuum in the rotating frames is "c". A second interesting consequence is that rotation induces aberration.

4. Electrodynamics in Uniformly Rotating Frames as Viewed from an Inertial Frame

4.1. Introduction

Real life applications include accelerating and rotating frames more often than the idealized case of inertial frames. Our daily experiments happen in the laboratories attached to the rotating, continuously accelerating Earth. Many books and papers have been dedicated to transformations between particular cases of rectilinear acceleration and/or rotation [42] and to the applications of such formulas [44-52]. The main idea of this chapter is to generate a standard blueprint for a general solution that gives equivalent of the Lorentz transforms for the case of the transforms between an inertial frame and a rotating frame.

4.2. Uniformly Rotating Motion – the Transforms of the Electromagnetic Field

In this section we discuss the case of the particle moving in an arbitrary plane, with the normal given by the constant angular velocity $\boldsymbol{\omega}(a,b,c)$. According to Moller [1], the simpler case when $\boldsymbol{\omega}$ is aligned with the z-axis produces the transformation between the rotating frame $S'(\tau)$ attached to the particle and an inertial, non-rotating frame S attached to the center of rotation:

$$\begin{pmatrix} x \\ y \\ z \\ t \end{pmatrix} = Phy\_rotation \begin{pmatrix} x' \\ y' \\ z' \\ t' \end{pmatrix} \tag{2.1}$$

where:

$$Phy\_rotation = \begin{bmatrix} \cos\alpha\cos\beta + \gamma\sin\alpha\sin\beta & \sin\alpha\cos\beta - \gamma\cos\alpha\sin\beta & 0 & u\gamma\sin\beta \\ \cos\alpha\sin\beta - \gamma\sin\alpha\cos\beta & \sin\alpha\sin\beta + \gamma\cos\alpha\cos\beta & 0 & -u\gamma\cos\beta \\ 0 & 0 & 1 & 0 \\ \dfrac{u\gamma\sin\alpha}{c^2} & -\dfrac{u\gamma\cos\alpha}{c^2} & 0 & \gamma \end{bmatrix}$$

$$\gamma = \dfrac{1}{\sqrt{1-\dfrac{u^2}{c^2}}} \tag{2.2}$$

$$u = r\omega$$
$$\alpha = \omega\gamma\tau$$
$$\beta = \omega\gamma^2\tau$$

In previous papers [53,54] we have derived the transformations from the rotating frame S' into the inertial frame S:

$$E_x = (\gamma\cos\alpha\cos\beta + \sin\alpha\sin\beta)E'_x + (\gamma\sin\alpha\cos\beta - \cos\alpha\sin\beta)E'_y - u\gamma\cos\beta B'_z$$
$$E_y = (\gamma\cos\alpha\sin\beta - \sin\alpha\cos\beta)E'_x + (\gamma\sin\alpha\sin\beta + \cos\alpha\cos\beta)E'_y - u\gamma\sin\beta B'_z$$
$$B_z = -\dfrac{u\gamma\cos\alpha}{c^2}E'_x - \dfrac{u\gamma\sin\alpha}{c^2}E'_y + \gamma B'_z \tag{2.3}$$

$$B_x = (\gamma\cos\alpha\cos\beta + \sin\alpha\sin\beta)B'_x + (\gamma\sin\alpha\cos\beta - \cos\alpha\sin\beta)B'_y - \dfrac{u\gamma\cos\beta}{c^2}E'_z$$
$$B_y = (\gamma\cos\alpha\sin\beta - \sin\alpha\cos\beta)B'_x + (\gamma\sin\alpha\sin\beta + \cos\alpha\cos\beta)B'_y - \dfrac{u\gamma\sin\beta}{c^2}E'_z$$
$$E_z = -u\gamma\cos\alpha B'_x - u\gamma\sin\alpha B'_y + \gamma E'_z$$

The above can be re-cast in matrix form:

$$\begin{bmatrix} E_x \\ E_y \\ B_z \end{bmatrix} = \begin{bmatrix} \gamma \cos\alpha \cos\beta + \sin\alpha \sin\beta & \gamma \sin\alpha \cos\beta - \cos\alpha \sin\beta & -u\gamma \cos\beta \\ \gamma \cos\alpha \sin\beta - \sin\alpha \cos\beta & \gamma \sin\alpha \sin\beta + \cos\alpha \cos\beta & -u\gamma \sin\beta \\ -\dfrac{u\gamma \cos\alpha}{c^2} & -\dfrac{u\gamma \sin\alpha}{c^2} & \gamma \end{bmatrix} \begin{bmatrix} E'_x \\ E'_y \\ B'_z \end{bmatrix} \quad (2.4)$$

$$\begin{bmatrix} B_x \\ B_y \\ E_z \end{bmatrix} = \begin{bmatrix} \gamma \cos\alpha \cos\beta + \sin\alpha \sin\beta & \gamma \sin\alpha \cos\beta - \cos\alpha \sin\beta & -\dfrac{u\gamma \cos\beta}{c^2} \\ \gamma \cos\alpha \sin\beta - \sin\alpha \cos\beta & \gamma \sin\alpha \sin\beta + \cos\alpha \cos\beta & -\dfrac{u\gamma \sin\beta}{c^2} \\ -u\gamma \cos\alpha & -u\gamma \sin\alpha & \gamma \end{bmatrix} \begin{bmatrix} B'_x \\ B'_y \\ E'_z \end{bmatrix} \quad (2.5)$$

This observation allows us an easy way of inverting the transforms in order to obtain the transforms from the inertial frame S into the rotating frame S':

$$\begin{bmatrix} E'_x \\ E'_y \\ B'_z \end{bmatrix} = \begin{bmatrix} \gamma \cos\alpha \cos\beta + \sin\alpha \sin\beta & \gamma \cos\alpha \sin\beta - \sin\alpha \cos\beta & u\gamma \cos\alpha \\ \gamma \sin\alpha \cos\beta - \cos\alpha \sin\beta & \gamma \sin\alpha \sin\beta + \cos\alpha \cos\beta & u\gamma \sin\alpha \\ \dfrac{u\gamma \cos\beta}{c^2} & \dfrac{u\gamma \sin\beta}{c^2} & \gamma \end{bmatrix} \begin{bmatrix} E_x \\ E_y \\ B_z \end{bmatrix} \quad (2.6)$$

$$\begin{bmatrix} B'_x \\ B'_y \\ E'_z \end{bmatrix} = \begin{bmatrix} \gamma \cos\alpha \cos\beta + \sin\alpha \sin\beta & \gamma \cos\alpha \sin\beta - \sin\alpha \cos\beta & \dfrac{u\gamma \cos\alpha}{c^2} \\ \gamma \sin\alpha \cos\beta - \cos\alpha \sin\beta & \gamma \sin\alpha \sin\beta + \cos\alpha \cos\beta & \dfrac{u\gamma \sin\alpha}{c^2} \\ u\gamma \cos\beta & u\gamma \sin\beta & \gamma \end{bmatrix} \begin{bmatrix} B_x \\ B_y \\ E_z \end{bmatrix} \quad (2.7)$$

A very nice consequence of (2.6)-(2.7) is that:

$$\mathbf{E'}^2 - c^2 \mathbf{B'}^2 = \mathbf{E}^2 - c^2 \mathbf{B}^2 \quad (2.8)$$

The rotation "mixes" the components of the electromagnetic tensor in a way that is different from the cases of inertial motion or uniformly accelerated motion.

4.3. Planar Wave Transformation and Speed of Light in a Uniformly Rotating Frame

In this section we apply the formalism derived in the previous paragraph in order to obtain the transform of a planar wave. Assume that a planar wave is propagating along the y axis in the inertial frame S. The wave has the electric component $\mathbf{E}_x$ and the magnetic component $\mathbf{B}_z$ along the x and z axes, respectively. The components equations are:

$$\mathbf{E}_x = E_{0x}\cos(\Omega t - k_y y + \theta)\mathbf{e}_x$$
$$\mathbf{B}_z = B_{0z}\cos(\Omega t - k_y y + \theta)\mathbf{e}_z \qquad (3.1)$$

On the other hand:

$$\begin{pmatrix} x \\ y \\ z \\ t \end{pmatrix} = \begin{bmatrix} b_{11} & b_{12} & 0 & b_{14} \\ b_{21} & b_{22} & 0 & b_{24} \\ 0 & 0 & 1 & 0 \\ b_{41} & b_{42} & 0 & b_{44} \end{bmatrix} \begin{pmatrix} x' \\ y' \\ z' \\ t' \end{pmatrix} \qquad (3.2)$$

where:

$$\begin{bmatrix} b_{11} & b_{12} & 0 & b_{14} \\ b_{21} & b_{22} & 0 & b_{24} \\ 0 & 0 & 1 & 0 \\ b_{41} & b_{42} & 0 & b_{44} \end{bmatrix} = \begin{bmatrix} \cos\alpha\cos\beta + \gamma\sin\alpha\sin\beta & \sin\alpha\cos\beta - \gamma\cos\alpha\sin\beta & 0 & u\gamma\sin\beta \\ \cos\alpha\sin\beta - \gamma\sin\alpha\cos\beta & \sin\alpha\sin\beta + \gamma\cos\alpha\cos\beta & 0 & -u\gamma\cos\beta \\ 0 & 0 & 1 & 0 \\ \dfrac{u\gamma\sin\alpha}{c^2} & -\dfrac{u\gamma\cos\alpha}{c^2} & 0 & \gamma \end{bmatrix} \qquad (3.3)$$

Substituting (3.2) into (3.1) we obtain:

$$\begin{aligned}
E_x &= E_{0x}\cos[\Omega(b_{41}x' + b_{42}y' + b_{44}t') - k_y(b_{21}x' + b_{22}y' + b_{24}t') + \theta] = \\
&= E_{0x}\cos[(\Omega b_{44} - k_y b_{24})t' - (k_y b_{22} - \Omega b_{42})y' - (k_y b_{21} - \Omega b_{41})x' + \theta] \\
E_y &= E_z = 0 \\
B_z &= B_{0z}\cos[(\Omega b_{44} - k_y b_{24})t' - (k_y b_{22} - \Omega b_{42})y' - (k_y b_{21} - \Omega b_{41})x' + \theta] \\
B_x &= B_y = 0
\end{aligned} \qquad (3.4)$$

On the other hand, in frame S', the wave equation has two alternative forms, First one:

$$\begin{aligned}
E'_x &= (\gamma\cos\alpha\cos\beta + \sin\alpha\sin\beta)E_x + (u\gamma\cos\alpha)B_z\} = \\
&= [(\gamma\cos\alpha\cos\beta + \sin\alpha\sin\beta)E_{0x} + (u\gamma\cos\alpha)B_{0z}]^* \\
&\quad * \cos[(\Omega b_{44} - k_y b_{24})t' - (k_y b_{22} - \Omega b_{42})y' - (k_y b_{21} - \Omega b_{41})x' + \theta] \\
E'_y &= [(\gamma\sin\alpha\cos\beta - \cos\alpha\sin\beta)E_{0x} + (u\gamma\sin\alpha)B_{0z}]^* \\
&\quad * \cos[(\Omega b_{44} - k_y b_{24})t' - (k_y b_{22} - \Omega b_{42})y' - (k_y b_{21} - \Omega b_{41})x' + \theta] \\
B'_z &= \gamma(B_{0z} + \dfrac{u\cos\beta}{c^2}E_{0x})^* \\
&\quad * \cos[(\Omega b_{44} - k_y b_{24})t' - (k_y b_{22} - \Omega b_{42})y' - (k_y b_{21} - \Omega b_{41})x' + \theta]
\end{aligned} \qquad (3.5)$$

Second one:

$$\mathbf{E}'_x = E'_{0x} \cos(\Omega' t' - k'_x x' - k'_y y' - k'_z z' + \theta') \mathbf{e}_x$$
$$\mathbf{E}'_y = E'_{0y} \cos(\Omega' t' - k'_x x' - k'_y y' - k'_z z' + \theta') \mathbf{e}_y \quad (3.6)$$
$$\mathbf{B}'_z = B'_{0z} \cos(\Omega' t' - k'_x x' - k'_y y' - k'_z z' + \theta') \mathbf{e}_z$$

Comparing (3.5) and (3.6) we obtain:

$$E'_{0x} = (\gamma \cos \alpha \cos \beta + \sin \alpha \sin \beta) E_{0x} + (u\gamma \cos \alpha) B_{0z}$$
$$\Omega' = \Omega b_{44} - k_y b_{24}$$
$$k'_x = k_y b_{21} - \Omega b_{41} \quad (3.7)$$
$$k'_y = k_y b_{22} - \Omega b_{42}$$
$$k'_z = 0$$
$$\theta' = \theta$$

The phase light speed v'_p in the rotating frame S' is:

$$v'_p = \frac{\Omega'}{\sqrt{k'^2_x + k'^2_y}} = \frac{\Omega b_{44} - k_y b_{24}}{\sqrt{(k_y b_{21} - \Omega b_{41})^2 + (k_y b_{22} - \Omega b_{42})^2}} = c \quad (3.7)$$

So, the light speed in the rotating frame equals the light speed in the inertial frame, c.

We can now proceed to calculating the amplitude and the phase transformation between the inertial and the rotating frame:

$$\theta' = \theta \quad (3.8)$$

The general equation of the Doppler effect is:

$$\Omega' = \Omega b_{44} - k_y b_{24} = \Omega(b_{44} - \frac{b_{24}}{c}) = \Omega(\gamma - \frac{u\gamma \cos \alpha}{c}) = \gamma(1 - \frac{u \cos \alpha}{c})\Omega$$
$$\gamma = \frac{1}{\sqrt{1 - \frac{u^2}{c^2}}}$$
$$u = r\omega \quad (3.9)$$
$$\alpha = \omega \gamma \tau$$
$$\beta = \omega \gamma^2 \tau$$

The general equations for aberration are:

$$k'_x = k(\sin\alpha\cos\beta - \gamma\cos\alpha\sin\beta - \frac{u\gamma\sin\beta}{c})$$

$$k'_y = k(\sin\alpha\sin\beta + \gamma\cos\alpha\cos\beta + \frac{u\gamma\cos\beta}{c}) \qquad (3.10)$$

$$k'_z = 0$$

In the rotating frame, the wave follows a curved trajectory. It is interesting to observe that for $\tau = 0$ we have so $\alpha = \beta = 0$ and:

$$k'_x = 0$$
$$k'_y = k\gamma(1+\frac{u}{c}) \qquad (3.11)$$
$$k'_z = 0$$

meaning that the wave vector in frame S' starts in the same direction as the wave vector in frame S before it starts to progressively curve away as the proper time τ increases.

4.4. General Case of Rotation about an Arbitrary Axis

In a prior paper we have shown [54] that the particular transformation (2.1) can be generalized for the case of arbitrary direction. The general case is treated by transforming the problem into the particular case treated in [42] through a transformation into the "canonical case," followed by an application of the transformation from the rotating frame into the inertial frame, ending with the inverse of the first transformation, as shown below:

$$\begin{pmatrix} x \\ y \\ z \\ t \end{pmatrix} = (Rr^{-1} * Phy\_rotation * Rr) \begin{pmatrix} x' \\ y' \\ z' \\ t' \end{pmatrix} \qquad (4.1)$$

$$Rr = Rot(\mathbf{e}_x)_{90^0} * Rot(\mathbf{e}_y)_{90^0-\varphi} * Rot_y \qquad (4.2)$$

$Rot(\mathbf{e}_y)_{90^0-\varphi} * Rot_y$ aligns $\boldsymbol{\omega}$ with $\mathbf{e}_y$. The second step is comprised by another rotation around the x-axis by 90^0 that aligns $\boldsymbol{\omega}$ with $\mathbf{e}_z$:

$$Rot(\mathbf{e}_x)_{90^0} = \begin{bmatrix} 1 & 0 & 0 & 0 \\ 0 & 0 & -1 & 0 \\ 0 & 1 & 0 & 0 \\ 0 & 0 & 0 & 1 \end{bmatrix} \quad (4.3)$$

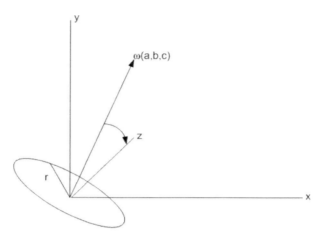

Figure 4.1. Uniform rotation with arbitrary direction of angular velocity.

Expression (4.1) gives the solution for the general case, of arbitrary angular velocity direction. Re-writing (2.6)-(2.7) in tensor form we get:

$$\begin{bmatrix} E'_x & B'_x & * & * \\ E'_y & * & B'_y & * \\ * & E'_z & * & * \\ * & * & * & B'_z \end{bmatrix} = \begin{bmatrix} \gamma\cos\alpha\cos\beta+\sin\alpha\sin\beta & \gamma\sin\alpha\cos\beta-\cos\alpha\sin\beta & u\gamma\cos\alpha & 0 \\ \gamma\cos\alpha\sin\beta-\sin\alpha\cos\beta & \gamma\sin\alpha\sin\beta+\cos\alpha\cos\beta & u\gamma\sin\alpha & 0 \\ u\gamma\cos\beta & u\gamma\sin\beta & 0 & \gamma \\ 0 & \dfrac{u\gamma\cos\beta}{c^2} & \dfrac{u\gamma\sin\beta}{c^2} & \gamma \end{bmatrix} \begin{bmatrix} E_x & B_x & B_y & 0 \\ E_y & B_y & B_x & E_x \\ B_z & \dfrac{E_z}{c^2} & \dfrac{E_z}{c^2} & E_y \\ 0 & E_z & 0 & B_z \end{bmatrix} \quad (4.4)$$

The asterisks represent entries with no particular physical meaning, we do not care about them. Then, the general transform is:

$$Rr * \begin{bmatrix} \gamma\cos\alpha\cos\beta+\sin\alpha\sin\beta & \gamma\sin\alpha\cos\beta-\cos\alpha\sin\beta & u\gamma\cos\alpha & 0 \\ \gamma\cos\alpha\sin\beta-\sin\alpha\cos\beta & \gamma\sin\alpha\sin\beta+\cos\alpha\cos\beta & u\gamma\sin\alpha & 0 \\ u\gamma\cos\beta & u\gamma\sin\beta & 0 & \gamma \\ 0 & \dfrac{u\gamma\cos\beta}{c^2} & \dfrac{u\gamma\sin\beta}{c^2} & \gamma \end{bmatrix} * Rr^{-1} \quad (4.5)$$

4.5. Application I: The General Expressions for Aberration and for the Doppler Effect

We have seen in section 3 that the relativistic Doppler effect can be derived from the frame invariance of the expression:

$$\Psi = \Omega t - (k_x x + k_y y + k_z z) + \varphi \qquad (5.1)$$

We have seen in section 4 that the general coordinate transformation from an accelerated frame S' into an inertial frame S is:

$$\begin{pmatrix} x \\ y \\ z \\ t \end{pmatrix} = \begin{bmatrix} a_{11} & a_{12} & a_{13} & a_{14} \\ a_{21} & a_{22} & a_{23} & a_{24} \\ a_{31} & a_{32} & a_{33} & a_{34} \\ a_{41} & a_{42} & a_{43} & a_{44} \end{bmatrix}_\omega \begin{pmatrix} x' \\ y' \\ z' \\ t' \end{pmatrix} \qquad (5.2)$$

The subscript ω represents the dependence of the matrix elements $a_{ij} = a_{ij}(\omega)$ of the angular velocity $\omega = (\omega_x, \omega_y, \omega_z)$ between frames S and S'. Substituting (5.2) into (5.1) we obtain:

$$\begin{aligned}\Psi &= \Omega(\mathbf{a_4}.\mathbf{x'}) - k_x(\mathbf{a_1}.\mathbf{x'}) - k_y(\mathbf{a_2}.\mathbf{x'}) - k_z(\mathbf{a_3}.\mathbf{x'}) + \varphi = \\ &= (\omega a_{44} - k_x a_{14} - k_y a_{24} - k_z a_{34})t' - (k_x a_{11} + k_y a_{21} + k_z a_{31} - \omega a_{41})x' - \\ &\quad -(k_x a_{12} + k_y a_{22} + k_z a_{32} - \omega a_{42})y' - (k_x a_{13} + k_y a_{23} + k_z a_{33} - \omega a_{43})z' + \varphi \end{aligned} \qquad (5.3)$$

On the other hand, in the rotating frame S':

$$\Psi' = \Omega' t' - (k'_x x' + k'_y y' + k'_z z') + \varphi' = \Psi \qquad (5.4)$$

Comparing (5.3) and (5.4) we obtain the general expressions of the relativistic Doppler effect between the inertial frame S and the accelerated frame S':

$$\begin{aligned} \Omega' &= \Omega a_{44} - k_x a_{14} - k_y a_{24} - k_z a_{34} \\ k'_x &= k_x a_{11} + k_y a_{21} + k_z a_{31} - \Omega a_{41} \\ k'_y &= k_x a_{12} + k_y a_{22} + k_z a_{32} - \Omega a_{42} \\ k'_z &= k_x a_{13} + k_y a_{23} + k_z a_{33} - \Omega a_{43} \\ \varphi' &= \varphi \end{aligned} \qquad (5.5)$$

In matrix form:

$$\begin{bmatrix} \Omega' \\ k_x' \\ k_y' \\ k_z' \end{bmatrix} = \begin{bmatrix} a_{44} & -a_{14} & -a_{24} & -a_{34} \\ -a_{41} & a_{11} & a_{21} & a_{31} \\ -a_{42} & a_{12} & a_{22} & a_{32} \\ -a_{43} & a_{13} & a_{23} & a_{33} \end{bmatrix}_\omega \begin{bmatrix} \Omega \\ k_x \\ k_y \\ k_z \end{bmatrix} \quad (5.6)$$

For the reverse transformations we start with:

$$\begin{pmatrix} x' \\ y' \\ z' \\ t' \end{pmatrix} = \begin{bmatrix} a_{11} & a_{12} & a_{13} & a_{14} \\ a_{21} & a_{22} & a_{23} & a_{24} \\ a_{31} & a_{32} & a_{33} & a_{34} \\ a_{41} & a_{42} & a_{43} & a_{44} \end{bmatrix}^{-1} \begin{pmatrix} x \\ y \\ z \\ t \end{pmatrix} \quad (5.7)$$

For simplicity, we re-write (5.7) as:

$$\begin{pmatrix} x' \\ y' \\ z' \\ t' \end{pmatrix} = \begin{bmatrix} b_{11} & b_{12} & b_{13} & b_{14} \\ b_{21} & b_{22} & b_{23} & b_{24} \\ b_{31} & b_{32} & b_{33} & b_{34} \\ b_{41} & b_{42} & b_{43} & b_{44} \end{bmatrix}_\omega \begin{pmatrix} x \\ y \\ z \\ t \end{pmatrix} \quad (5.8)$$

In the rotating frame S':

$$\Psi' = \Omega' t' - (k_x' x' + k_y' y' + k_z' z') + \varphi' \quad (5.9)$$

Substituting (5.8) into (5.9):

$$\Psi' = \Omega'(\mathbf{b_4} \cdot \mathbf{x}) - k_x'(\mathbf{b_1} \cdot \mathbf{x}) - k_y'(\mathbf{b_2} \cdot \mathbf{x}) - k_z'(\mathbf{b_3} \cdot \mathbf{x}) + \varphi' = \\ = (\Omega' b_{44} - k_x' b_{14} - k_y' b_{24} - k_z' b_{34})t - (k_x' b_{11} + k_y' b_{21} + k_z' b_{31} - \Omega' b_{41})x - \\ -(k_x' b_{12} + k_y' b_{22} + k_z' b_{32} - \Omega' b_{42})y - (k_x' b_{13} + k_y' b_{23} + k_z' b_{33} - \Omega' b_{43})z + \varphi' \quad (5.10)$$

On the other hand, in frame S:

$$\Psi = \Omega t - (k_x x + k_y y + k_z z) + \varphi = \Psi' \quad (5.11)$$

Comparing (5.10) and (5.11) we obtain the general expressions of the relativistic Doppler effect between the inertial frame S' and the accelerated frame S:

$$\Omega = \Omega' b_{44} - k_x' b_{14} - k_y' b_{24} - k_z' b_{34}$$
$$k_x = k_x' b_{11} + k_y' b_{21} + k_z' b_{31} - \Omega' b_{41}$$
$$k_y = k_x' b_{12} + k_y' b_{22} + k_z' b_{32} - \Omega' b_{42} \qquad (5.12)$$
$$k_z = k_x' b_{13} + k_y' b_{23} + k_z' b_{33} - \Omega' b_{43}$$
$$\varphi = \varphi'$$

In matrix form:

$$\begin{bmatrix} \Omega \\ k_x \\ k_y \\ k_z \end{bmatrix} = \begin{bmatrix} b_{44} & -b_{14} & -b_{24} & -b_{34} \\ -b_{41} & b_{11} & b_{21} & b_{31} \\ -b_{42} & b_{12} & b_{22} & b_{32} \\ -b_{43} & b_{13} & b_{23} & b_{33} \end{bmatrix}_\omega \begin{bmatrix} \Omega' \\ k_x' \\ k_y' \\ k_z' \end{bmatrix} \qquad (5.13)$$

4.6. Application II: The Doppler Effect for Rotating Emitter and Receiver

In this section we treat the general case, the emitter moves with angular velocity $\omega 1$ and the receiver moves with arbitrary angular velocity $\omega 2$, both with respect to the inertial frame S.
According to (5.13):

$$\begin{bmatrix} \Omega \\ k_x \\ k_y \\ k_z \end{bmatrix} = \begin{bmatrix} b_{44} & -b_{14} & -b_{24} & -b_{34} \\ -b_{41} & b_{11} & b_{21} & b_{31} \\ -b_{42} & b_{12} & b_{22} & b_{32} \\ -b_{43} & b_{13} & b_{23} & b_{33} \end{bmatrix}_{\omega 1} \begin{bmatrix} \Omega' \\ k_x' \\ k_y' \\ k_z' \end{bmatrix}_{emitter} \qquad (6.1)$$

According to (5.6):

$$\begin{bmatrix} \Omega' \\ k_x' \\ k_y' \\ k_z' \end{bmatrix}_{receiver} = \begin{bmatrix} a_{44} & -a_{14} & -a_{24} & -a_{34} \\ -a_{41} & a_{11} & a_{21} & a_{31} \\ -a_{42} & a_{12} & a_{22} & a_{32} \\ -a_{43} & a_{13} & a_{23} & a_{33} \end{bmatrix}_{\omega 2} \begin{bmatrix} \Omega \\ k_x \\ k_y \\ k_z \end{bmatrix} \qquad (6.2)$$

From (6.1) and (6.2) we obtain the general form of Doppler effect and aberration for the case of the emitter and the receiver moving with arbitrary angular velocities $\omega 1$ and $\omega 2$ with respect to the same inertial reference frame:

Electrodynamics in Uniformly Accelerated/Rotating Frames 267

$$\begin{bmatrix} \Omega' \\ k'_x \\ k'_y \\ k'_z \end{bmatrix}_{receiver} = \begin{bmatrix} a_{44} & -a_{14} & -a_{24} & -a_{34} \\ -a_{41} & a_{11} & a_{21} & a_{31} \\ -a_{42} & a_{12} & a_{22} & a_{32} \\ -a_{43} & a_{13} & a_{23} & a_{33} \end{bmatrix}_{\omega 2} \begin{bmatrix} b_{44} & -b_{14} & -b_{24} & -b_{34} \\ -b_{41} & b_{11} & b_{21} & b_{31} \\ -b_{42} & b_{12} & b_{22} & b_{32} \\ -b_{43} & b_{13} & b_{23} & b_{33} \end{bmatrix}_{\omega 1} \begin{bmatrix} \Omega' \\ k'_x \\ k'_y \\ k'_z \end{bmatrix}_{emitter} \quad (6.3)$$

A quick sanity check shows that for $\omega 1 = \omega 2$:

$$\begin{bmatrix} \Omega' \\ k'_x \\ k'_y \\ k'_z \end{bmatrix}_{receiver} = \begin{bmatrix} 1 & 0 & 0 & 0 \\ 0 & * & * & * \\ 0 & * & * & * \\ 0 & * & * & * \end{bmatrix} \begin{bmatrix} \Omega' \\ k'_x \\ k'_y \\ k'_z \end{bmatrix}_{emitter} \quad (6.4)$$

In this case, there is no Doppler effect but there is obviously aberration since the terms denoted by asterisks are not null.

5. Electrodynamics in Uniformly Rotating Frames: The Central Observer Point of View

5.1. Introduction

Real life applications include accelerating and rotating frames more often than the idealized case of inertial frames. Our daily experiments happen in the laboratories attached to the rotating, continuously accelerating Earth. Many books and papers have been dedicated to transformations between particular cases of rectilinear acceleration and/or rotation [55] and to the applications of such formulas [56-66]. The main idea of this chapter is to generate a standard blueprint for a general solution that gives equivalent of the Lorentz transforms for the case of the transforms between an inertial frame and a uniformly rotating frame.

5.2. Uniformly Rotating Motion – The Transforms of the Electromagnetic Field

According to Moller [55], the transformation between the rotating frame S' attached to a uniformly rotating observer located along the rotation axis and an inertial, non-rotating frame S attached to the center of rotation is:

$$\begin{pmatrix} x \\ y \\ z \\ t \end{pmatrix} = Phy\_rotation \begin{pmatrix} x' \\ y' \\ z' \\ t' \end{pmatrix} \quad (2.1)$$

where:

$$Phy\_rotation = \begin{bmatrix} \cos\Omega t & \sin\Omega t & 0 & 0 \\ -\sin\Omega t & \cos\Omega t & 0 & 0 \\ 0 & 0 & 1 & 0 \\ 0 & 0 & 0 & 1 \end{bmatrix} \quad (2.2)$$

In order to simplify things, we will use the notation:

$$Phy\_rotation = \begin{bmatrix} a_{11} & a_{12} & 0 & 0 \\ a_{21} & a_{22} & 0 & 0 \\ 0 & 0 & 1 & 0 \\ 0 & 0 & 0 & 1 \end{bmatrix} \quad (2.3)$$

We also need the following:

$$\begin{aligned} \frac{\partial}{\partial x'} &= \frac{\partial}{\partial x}\frac{\partial x}{\partial x'} + \frac{\partial}{\partial y}\frac{\partial y}{\partial x'} \\ \frac{\partial}{\partial y'} &= \frac{\partial}{\partial x}\frac{\partial x}{\partial y'} + \frac{\partial}{\partial y}\frac{\partial y}{\partial y'} \\ \frac{\partial}{\partial z'} &= \frac{\partial}{\partial z} \\ \frac{\partial}{\partial t'} &= \frac{\partial}{\partial t} \end{aligned} \quad (2.4)$$

$$\begin{bmatrix} \frac{\partial x}{\partial x'} & \frac{\partial x}{\partial y'} & \frac{\partial x}{\partial z'} & \frac{\partial x}{\partial t'} \\ \frac{\partial y}{\partial x'} & \frac{\partial y}{\partial y'} & \frac{\partial y}{\partial z'} & \frac{\partial y}{\partial t'} \\ \frac{\partial z}{\partial x'} & \frac{\partial z}{\partial y'} & \frac{\partial z}{\partial z'} & \frac{\partial z}{\partial t'} \\ \frac{\partial t}{\partial x'} & \frac{\partial t}{\partial y'} & \frac{\partial t}{\partial z'} & \frac{\partial t}{\partial t'} \end{bmatrix} = \begin{bmatrix} a_{11} & a_{12} & 0 & 0 \\ a_{21} & a_{22} & 0 & 0 \\ 0 & 0 & 1 & 0 \\ 0 & 0 & 0 & 1 \end{bmatrix} \quad (2.5)$$

We also need the inverse transformation, from frame $S'(\tau)$ into frame S because it gives us:

$$\begin{bmatrix} \dfrac{\partial x'}{\partial x} & \dfrac{\partial x'}{\partial y} & \dfrac{\partial x'}{\partial z} & \dfrac{\partial x'}{\partial t} \\ \dfrac{\partial y'}{\partial x} & \dfrac{\partial y'}{\partial y} & \dfrac{\partial y'}{\partial z} & \dfrac{\partial y'}{\partial t} \\ \dfrac{\partial z'}{\partial x} & \dfrac{\partial z'}{\partial y} & \dfrac{\partial z'}{\partial z} & \dfrac{\partial z'}{\partial t} \\ \dfrac{\partial t'}{\partial x} & \dfrac{\partial t'}{\partial y} & \dfrac{\partial t'}{\partial z} & \dfrac{\partial t'}{\partial t} \end{bmatrix} = \begin{bmatrix} b_{11} & b_{12} & 0 & 0 \\ b_{21} & b_{22} & 0 & 0 \\ 0 & 0 & 1 & 0 \\ 0 & 0 & 0 & 1 \end{bmatrix} = \begin{bmatrix} a_{11} & -a_{12} & 0 & 0 \\ -a_{21} & a_{22} & 0 & 0 \\ 0 & 0 & 1 & 0 \\ 0 & 0 & 0 & 1 \end{bmatrix} \quad (2.6)$$

The electromagnetic potential, by virtue of being a 4-vector transforms the same way as described by (2.1):

$$\begin{pmatrix} A_x \\ A_y \\ A_z \\ \varphi \end{pmatrix} = Phy\_rotation \begin{pmatrix} A'_x \\ A'_y \\ A'_z \\ \varphi' \end{pmatrix} \quad (2.7)$$

In the inertial frame, the differential Maxwell equations in vacuum, in the absence of electric charge, are [55]:

$$-E_x = \frac{\partial A_x}{\partial t} + c^2 \frac{\partial \varphi}{\partial x} \quad (2.8)$$

$$-E_y = \frac{\partial A_y}{\partial t} + c^2 \frac{\partial \varphi}{\partial y} \quad (2.9)$$

$$-E_z = \frac{\partial A_z}{\partial t} + c^2 \frac{\partial \varphi}{\partial z} \quad (2.10)$$

$$\mathbf{B} = curl\mathbf{A} \quad (2.11)$$

Let's start with (2.8):

$$A_x = a_{11} A'_x + a_{12} A'_y$$
$$\frac{\partial A_x}{\partial t} = a_{11} \frac{\partial A'_x}{\partial t'} + a_{12} \frac{\partial A'_y}{\partial t'} \quad (2.12)$$

$$\frac{\partial \varphi}{\partial x} = \frac{\partial \varphi}{\partial x'}\frac{\partial x'}{\partial x} + \frac{\partial \varphi}{\partial y'}\frac{\partial y'}{\partial x} + \frac{\partial \varphi}{\partial z'}\frac{\partial z'}{\partial x} = a_{11}\frac{\partial \varphi'}{\partial x'} - a_{21}\frac{\partial \varphi'}{\partial y'} = a_{11}\frac{\partial \varphi'}{\partial x'} + a_{12}\frac{\partial \varphi'}{\partial y'} \quad (2.13)$$

$$-E_x = \frac{\partial A_x}{\partial t} + c^2 \frac{\partial \varphi}{\partial x} = a_{11}\frac{\partial A_x'}{\partial t'} + a_{12}\frac{\partial A_y'}{\partial t'} + c^2 a_{11}\frac{\partial \varphi'}{\partial x'} + c^2 a_{12}\frac{\partial \varphi'}{\partial y'} = -a_{11}E_x' - a_{12}E_y' \qquad (2.14)$$

So:
$$E_x = a_{11}E_x' + a_{12}E_y' \qquad (2.15)$$

Moving on to the second Maxwell equation, (2.9):

$$\frac{\partial A_y}{\partial t} = a_{21}\frac{\partial A_x'}{\partial t'} + a_{22}\frac{\partial A_y'}{\partial t'} \qquad (2.16)$$

$$\frac{\partial \varphi}{\partial y} = a_{44}b_{12}\frac{\partial \varphi'}{\partial x'} + a_{44}b_{22}\frac{\partial \varphi'}{\partial y'} = -a_{12}\frac{\partial \varphi'}{\partial x'} + a_{22}\frac{\partial \varphi'}{\partial y'} = a_{21}\frac{\partial \varphi'}{\partial x'} + a_{22}\frac{\partial \varphi'}{\partial y'} \qquad (2.17)$$

$$E_y = a_{21}E_x' + a_{22}E_y' \qquad (2.18)$$

Moving on to (2.10):

$$\frac{\partial A_z}{\partial t} = \frac{\partial A_z'}{\partial t} = \frac{\partial A_z'}{\partial t'} \qquad (2.19)$$

$$\frac{\partial \varphi}{\partial z} = \frac{\partial \varphi}{\partial z'} = \frac{\partial \varphi'}{\partial z'} \qquad (2.20)$$

$$-E_z = \frac{\partial A_z}{\partial t} + c^2\frac{\partial \varphi}{\partial z} = \frac{\partial A_z'}{\partial t'} + c^2\frac{\partial \varphi'}{\partial z'} = -E_z' \qquad (2.21)$$

Therefore:
$$E_z = E_z' \qquad (2.22)$$

$$B_z = \frac{\partial A_y}{\partial x} - \frac{\partial A_x}{\partial y} \qquad (2.23)$$

$$\frac{\partial A_y}{\partial x} = \frac{\partial}{\partial x}(A_x' a_{21} + A_y' a_{22}) \qquad (2.24)$$

$$\frac{\partial A_x^{'}}{\partial x} = \frac{\partial A_x^{'}}{\partial x'}b_{11} + \frac{\partial A_x^{'}}{\partial y'}b_{21} \tag{2.25}$$

$$\frac{\partial A_y^{'}}{\partial x} = \frac{\partial A_y^{'}}{\partial x'}b_{11} + \frac{\partial A_y^{'}}{\partial y'}b_{21} \tag{2.26}$$

$$\frac{\partial A_y}{\partial x} = a_{21}(\frac{\partial A_x^{'}}{\partial x'}b_{11} + \frac{\partial A_x^{'}}{\partial y'}b_{21}) + a_{22}(\frac{\partial A_y^{'}}{\partial x'}b_{11} + \frac{\partial A_y^{'}}{\partial y'}b_{21}) \tag{2.27}$$

$$\frac{\partial A_x}{\partial y} = \frac{\partial}{\partial y}(A_x^{'}a_{11} + A_y^{'}a_{12}) \tag{2.28}$$

$$\frac{\partial A_x^{'}}{\partial y} = \frac{\partial A_x^{'}}{\partial x'}b_{12} + \frac{\partial A_x^{'}}{\partial y'}b_{22} \tag{2.29}$$

$$\frac{\partial A_y^{'}}{\partial y} = \frac{\partial A_y^{'}}{\partial x'}b_{12} + \frac{\partial A_y^{'}}{\partial y'}b_{22} + \frac{\partial A_y^{'}}{\partial t'}b_{42} \tag{2.30}$$

$$\frac{\partial A_x}{\partial y} = a_{11}(\frac{\partial A_x^{'}}{\partial x'}b_{12} + \frac{\partial A_x^{'}}{\partial y'}b_{22}) + a_{12}(\frac{\partial A_y^{'}}{\partial x'}b_{12} + \frac{\partial A_y^{'}}{\partial y'}b_{22}) \tag{2.31}$$

$$B_z = \frac{\partial A_y}{\partial x} - \frac{\partial A_x}{\partial y} = (a_{21}b_{11} - a_{11}b_{12})\frac{\partial A_x^{'}}{\partial x'} + (a_{22}b_{21} - a_{12}b_{22})\frac{\partial A_y^{'}}{\partial y'} +$$
$$+(a_{21}b_{21}\frac{\partial A_x^{'}}{\partial y'} - a_{12}b_{12}\frac{\partial A_y^{'}}{\partial x'}) - (a_{11}b_{22}\frac{\partial A_x^{'}}{\partial y'} - a_{22}b_{11}\frac{\partial A_y^{'}}{\partial x'}) = \tag{2.32}$$
$$= (a_{21}b_{21}\frac{\partial A_x^{'}}{\partial y'} - a_{12}b_{12}\frac{\partial A_y^{'}}{\partial x'}) - (a_{11}b_{22}\frac{\partial A_x^{'}}{\partial y'} - a_{22}b_{11}\frac{\partial A_y^{'}}{\partial x'}) = (a_{21}^2 + a_{11}^2)(\frac{\partial A_y^{'}}{\partial x'} - \frac{\partial A_x^{'}}{\partial y'}) = B_z^{'}$$

$$B_z = B_z^{'} \tag{2.33}$$

$$B_x = \frac{\partial A_z}{\partial y} - \frac{\partial A_y}{\partial z} \tag{2.34}$$

$$\frac{\partial A_z}{\partial y} = \frac{\partial A_z^{'}}{\partial y} = \frac{\partial A_z^{'}}{\partial x'}\frac{\partial x'}{\partial y} + \frac{\partial A_z^{'}}{\partial y'}\frac{\partial y'}{\partial y} + \frac{\partial A_z^{'}}{\partial t'}\frac{\partial t'}{\partial y} = \frac{\partial A_z^{'}}{\partial x'}b_{12} + \frac{\partial A_z^{'}}{\partial y'}b_{22} \tag{2.35}$$

$$\frac{\partial A_y}{\partial z} = \frac{\partial A_y}{\partial z'} = \frac{\partial}{\partial z'}(A'_x a_{21} + A'_z a_{22} + \varphi' a_{24}) = \frac{\partial A'_x}{\partial z'} a_{21} + \frac{\partial A'_y}{\partial z'} a_{22} \qquad (2.36)$$

$$\begin{aligned}B_x &= \frac{\partial A'_z}{\partial x'} b_{12} + \frac{\partial A'_z}{\partial y'} b_{22} - \frac{\partial A'_x}{\partial z'} a_{21} - \frac{\partial A'_y}{\partial z'} a_{22} = \\ &= (\frac{\partial A'_z}{\partial y'} b_{22} - \frac{\partial A'_y}{\partial z'} a_{22}) + (\frac{\partial A'_z}{\partial x'} b_{12} - \frac{\partial A'_x}{\partial z'} a_{21}) = a_{22}(\frac{\partial A'_z}{\partial y'} - \frac{\partial A'_y}{\partial z'}) + a_{12}(\frac{\partial A'_x}{\partial z'} - \frac{\partial A'_z}{\partial x'}) = \\ &= a_{22} B'_x + a_{12} B'_y\end{aligned} \qquad (2.37)$$

$$B_x = a_{22} B'_x + a_{12} B'_y \qquad (2.38)$$

$$B_y = \frac{\partial A_x}{\partial z} - \frac{\partial A_z}{\partial x} \qquad (2.39)$$

$$\frac{\partial A_x}{\partial z} = \frac{\partial A_x}{\partial z'} = \frac{\partial}{\partial z'}(A'_x a_{11} + A'_z a_{12}) = \frac{\partial A'_x}{\partial z'} a_{11} + \frac{\partial A'_y}{\partial z'} a_{12} \qquad (2.40)$$

$$\frac{\partial A_z}{\partial x} = \frac{\partial A'_z}{\partial x} = \frac{\partial A'_z}{\partial x'}\frac{\partial x'}{\partial x} + \frac{\partial A'_z}{\partial y'}\frac{\partial y'}{\partial x} + \frac{\partial A'_z}{\partial t'}\frac{\partial t'}{\partial x} = \frac{\partial A'_z}{\partial x'} b_{11} + \frac{\partial A'_z}{\partial y'} b_{21} \qquad (2.41)$$

$$\begin{aligned}\frac{\partial A_x}{\partial z} - \frac{\partial A_z}{\partial x} &= (\frac{\partial A'_x}{\partial z'} a_{11} - \frac{\partial A'_z}{\partial x'} b_{11}) + (\frac{\partial A'_y}{\partial z'} a_{12} - \frac{\partial A'_z}{\partial y'} b_{21}) = \\ &= a_{11}(\frac{\partial A'_x}{\partial z'} - \frac{\partial A'_z}{\partial x'}) + a_{12}(\frac{\partial A'_y}{\partial z'} - \frac{\partial A'_z}{\partial y'}) = a_{11} B'_y - a_{12} B'_x\end{aligned} \qquad (2.42)$$

$$B_y = a_{11} B'_y - a_{12} B'_x \qquad (2.43)$$

To summarize:

$$\begin{aligned} E_x &= a_{11} E'_x + a_{12} E'_y \\ B_x &= a_{22} B'_x + a_{12} B'_y = a_{11} B'_x + a_{12} B'_y \\ E_y &= a_{21} E'_x + a_{22} E'_y = a_{11} E'_y - a_{12} E'_x \\ B_y &= a_{12} B'_x + a_{11} B'_y = a_{11} B'_y - a_{12} B'_x \\ E_z &= E'_z \\ B_z &= B'_z \end{aligned} \qquad (2.44)$$

An immediate consequence is:

$$E = E'$$
$$B = B' \qquad (2.45)$$

Obviously, the quantity $E^2 - (Bc)^2$ is invariant with respect to the transforms (2.1). Other invariants will be studied in the next section.

Recasting (2.44) in matrix form allows us to notice another important property:

$$\begin{pmatrix} E_x \\ E_y \\ E_z \\ B_z \end{pmatrix} = Phy\_rotation \begin{pmatrix} E'_x \\ E'_y \\ E'_z \\ B'_z \end{pmatrix}$$

$$\begin{pmatrix} B_x \\ B_y \\ E_z \\ B_z \end{pmatrix} = Phy\_rotation \begin{pmatrix} B'_x \\ B'_y \\ E'_z \\ B'_z \end{pmatrix} \qquad (2.46)$$

The above produces the inverse transforms:

$$\begin{pmatrix} E'_x \\ E'_y \\ E'_z \\ B'_z \end{pmatrix} = Phy\_rotation^{-1} \begin{pmatrix} E_x \\ E_y \\ E_z \\ B_z \end{pmatrix}$$

$$\begin{pmatrix} B'_x \\ B'_y \\ E'_z \\ B'_z \end{pmatrix} = Phy\_rotation^{-1} \begin{pmatrix} B_x \\ B_y \\ E_z \\ B_z \end{pmatrix} \qquad (2.47)$$

5.2.2. Consequences

5.2.2.1. Maxwell Laws in a Uniformly Rotating Frame

From the calculations in the previous section we deduce immediately that:

$$\mathbf{E}' = -\frac{\partial \mathbf{A}'}{\partial t'} - c^2 \nabla \varphi'$$
$$\mathbf{B}' = curl \mathbf{A}' \qquad (2.48)$$

Indeed:

$$\frac{\partial \mathbf{A}'}{\partial t'} = \frac{\partial}{\partial t}(\mathbf{i}A'_x + \mathbf{j}A'_y + \mathbf{k}A'_z) = \mathbf{i}\frac{\partial}{\partial t}(b_{11}A_x + b_{12}A_y) + \mathbf{j}\frac{\partial}{\partial t}(b_{21}A_x + b_{22}A_y) + \mathbf{k}\frac{\partial A_z}{\partial t} = $$
$$= \mathbf{i}(b_{11}\frac{\partial A_x}{\partial t} + b_{12}\frac{\partial A_y}{\partial t}) + \mathbf{j}(b_{21}\frac{\partial A_x}{\partial t} + b_{22}\frac{\partial A_y}{\partial t}) + \mathbf{k}\frac{\partial A_z}{\partial t} \quad (2.49)$$

$$\nabla\varphi' = \mathbf{i}\frac{\partial\varphi'}{\partial x'} + \mathbf{j}\frac{\partial\varphi'}{\partial y'} + \mathbf{k}\frac{\partial\varphi'}{\partial z'} = \mathbf{i}(\frac{\partial\varphi}{\partial x}\frac{\partial x}{\partial x'} + \frac{\partial\varphi}{\partial y}\frac{\partial y}{\partial x'} + \frac{\partial\varphi}{\partial z}\frac{\partial z}{\partial x'}) +$$
$$+ \mathbf{j}(\frac{\partial\varphi}{\partial x}\frac{\partial x}{\partial y'} + \frac{\partial\varphi}{\partial y}\frac{\partial y}{\partial y'} + \frac{\partial\varphi}{\partial z}\frac{\partial z}{\partial y'}) + \mathbf{k}\frac{\partial\varphi}{\partial z} = \quad (2.50)$$
$$= \mathbf{i}(a_{11}\frac{\partial\varphi}{\partial x} + a_{21}\frac{\partial\varphi}{\partial y}) + \mathbf{j}(a_{12}\frac{\partial\varphi}{\partial x} + a_{22}\frac{\partial\varphi}{\partial y}) + \mathbf{k}\frac{\partial\varphi}{\partial z}$$

$$\frac{\partial \mathbf{A}'}{\partial t'} + c^2\nabla\varphi' = \mathbf{i}(b_{11}\frac{\partial A_x}{\partial t} + b_{12}\frac{\partial A_y}{\partial t}) + \mathbf{j}(b_{21}\frac{\partial A_x}{\partial t} + b_{22}\frac{\partial A_y}{\partial t}) + \mathbf{k}\frac{\partial A_z}{\partial t} +$$
$$+ \mathbf{i}c^2(a_{11}\frac{\partial\varphi}{\partial x} + a_{21}\frac{\partial\varphi}{\partial y}) + \mathbf{j}c^2(a_{12}\frac{\partial\varphi}{\partial x} + a_{22}\frac{\partial\varphi}{\partial y}) + \mathbf{k}c^2\frac{\partial\varphi}{\partial z} =$$
$$= \mathbf{i}(b_{11}\frac{\partial A_x}{\partial t} + b_{12}\frac{\partial A_y}{\partial t} + a_{11}c^2\frac{\partial\varphi}{\partial x} + a_{21}c^2\frac{\partial\varphi}{\partial y}) + \quad (2.51)$$
$$+ \mathbf{j}(b_{21}\frac{\partial A_x}{\partial t} + b_{22}\frac{\partial A_y}{\partial t} + c^2 a_{12}\frac{\partial\varphi}{\partial x} + c^2 a_{22}\frac{\partial\varphi}{\partial y}) + \mathbf{k}(\frac{\partial A_z}{\partial t} + c^2\frac{\partial\varphi}{\partial z}) =$$
$$= \mathbf{i}(a_{11}E_x + a_{21}E_y) + \mathbf{j}(a_{12}E_x + a_{22}E_y) + \mathbf{k}(\frac{\partial A_z}{\partial t} + c^2\frac{\partial\varphi}{\partial z}) =$$
$$= -(\mathbf{i}E'_x + \mathbf{j}E'_y + \mathbf{k}E'_z) = -\mathbf{E}'$$

$$\mathrm{curl}\mathbf{A}' = \begin{bmatrix} \mathbf{i} & \mathbf{j} & \mathbf{k} \\ \frac{\partial}{\partial x'} & \frac{\partial}{\partial y'} & \frac{\partial}{\partial z'} \\ A'_x & A'_y & A'_z \end{bmatrix} = \mathbf{i}(\frac{\partial A'_z}{\partial y'} - \frac{\partial A'_y}{\partial z'}) + \mathbf{j}(\frac{\partial A'_x}{\partial z'} - \frac{\partial A'_z}{\partial x'}) + \mathbf{k}(\frac{\partial A'_y}{\partial x'} - \frac{\partial A'_x}{\partial y'}) \quad (2.52)$$

$$\frac{\partial A'_z}{\partial y'} = \frac{\partial A_z}{\partial y} \quad (2.53)$$

$$\frac{\partial A'_y}{\partial z'} = \frac{\partial A_y}{\partial z} \quad (2.54)$$

$$\frac{\partial A'_x}{\partial z'} = \frac{\partial A'_x}{\partial z} = \frac{\partial}{\partial z}(b_{11}A_x + b_{12}A_y) = b_{11}\frac{\partial A_x}{\partial z} + b_{12}\frac{\partial A_y}{\partial z} = a_{11}\frac{\partial A_x}{\partial z} - a_{12}\frac{\partial A_y}{\partial z} \quad (2.55)$$

$$\frac{\partial A'_z}{\partial x'} = \frac{\partial A_z}{\partial x'} = \frac{\partial A_z}{\partial x}\frac{\partial x}{\partial x'} + \frac{\partial A_z}{\partial y}\frac{\partial y}{\partial x'} + \frac{\partial A_z}{\partial z}\frac{\partial z}{\partial x'} = a_{11}\frac{\partial A_z}{\partial x} + a_{21}\frac{\partial A_z}{\partial y} = a_{11}\frac{\partial A_z}{\partial x} - a_{12}\frac{\partial A_z}{\partial y} \quad (2.56)$$

$$\frac{\partial A'_y}{\partial x'} = \frac{\partial A_y}{\partial x'} = \frac{\partial A_y}{\partial x}\frac{\partial x}{\partial x'} + \frac{\partial A_y}{\partial y}\frac{\partial y}{\partial x'} + \frac{\partial A_y}{\partial z}\frac{\partial z}{\partial x'} = a_{11}\frac{\partial A_y}{\partial x} + a_{21}\frac{\partial A_y}{\partial y} = a_{11}\frac{\partial A_y}{\partial x} - a_{12}\frac{\partial A_y}{\partial y} \quad (2.57)$$

$$\frac{\partial A'_x}{\partial y'} = \frac{\partial A'_x}{\partial y} = \frac{\partial}{\partial y}(b_{11}A_x + b_{12}A_y) = b_{11}\frac{\partial A_x}{\partial y} + b_{12}\frac{\partial A_y}{\partial y} = a_{11}\frac{\partial A_x}{\partial y} - a_{12}\frac{\partial A_y}{\partial y} \quad (2.58)$$

Therefore:

$$curl\mathbf{A}' = \mathbf{i}(\frac{\partial A_z}{\partial y} - \frac{\partial A_y}{\partial z}) + \mathbf{j}[a_{11}(\frac{\partial A_x}{\partial z} - \frac{\partial A_z}{\partial x}) + a_{12}(\frac{\partial A_z}{\partial y} - \frac{\partial A_y}{\partial z})] +$$
$$+ \mathbf{k}[a_{11}(\frac{\partial A_y}{\partial x} - \frac{\partial A_x}{\partial y}) + a_{12}(\frac{\partial A_y}{\partial y} - \frac{\partial A_y}{\partial y})] = \mathbf{i}B'_x + \mathbf{j}B'_y + \mathbf{k}B'_z = \mathbf{B'} \quad (2.59)$$

5.2.2.2. The Gauge Invariance Condition in a Uniformly Rotating Frame

$$div\mathbf{A}' + \frac{\partial \varphi'}{\partial t'} = div\mathbf{A} + \frac{\partial \varphi}{\partial t} = 0 \quad (2.60)$$

$$div\mathbf{A}' = \frac{\partial A'_x}{\partial x'} + \frac{\partial A'_y}{\partial y'} + \frac{\partial A'_z}{\partial z'} = \frac{\partial A'_x}{\partial x'} + \frac{\partial A'_y}{\partial y'} + \frac{\partial A_z}{\partial z} \quad (2.61)$$

$$\frac{\partial A'_x}{\partial x'} = b_{11}(\frac{\partial A_x}{\partial x}a_{11} + \frac{\partial A_x}{\partial y}a_{21}) + b_{12}(\frac{\partial A_y}{\partial x}a_{11} + \frac{\partial A_y}{\partial y}a_{21}) + b_{14}(\frac{\partial \varphi}{\partial x}a_{11} + \frac{\partial \varphi}{\partial y}a_{21}) \quad (2.62)$$

$$\frac{\partial A'_y}{\partial y'} = b_{21}(\frac{\partial A_x}{\partial x}a_{12} + \frac{\partial A_x}{\partial y}a_{22}) + b_{22}(\frac{\partial A_y}{\partial x}a_{12} + \frac{\partial A_y}{\partial y}a_{22}) + b_{24}(\frac{\partial \varphi}{\partial x}a_{12} + \frac{\partial \varphi}{\partial y}a_{22}) \quad (2.63)$$

$$\frac{\partial \varphi'}{\partial t'} = b_{44}(\frac{\partial \varphi}{\partial x}a_{14} + \frac{\partial \varphi}{\partial y}a_{24} + \frac{\partial \varphi}{\partial t}a_{44}) \quad (2.64)$$

$$div\mathbf{A}' + \frac{\partial \varphi'}{\partial t'} = (b_{11}a_{11} + b_{21}a_{12})\frac{\partial A_x}{\partial x} + (b_{11}a_{21} + b_{21}a_{22})\frac{\partial A_x}{\partial y} + (b_{12}a_{11} + b_{22}a_{12})\frac{\partial A_y}{\partial x} + (b_{12}a_{21} + b_{22}a_{22})\frac{\partial A_y}{\partial y} +$$
$$+ (b_{14}a_{11} + b_{24}a_{12} + b_{44}a_{14})\frac{\partial \varphi}{\partial x} + (b_{14}a_{21} + b_{24}a_{22} + b_{44}a_{24})\frac{\partial \varphi}{\partial y} + a_{44}b_{44}\frac{\partial \varphi}{\partial t} + \frac{\partial A_z}{\partial z} = \frac{\partial A_x}{\partial x} + \frac{\partial A_y}{\partial y} + \frac{\partial A_z}{\partial z} + \frac{\partial \varphi}{\partial t} \quad (2.65)$$

Equality (2.65) results into:

$$\text{div}\mathbf{A}' + \frac{\partial \varphi'}{\partial t'} = \text{div}\mathbf{A} + \frac{\partial \varphi}{\partial t} = 0 \tag{2.66}$$

Equalities (2.48) and (2.66) result into Maxwell's equations having the same exact form in the uniformly rotating frame as the equations in the inertial frames with the immediate consequence that light speed in vacuum in a uniformly a rotating frame is "c". Indeed, (2.48) results into:

$$\frac{1}{c^2} \frac{\partial^2 \mathbf{E}'}{\partial t'^2} - \nabla^2 \mathbf{E}' = 0$$
$$\frac{1}{c^2} \frac{\partial^2 \mathbf{B}'}{\partial t'^2} - \nabla^2 \mathbf{B}' = 0 \tag{2.67}$$

The above means that electromagnetic waves propagate in vacuum, in uniformly rotating frames, at the same speed as they propagate in inertial frames. In the next section we will derive this plus some other very interesting facts through a different approach.

5.3. Planar Wave Transformation and Speed of Light in a Uniformly Rotating Frame

In this section we apply the formalism derived in the previous paragraph in order to obtain the transform of a planar wave. Assume that a planar wave is propagating along the y' axis in the accelerated frame $S'(\tau)$. The wave has the electric component $\mathbf{E}'_x$ and the magnetic component $\mathbf{B}'_z$ along the x' and z' axes, respectively. The components equations are:

$$\mathbf{E}'_x = E'_{0x} \cos(\Omega' t' - k'_y y' + \theta') \mathbf{e}_x$$
$$\mathbf{B}'_z = B'_{0z} \cos(\Omega' t' - k'_y y' + \theta') \mathbf{e}_z \tag{3.1}$$

On the other hand:

$$\begin{pmatrix} x' \\ y' \\ z' \\ t' \end{pmatrix} = \begin{bmatrix} b_{11} & b_{12} & 0 & 0 \\ b_{21} & b_{22} & 0 & 0 \\ 0 & 0 & 1 & 0 \\ 0 & 0 & 0 & 1 \end{bmatrix} \begin{pmatrix} x \\ y \\ z \\ t \end{pmatrix} \tag{3.2}$$

In frame S, the wave equation is either:

$$E_x = (\gamma \cos\alpha \cos\beta + \sin\alpha \sin\beta)E'_x - (u\gamma \cos\beta)B'_z =$$
$$= [(\gamma \cos\alpha \cos\beta + \sin\alpha \sin\beta)E'_{0x} - (u\gamma \cos\beta)B'_{0z}]*$$
$$*\cos[(\Omega' b_{44} - k'_y b_{24})t - (k'_y b_{22} - \Omega' b_{42})y - (k'_y b_{21} - \Omega' b_{41})x + \theta'] \quad (3.3)$$
$$E_y = [(\gamma \cos\alpha \sin\beta - \sin\alpha \cos\beta)E'_{0x} - (u\gamma \sin\beta)B'_{0z}]*$$
$$*\cos[(\Omega' b_{44} - k'_y b_{24})t - (k'_y b_{22} - \Omega' b_{42})y - (k'_y b_{21} - \Omega' b_{41})x + \theta']$$
$$B_z = \gamma(B'_{0z} - \frac{u\cos\alpha}{c^2}E'_{0x})*$$
$$*\cos[(\Omega' b_{44} - k'_y b_{24})t - (k'_y b_{22} - \Omega' b_{42})y - (k'_y b_{21} - \Omega' b_{41})x + \theta']$$

or:

$$\mathbf{E}_x = E_{0x}\cos(\Omega t - k_x x - k_y y - k_z z + \theta)\mathbf{e}_x$$
$$\mathbf{E}_y = E_{0y}\cos(\Omega t - k_x x - k_y y - k_z z + \theta)\mathbf{e}_y \quad (3.4)$$
$$\mathbf{B}_z = B_{0z}\cos(\Omega t - k_x x - k_y y - k_z z + \theta)\mathbf{e}_z$$

Comparing (3.4) and (3.3) we obtain:

$$E_{0x} = (\gamma \cos\alpha \cos\beta + \sin\alpha \sin\beta)E'_{0x} - (u\cos\beta)B'_{0z}$$
$$\Omega = \Omega' b_{44} - k'_y b_{24}$$
$$k_x = k'_y b_{21} - \Omega' b_{41} \quad (3.5)$$
$$k_y = k'_y b_{22} - \Omega' b_{42}$$
$$k_z = 0$$
$$\theta = \theta'$$

$$c = \frac{\Omega}{\sqrt{k_x^2 + k_y^2}} = \frac{\Omega' b_{44} - k'_y b_{24}}{\sqrt{(k'_y b_{21} - \Omega' b_{41})^2 + (k'_y b_{22} - \Omega' b_{42})^2}}$$

$$\frac{\Omega'^2}{k'^2}(c^2 b_{41}^2 + c^2 b_{42}^2 - b_{44}^2) - 2\frac{\Omega'}{k'}(c^2 b_{21}b_{41} + c^2 b_{22}b_{42} - b_{24}b_{44}) + c^2 b_{21}^2 + c^2 b_{22}^2 - b_{24}^2 = 0 \quad (3.6)$$

$$c^2(b_{41}^2 + b_{42}^2) - b_{44}^2 = c^2((\frac{u\gamma \sin\beta}{c^2})^2 + (\frac{u\gamma \cos\beta}{c^2})^2) - \gamma^2 = \frac{u^2\gamma^2}{c^2} - \gamma^2 = -1$$

$$c^2(b_{21}b_{41} + b_{22}b_{42}) - b_{24}b_{44} = -u\gamma \sin\beta(\sin\alpha \cos\beta - \gamma \cos\alpha \sin\beta) +$$
$$+u\gamma \cos\beta(\sin\alpha \sin\beta + \gamma \cos\alpha \cos\beta) - u\gamma^2 \cos\alpha = 0$$

$$c^2(b_{21}^2 + b_{22}^2) - b_{24}^2 = c^2[(\sin\alpha \cos\beta - \gamma \cos\alpha \sin\beta)^2 + (\sin\alpha \sin\beta + \gamma \cos\alpha \cos\beta)^2] -$$
$$-(u\gamma \cos\alpha)^2 = c^2(\sin^2\alpha + \gamma^2 \cos^2\alpha) - u^2\gamma^2 \cos^2\alpha = c^2$$

$$-\frac{\Omega'^2}{k'^2} + c^2 = 0$$

$$\frac{\Omega'}{k'} = \pm c$$

We can now calculate the phase light speed in the rotating frame:

$$v'_p = \frac{\Omega'}{k'} = \frac{\Omega}{k} = \pm c \quad (3.7)$$

So, the light speed in the rotating frame equals the light speed in the inertial frame, c.

We can now proceed to calculating the amplitude and the phase transformation between the inertial and the rotating frame:

$$\theta = \theta' \tag{3.8}$$

The general equation of the Doppler effect is:

$$\Omega = \Omega' b_{44} - k'_y b_{24} = \Omega'(b_{44} - \frac{b_{24}}{c}) = \Omega'(\gamma - \frac{u\gamma \cos\alpha}{c}) = \gamma(1 - \frac{u\cos\alpha}{c})\Omega'$$

$$\gamma = \frac{1}{\sqrt{1 - \frac{u^2}{c^2}}} \tag{3.9}$$

$$u = r\omega$$

$$\alpha = \omega\gamma\tau$$

The general equations for aberration are:

$$k_x = k'_y b_{21} - \Omega' b_{41} = k' b_{21} - \Omega' b_{41} = k'(b_{21} - b_{41}c) = k'(\sin\alpha\cos\beta - \gamma\cos\alpha\sin\beta - \frac{u\gamma\sin\beta}{c})$$

$$k_y = k'_y b_{22} - \Omega' b_{42} = k' b_{22} - \Omega' b_{42} = k'(b_{22} - b_{42}c) = k'(\sin\alpha\sin\beta + \gamma\cos\alpha\cos\beta + \frac{u\gamma\cos\beta}{c}) \tag{3.10}$$

$$k_z = 0$$

$$\beta = \omega\gamma^2\tau$$

5.4. General Case of Rotation about an Arbitrary Axis

In a prior paper we have shown [66] that the particular transformation (2.1) can be generalized for the case of arbitrary direction. The general case is treated by transforming the problem into the particular case treated in [55] through a transformation into the "canonical case", followed by an application of the transformation from the rotating frame into the inertial frame, ending with the inverse of the first transformation, as shown below:

$$\begin{pmatrix} x \\ y \\ z \\ t \end{pmatrix} = (Rr^{-1} * Phy\_rotation * Rr) \begin{pmatrix} x' \\ y' \\ z' \\ t' \end{pmatrix} \tag{4.1}$$

$$Rr = Rot(\mathbf{e}_x)_{90°} * Rot(\mathbf{e}_y)_{90°-\varphi} * Rot_y \tag{4.2}$$

$Rot(\mathbf{e}_y)_{90°-\varphi} * Rot_y$ aligns $\boldsymbol{\omega}$ with $\mathbf{e}_y$. The second step is comprised by another rotation around the x-axis by $90°$ that aligns $\boldsymbol{\omega}$ with $\mathbf{e}_z$:

$$Rot(\mathbf{e}_x)_{90°} = \begin{bmatrix} 1 & 0 & 0 & 0 \\ 0 & 0 & -1 & 0 \\ 0 & 1 & 0 & 0 \\ 0 & 0 & 0 & 1 \end{bmatrix} \quad (4.3)$$

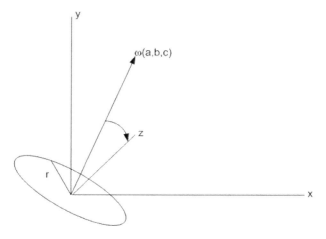

Figure 5.1. Uniform rotation with arbitrary direction of angular velocity.

Expression (4.1) gives the solution for the general case, of arbitrary angular velocity direction. Re-writing (2.45) in tensor form we get:

$$\begin{bmatrix} E_x & B_x & * & * \\ E_y & * & B_y & * \\ * & E_z & * & * \\ * & * & * & B_z \end{bmatrix} = \begin{bmatrix} \gamma\cos\alpha\cos\beta+\sin\alpha\sin\beta & \gamma\sin\alpha\cos\beta-\cos\alpha\sin\beta & -u\gamma\cos\beta & 0 \\ \gamma\cos\alpha\sin\beta-\sin\alpha\cos\beta & \gamma\sin\alpha\sin\beta+\cos\alpha\cos\beta & -u\gamma\sin\beta & 0 \\ u\gamma\cos\alpha & -u\gamma\sin\alpha & 0 & \gamma \\ 0 & -\dfrac{u\gamma\cos\alpha}{c^2} & -\dfrac{u\gamma\sin\alpha}{c^2} & \gamma \end{bmatrix} \begin{bmatrix} E'_x & B'_x & B'_y & 0 \\ E'_y & B'_y & B'_x & E'_x \\ B'_z & \dfrac{E'_z}{c^2} & \dfrac{E'_z}{c^2} & E'_y \\ 0 & E'_z & 0 & B'_z \end{bmatrix} \quad (4.4)$$

The asterisks represent entries with no particular physical meaning, we do not care about them. Then, the general transform is:

$$Rr^{-1} * \begin{bmatrix} \gamma\cos\alpha\cos\beta+\sin\alpha\sin\beta & \gamma\sin\alpha\cos\beta-\cos\alpha\sin\beta & -u\gamma\cos\beta & 0 \\ \gamma\cos\alpha\sin\beta-\sin\alpha\cos\beta & \gamma\sin\alpha\sin\beta+\cos\alpha\cos\beta & -u\sin\beta & 0 \\ u\gamma\cos\alpha & -u\gamma\sin\alpha & 0 & \gamma \\ 0 & -\dfrac{u\gamma\cos\alpha}{c^2} & -\dfrac{u\gamma\sin\alpha}{c^2} & \gamma \end{bmatrix} * Rr \quad (4.5)$$

6. The Mossbauer Rotor Effect - Relativistic Electrodynamics in Uniformly Rotating Frames

6.1. Introduction – The Mossbauer Rotor Experiment

A confirmation of the relativistic Doppler effect was achieved by the Mössbauer rotor experiment [67]. Gamma rays are sent from a source in the middle of a rotating disk (see

Figure 6) to an absorber at the rim and a stationary counter is placed beyond the absorber. The characteristic resonance absorption frequency of the moving absorber at the rim should decrease due to time dilation, so the transmission of gamma rays through the absorber increases, which is subsequently measured by the stationary counter beyond the absorber. The maximal deviation from time dilation was 10^{-5}. Such experiments were performed by Hay et al. [68,69], Champeneyet al.[70,72] and by Kündig [71].

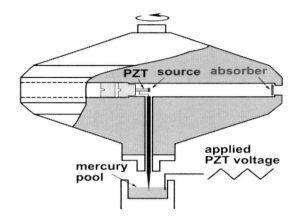

Figure 6.1. The Mossbauer rotor experiment.

6.2. The Mossbauer Experiment Theory Explained in Rotating Frames

In this section we apply the formalism derived in [74,75] in order to obtain the explanation of the Mossbauer experiment Assume that a planar wave is propagating along the y' axis in the accelerated frame $S'(\tau)$. The wave has the electric component $\mathbf{E}'_x$ and the magnetic component $\mathbf{B}'_z$ along the x' and z' axes, respectively. The components equations are:

$$\mathbf{E}'_x = E'_{0x} \cos(\Omega' t' - k'_y y' + \theta') \mathbf{e}_x$$
$$\mathbf{B}'_z = B'_{0z} \cos(\Omega' t' - k'_y y' + \theta') \mathbf{e}_z$$
(2.1)

On the other hand, from [8,9]:

$$\begin{pmatrix} x' \\ y' \\ z' \\ t' \end{pmatrix} = \begin{bmatrix} b_{11} & b_{12} & 0 & 0 \\ b_{21} & b_{22} & 0 & 0 \\ 0 & 0 & 1 & 0 \\ 0 & 0 & 0 & 1 \end{bmatrix} \begin{pmatrix} x \\ y \\ z \\ t \end{pmatrix}$$
(2.2)

In frame S, the wave equation is [8,9] either:

$$E_x = (\gamma \cos\alpha \cos\beta + \sin\alpha \sin\beta)E'_x - (u\gamma \cos\beta)B'_z =$$
$$= [(\gamma \cos\alpha \cos\beta + \sin\alpha \sin\beta)E'_{0x} - (u\gamma \cos\beta)B'_{0z}]*$$
$$*\cos[(\Omega'b_{44} - k'_y b_{24})t - (k'_y b_{22} - \Omega'b_{42})y - (k'_y b_{21} - \Omega'b_{41})x + \theta'] \quad (2.3)$$
$$E_y = [(\gamma \cos\alpha \sin\beta - \sin\alpha \cos\beta)E'_{0x} - (u\gamma \sin\beta)B'_{0z}]*$$
$$*\cos[(\Omega'b_{44} - k'_y b_{24})t - (k'_y b_{22} - \Omega'b_{42})y - (k'_y b_{21} - \Omega'b_{41})x + \theta']$$
$$B_z = \gamma(B'_{0z} - \frac{u\cos\alpha}{c^2}E'_{0x})*$$
$$*\cos[(\Omega'b_{44} - k'_y b_{24})t - (k'_y b_{22} - \Omega'b_{42})y - (k'_y b_{21} - \Omega'b_{41})x + \theta']$$

or:

$$\mathbf{E}_x = E_{0x}\cos(\Omega t - k_x x - k_y y - k_z z + \theta)\mathbf{e}_x$$
$$\mathbf{E}_y = E_{0y}\cos(\Omega t - k_x x - k_y y - k_z z + \theta)\mathbf{e}_y \quad (2.4)$$
$$\mathbf{B}_z = B_{0z}\cos(\Omega t - k_x x - k_y y - k_z z + \theta)\mathbf{e}_z$$

Comparing (2.4) and (2.3) we obtain:

$$E_{0x} = (\gamma \cos\alpha \cos\beta + \sin\alpha \sin\beta)E'_{0x} - (u\cos\beta)B'_{0z}$$
$$\Omega = \Omega'b_{44} - k'_y b_{24}$$
$$k_x = k'_y b_{21} - \Omega'b_{41} \quad (2.5)$$
$$k_y = k'_y b_{22} - \Omega'b_{42}$$
$$k_z = 0$$
$$\theta = \theta'$$

$$c = \frac{\Omega}{\sqrt{k_x^2 + k_y^2}} = \frac{\Omega'b_{44} - k'_y b_{24}}{\sqrt{(k'_y b_{21} - \Omega'b_{41})^2 + (k'_y b_{22} - \Omega'b_{42})^2}}$$

$$\frac{\Omega'^2}{k'^2}(c^2 b_{41}^2 + c^2 b_{42}^2 - b_{44}^2) - 2\frac{\Omega'}{k'}(c^2 b_{21} b_{41} + c^2 b_{22} b_{42} - b_{24} b_{44}) + c^2 b_{21}^2 + c^2 b_{22}^2 - b_{24}^2 = 0$$
$$c^2(b_{41}^2 + b_{42}^2) - b_{44}^2 = c^2((\frac{u\gamma \sin\beta}{c^2})^2 + (\frac{u\gamma \cos\beta}{c^2})^2) - \gamma^2 = \frac{u^2\gamma^2}{c^2} - \gamma^2 = -1$$
$$c^2(b_{21} b_{41} + b_{22} b_{42}) - b_{24} b_{44} = -u\gamma \sin\beta(\sin\alpha \cos\beta - \gamma \cos\alpha \sin\beta) +$$
$$+ u\gamma \cos\beta(\sin\alpha \sin\beta + \gamma \cos\alpha \cos\beta) - u\gamma^2 \cos\alpha = 0$$
$$c^2(b_{21}^2 + b_{22}^2) - b_{24}^2 = c^2[(\sin\alpha \cos\beta - \gamma \cos\alpha \sin\beta)^2 + (\sin\alpha \sin\beta + \gamma \cos\alpha \cos\beta)^2] -$$
$$- (u\gamma \cos\alpha)^2 = c^2(\sin^2\alpha + \gamma^2 \cos^2\alpha) - u^2\gamma^2 \cos^2\alpha = c^2$$
$$-\frac{\Omega'^2}{k'^2} + c^2 = 0$$
$$\frac{\Omega'}{k'} = \pm c \quad (2.6)$$

We can now calculate the phase light speed in the rotating frame:

$$v'_p = \frac{\Omega'}{k'} = \frac{\Omega}{k} = \pm c \quad (2.7)$$

We can now proceed to calculating the amplitude and the phase transformation between the inertial and the rotating frame:

$$\theta = \theta' \tag{2.8}$$

This means that the frequency at the Mossbauer receiver is a periodic function of time: The general equation of the Doppler effect is:

$$\Omega = \Omega' b_{44} - k'_y b_{24} = \gamma(1 - \frac{u \cos \alpha}{c})\Omega'$$

$$\gamma = \frac{1}{\sqrt{1 - \frac{u^2}{c^2}}} \tag{2.9}$$

$$u = r\omega$$

$$\alpha = \omega\gamma\tau = \omega t$$

$$\Omega(t) = \gamma(1 - \frac{u \cos \omega t}{c})\Omega'_0 \tag{2.10}$$

Since the receiver integrates all the frequencies received during an integer number of full revolutions of the transmitter, the resultant frequency is:

$$\Omega = \gamma\Omega'_0 \frac{\int_0^{k\frac{2\pi}{\omega}} (1 - \frac{u \cos \omega t}{c})dt}{k\frac{2\pi}{\omega}} = \gamma\Omega'_0 \tag{2.11}$$

The above confirms that the Mossbauer effect measures the purest form of transverse Doppler effect. On an interesting note:

$$\gamma\Omega'_0 = \Omega'_0[1 + \frac{1}{2}\frac{u^2}{c^2} + R_n(\frac{u^2}{c^2})] \tag{2.12}$$

where $R_n(\frac{u^2}{c^2})$ is the Lagrange remainder of the Taylor series:

$$R_n(\frac{u^2}{c^2}) = \frac{3}{8}(\frac{u^2}{c^2})^2$$

$$\gamma\Omega'_0 \approx \Omega'_0[1 + \frac{1}{2}\frac{u^2}{c^2}(1 + \frac{3}{4}(\frac{u^2}{c^2}))] \tag{2.13}$$

The deviation from Kundig's approximation [5]:

$$\gamma\Omega_0' = \frac{\Omega_0'}{\sqrt{1-\frac{u^2}{c^2}}} \approx \Omega_0'(1+\frac{1}{2}\frac{u^2}{c^2}) \qquad (2.14)$$

is extremely small given that $u \approx 54 m/s$.

If the integration is not done over an integer number of full revolutions of the transmitter, the sinusoidal term can introduce some small errors as well, for example:

$$\gamma \frac{\int_0^{kT+T/4}(1-\frac{u\cos\omega t}{c})dt}{kT} = \gamma[1+\frac{1}{k}(\frac{1}{4}+\frac{\frac{u}{c}}{2\pi})] \qquad (2.15)$$

Only when the integration occurs over precisely an integer number of revolutions, will the departure from γ be null. But it is very difficult to ensure that the integration occurred over precisely an integer number of revolutions. We can see that the error will vary with the total number of revolutions, k and with the tangential speed of the emitter. It is also easy to see that the error can be minimized by extending the total number of revolutions towards a very large number. The error in this case can be much larger, as it can be gleaned from (2.15). This type of error motivated Kolmetskii [76,77] to develop a whole new theory of gravitation and ignited a controversy with Corda who proved that such a theory is not only not needed but also incorrect [78]. Lastly, from [74,75] we obtain the equations of aberration:

$$\begin{aligned}
k_x &= k_y' b_{21} - \Omega' b_{41} = k' b_{21} - \Omega' b_{41} = k'(b_{21} - b_{41}c) = k'(\sin\alpha\cos\beta - \gamma\cos\alpha\sin\beta - \frac{u\gamma\sin\beta}{c}) \\
k_y &= k_y' b_{22} - \Omega' b_{42} = k' b_{22} - \Omega' b_{42} = k'(b_{22} - b_{42}c) = k'(\sin\alpha\sin\beta + \gamma\cos\alpha\cos\beta + \frac{u\gamma\cos\beta}{c}) \\
k_z &= 0 \\
\alpha &= \omega\gamma\tau = \omega t \\
\beta &= \omega\gamma^2\tau = \gamma\omega t \\
u &= r\omega
\end{aligned} \qquad (2.16)$$

The above provides us with an interesting finding, the light ray that follows a straight line (radial) path in the rotating frame of the rotor, will follow a curve in the frame of the lab. The x and y directions of the curve are given by:

$$\begin{aligned}
k' &= k_y' = 1 \\
k_x &= \sin\alpha\cos\beta - \gamma\cos\alpha\sin\beta - \frac{u\gamma\sin\beta}{c} \\
k_y &= \sin\alpha\sin\beta + \gamma\cos\alpha\cos\beta + \frac{u\gamma\cos\beta}{c}
\end{aligned} \qquad (2.17)$$

The light ray directions are time varying. It is also interesting to notice that the reverse is also true, the light ray that follows a straight line (radial) path in lab, will follow a curve in the frame commoving with the rotor:

$$k'_x = k(\sin\alpha\cos\beta - \gamma\cos\alpha\sin\beta - \frac{u\gamma\sin\beta}{c})$$
$$k'_y = k(\sin\alpha\sin\beta + \gamma\cos\alpha\cos\beta + \frac{u\gamma\cos\beta}{c}) \qquad (2.18)$$
$$k'_z = 0$$

The behavior is similar to the Coriolis effect.

7. The Relativistic Cyclotron Radiation in the Circular Rotating Frame of the Moving Heavy Particle

7.1. Introduction - Bremsstrahlung

Bremsstrahlung is the electromagnetic radiation produced by the deceleration of a charged particle. The moving particle loses kinetic energy, which is converted into a photon, it is the process of producing the energy radiation [79]:

$$p = \frac{q^2\gamma^6}{6\pi\varepsilon_0 c}(\dot{\beta}^2 - (\vec{\beta}\times\dot{\vec{\beta}})^2)$$
$$\gamma = \frac{1}{\sqrt{1-\beta^2}}$$
$$\vec{\beta} = \frac{\vec{v}}{c} \qquad (1.1)$$
$$\dot{\vec{\beta}} = \frac{\vec{a}}{c}$$
$$\vec{a} = \frac{d\vec{v}}{d\tau}$$

For the case of acceleration perpendicular to the velocity (as in the case of synchrotrons), the formula simplifies to:

$$p = \frac{q^2 a^2 \gamma^4}{6\pi\varepsilon_0 c^3} \qquad (1.2)$$

where p is the power measured in the frame of the lab. In the current paper we will make the attempt of finding the power as expressed in the frame co-moving with the particle.

Electrodynamics in Uniformly Accelerated/Rotating Frames 285

7.2. Kinematics in Uniform Angular Velocity Rotation

In this section we introduce all the fundamental notions that will help discussing the case of the particle moving in an arbitrary plane, with the normal given by the constant angular velocity $\boldsymbol{\omega}(a,b,c)$, as in the case of a charged particle in circular motion in a synchrotron. According to Moller [2], the simpler case when $\boldsymbol{\omega}$ is aligned with the z-axis produces the transformation between the rotating frame $S'(\tau)$ attached to the particle and an inertial, non-rotating frame S attached to the center of rotation [80-85, 93, 94] (see Figure 7.1):

$$\begin{pmatrix} x \\ y \\ z \\ ct \end{pmatrix} = \mathbf{A} \begin{pmatrix} x' \\ y' \\ z' \\ ct' \end{pmatrix} \tag{2.1}$$

where [85]:

$$\mathbf{A} = \begin{bmatrix} \cos\alpha\cos\beta + \gamma\sin\alpha\sin\beta & \sin\alpha\cos\beta - \gamma\cos\alpha\sin\beta & 0 & -\dfrac{u\gamma}{c}\sin\beta \\ \cos\alpha\sin\beta - \gamma\sin\alpha\cos\beta & \sin\alpha\sin\beta + \gamma\cos\alpha\cos\beta & 0 & \dfrac{u\gamma}{c}\cos\beta \\ 0 & 0 & 1 & 0 \\ \dfrac{u\gamma}{c}\sin\alpha & -\dfrac{u\gamma}{c}\cos\alpha & 0 & \gamma \end{bmatrix} \tag{2.2}$$

$$\mathbf{A}^{-1} = \begin{bmatrix} \cos\alpha\cos\beta + \gamma\sin\alpha\sin\beta & \cos\alpha\sin\beta - \gamma\sin\alpha\cos\beta & 0 & -\dfrac{u\gamma}{c}\sin\alpha \\ \sin\alpha s\cos\beta - \gamma\cos\alpha\sin\beta & \sin\alpha\sin\beta + \gamma\cos\alpha\cos\beta & 0 & \dfrac{u\gamma}{c}\cos\alpha \\ 0 & 0 & 1 & 0 \\ -\dfrac{u\gamma}{c}\sin\beta & \dfrac{u\gamma}{c}\cos\beta & 0 & \gamma \end{bmatrix}$$

$$\gamma = \dfrac{1}{\sqrt{1 - \dfrac{u^2}{c^2}}}$$

$$u = r\omega$$

$$\alpha = \omega\gamma\tau$$

$$\beta = \omega\gamma^2\tau$$

(2.3)

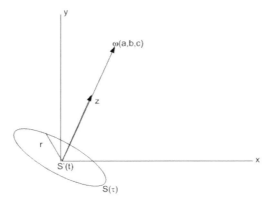

Figure 7.1. Relationship between rotating and inertial frames.

7.3. Bremsstrahlung in a Uniformly Rotating Frame

Assume that we have a particle of charge q and mass m moving in the x-y plane under the influence of a constant magnetic field **B** aligned with the z axis. The magnetic field is the only field present since the particle is to have a circular motion [4,8]. We know that in the frame of the lab, the expression of the Lorentz force acting on the particle is [8]:

$$\mathbf{F} = q\mathbf{v} \times \mathbf{B} = q \begin{bmatrix} \mathbf{i} & \mathbf{j} & \mathbf{k} \\ r\omega\cos(\omega t) & r\omega\sin(\omega t) & 0 \\ 0 & 0 & B \end{bmatrix} \quad (3.1)$$

We would like to find out the expression of the force in the frame co-rotating with the charged particle. For this purpose we will resort to the fact [9] that four-force transforms like four-coordinate (2.1)

$$\begin{pmatrix} \gamma'(u')F'_x \\ \gamma'(u')F'_y \\ \gamma'(u')F'_z \\ \gamma'(u')\dfrac{\mathbf{F'}.\mathbf{u'}}{c} \end{pmatrix} = \mathbf{A}^{-1} \begin{pmatrix} \gamma(u)F_x \\ \gamma(u)F_y \\ \gamma(u)F_z \\ \gamma(u)\dfrac{\mathbf{F}.\mathbf{u}}{c} \end{pmatrix} \quad (3.2)$$

The term $\gamma(u)\dfrac{\mathbf{F}.\mathbf{u}}{c}$ represents the power imparted by the magnetic field to the particle measured in the lab frame (divided by c) while the term $\gamma'(u')\dfrac{\mathbf{F'}.\mathbf{u'}}{c}$ represents the power imparted by the magnetic field to the particle measured in the frame commoving with the particle (divided by c). Transformation (3.2) gives the general formulas for transforming four-force (proper force) in rotating frames.

We know from [86] that:

Electrodynamics in Uniformly Accelerated/Rotating Frames

$$\begin{pmatrix} x \\ y \\ z \\ t \end{pmatrix} = \begin{pmatrix} r\cos(\omega t) \\ r\sin(\omega t) \\ 0 \\ t \end{pmatrix} \tag{3.3}$$

$$\begin{aligned} u_x &= -r\omega \sin \omega t \\ u_y &= r\omega \cos \omega t \\ u_z &= 0 \\ u &= \sqrt{u_x^2 + u_y^2 + u_z^2} = r\omega = u_0 \\ \gamma(u) &= \gamma(u_0) \end{aligned} \tag{3.4}$$

$$\begin{aligned} F_x &= -qBr\omega \sin \omega t \\ F_y &= qBr\omega \cos \omega t \\ F_z &= 0 \end{aligned} \tag{3.5}$$

Substituting (3.4), (3.5) into (3.2) we obtain:

$$\begin{pmatrix} \gamma'(u')F'_x \\ \gamma'(u')F'_y \\ \gamma'(u')F'_z \\ P'/c \end{pmatrix} = \mathbf{A}^{-1} \begin{pmatrix} \gamma(u_0)F_x \\ \gamma(u_0)F_y \\ \gamma(u_0)F_z \\ \gamma(u_0)\dfrac{\mathbf{F}\cdot\mathbf{u}}{c} \end{pmatrix} = \mathbf{A}^{-1} \begin{pmatrix} \gamma(u_0)qBu_0 \sin \omega t \\ -\gamma(u_0)qBu_0 \cos \omega t \\ 0 \\ \gamma(u_0)\dfrac{qBu_0^2}{c} \end{pmatrix} \tag{3.6}$$

Therefore:

$$\begin{pmatrix} \gamma'(u')F'_x \\ \gamma'(u')F'_y \\ \gamma'(u')F'_z \\ P'/c \end{pmatrix} = $$

$$= \begin{bmatrix} \cos\omega t \cos\gamma\omega t + \gamma \sin\omega t \sin\gamma\omega t & \sin\omega t \cos\gamma\omega t - \gamma \cos\omega t \sin\gamma\omega t & 0 & -\dfrac{u_0\gamma \sin\gamma\omega t}{c} \\ \cos\omega t \sin\gamma\omega t - \gamma \sin\omega t \cos\gamma\omega t & \sin\omega t \sin\gamma\omega t + \gamma \cos\omega t \cos\gamma\omega t & 0 & \dfrac{u_0\gamma \cos\gamma\omega t}{c} \\ 0 & 0 & 1 & 0 \\ -\dfrac{u_0\gamma \sin\omega t}{c} & \dfrac{u_0\gamma \cos\omega t}{c} & 0 & \gamma \end{bmatrix} \begin{pmatrix} \gamma qBu_0 \sin \omega t \\ -\gamma qBu_0 \cos \omega t \\ 0 \\ \gamma \dfrac{qBu_0^2}{c} \end{pmatrix} \tag{3.7}$$

The final formula for the power imparted by the magnetic field to the particle measured in the frame commoving with the particle is:

$$P' = 2qB\gamma^2(u_0)u_0^2 \tag{3.8}$$

Expression (3.7) provides us with the expression of the force acting on the particle as measured in the frame of the particle (the proper force). For example:

$$\begin{aligned}
\tilde{F}_x' &= \gamma'(u')F_x' = \gamma^2(u_0)qBu_0(1-\frac{u_0^2}{c^2})\sin\gamma\omega t = qBu_0\sin\gamma\omega t \\
\tilde{F}_y' &= \gamma'(u')F_y' = -\gamma^2(u_0)qBu_0(1-\frac{u_0^2}{c^2})\cos\gamma\omega t = qBu_0\cos\gamma\omega t \\
\tilde{F}_z' &= 0
\end{aligned} \tag{3.9}$$

The term $\gamma^2(u_0)qBu_0$ in (3.9) represents the non-fictitious component, the active Lorentz force while the term $-\gamma^2(u_0)qBu_0(\frac{u_0^2}{c^2})$ represents the fictitious component, the centrifugal "force" due to the calculations being done in the (uniformly) rotating frame.

We are now ready to derive the radiated power. From [82] we know that:

$$\omega = \frac{qB}{\gamma(v_0)m} \tag{3.10}$$

Substituting (3.4),(3.10) into (1.2) we obtain:

$$p = \frac{q^2 a^2 \gamma^4}{6\pi\varepsilon_0 c^3} = \frac{q^2 \gamma^6(u_0)u_0^2\omega^2}{6\pi\varepsilon_0 c^3} = \frac{q^2 B^2}{\gamma^2(u_0)m^2} \frac{q^2\gamma^6(u_0)u_0^2}{6\pi\varepsilon_0 c^3} = \frac{q^4\gamma^4(u_0)u_0^2 B^2}{6\pi\varepsilon_0 c^3 m^2} \tag{3.11}$$

The radiated power in this case is a constant that depends on the mass of the particle, m, its charge, q, its initial speed of injection into the synchrotron, u_0 and the magnitude of the magnetic field B. The constancy is due to the fact that $r\omega = u_0$. The braking force due to radiation acts in direct opposition to the direction of motion (direction of **u**):

$$\begin{aligned}
f_x &= +\frac{p}{u_0}\sin\omega t \\
f_y &= -\frac{p}{u_0}\cos\omega t \\
f_z &= 0 \\
\frac{f_x}{f_y} &= -\tan\omega t
\end{aligned} \tag{3.12}$$

We know that:

$$\begin{pmatrix} \gamma'(u')f'_x \\ \gamma'(u')f'_y \\ \gamma'(u')f'_z \\ p'/c \end{pmatrix} = \mathbf{A}^{-1} \begin{pmatrix} \gamma(u_0)f_x \\ \gamma(u_0)f_y \\ \gamma(u_0)f_z \\ \gamma(u_0)\dfrac{\mathbf{f}\cdot\mathbf{u}}{c} \end{pmatrix} = \mathbf{A}^{-1} \begin{pmatrix} \gamma(u_0)\dfrac{p}{u_0}\sin\omega t \\ -\gamma(u_0)\dfrac{p}{u_0}\cos\omega t \\ 0 \\ \gamma(u_0)\dfrac{p}{c} \end{pmatrix} = \gamma(u_0)p\mathbf{A}^{-1} \begin{pmatrix} \dfrac{\sin\omega t}{u_0} \\ -\dfrac{\cos\omega t}{u_0} \\ 0 \\ \dfrac{1}{c} \end{pmatrix} \quad (3.13)$$

Therefore:

$$\begin{pmatrix} \gamma'(u')f'_x \\ \gamma'(u')f'_y \\ \gamma'(u')f'_z \\ p'/c \end{pmatrix} =$$

$$= \gamma(u_0)p \begin{bmatrix} \cos\omega t\cos\gamma\omega t + \gamma\sin\omega t\sin\gamma\omega t & \sin\omega t\cos\gamma\omega t - \gamma\cos\omega t\sin\gamma\omega t & 0 & -\dfrac{u_0\gamma\sin\gamma\omega t}{c} \\ \cos\omega t\sin\gamma\omega t - \gamma\sin\omega t\cos\gamma\omega t & \sin\omega t\sin\gamma\omega t + \gamma\cos\omega t\cos\gamma\omega t & 0 & \dfrac{u_0\gamma\cos\gamma\omega t}{c} \\ 0 & 0 & 1 & 0 \\ -\dfrac{u_0\gamma\sin\omega t}{c} & \dfrac{u_0\gamma\cos\omega t}{c} & 0 & \gamma \end{bmatrix} \begin{pmatrix} \dfrac{\sin\omega t}{u_0} \\ -\dfrac{\cos\omega t}{u_0} \\ 0 \\ \dfrac{1}{c} \end{pmatrix} \quad (3.14)$$

From (3.14) we get the radiated power measured in the frame co-moving with the particle:

$$p' = 0 \quad (3.15)$$

This should come as no surprise since the particle is not accelerating in its own frame of reference. We also get the braking force due to radiation:

$$\tilde{f}'_x = \gamma'(u')f'_x = -\gamma^2(u_0)\dfrac{p}{u_0}\left(1-\dfrac{u_0^2}{c^2}\right)\sin\gamma\omega t = -\dfrac{p}{u_0}\sin\gamma\omega t$$

$$\tilde{f}'_y = \gamma'(u')f'_y = \gamma^2(u_0)\dfrac{p}{u_0}\left(1-\dfrac{u_0^2}{c^2}\right)\cos\gamma\omega t = \dfrac{p}{u_0}\cos\gamma\omega t \quad (3.16)$$

$$\tilde{f}'_z = 0$$

$$\dfrac{\tilde{f}'_x}{\tilde{f}'_y} = -\tan\gamma\omega t$$

Notice the similarity between expressions (3.16) and (3.13). Also notice how the ratio between the force components in the x and y directions has been increased by the contribution of the gamma factor. We can see that the particle experiences a braking force in the frame co-moving with it. This is extremely important for practical reasons: despite of the absence of radiating power as measured in the co-moving frame, the particle is still being slowed down, independent of the frame of reference used for calculations, thus requiring external energy to be imparted in order to maintain its speed [10-14].

7.4. Braking Force as a Percentage of the Lorentz Force

The total radiated power goes as m^{-4}, which accounts for why electrons lose energy due to bremsstrahlung radiation much more rapidly than heavier charged particles (e.g., muons, protons, alpha particles). This is the reason why the TeV energy electron-positron colliders cannot use circular tunnels (requiring constant acceleration), while a proton-proton colliders (such as LHC) can utilize circular tunnels (and, consequently, the acting Lorentz force is dependent only on the magnetic field, as explained in (3.1)). The electrons lose energy due to bremsstrahlung at a rate $(m_p/m_e)^4 \approx 10^{13}$ times higher than protons do [10-14].

Another way of looking at the issue is by calculating the ratio between the braking force and the active (Lorentz) force. From (3.12) we know that the braking force is:

$$f = \frac{p}{u_0} = \frac{q^4 \gamma^4(u_0) u_0 B^2}{6\pi\varepsilon_0 c^3 m^2} \tag{4.1}$$

The Lorentz force is:

$$F = qBu_0 \tag{4.2}$$

Therefore, their ratio is:

$$r = \frac{f}{F} = \frac{q^3 \gamma^4(u_0) B}{6\pi\varepsilon_0 c^3 m^2} \tag{4.3}$$

When comparing the ratios for the cases of an electron vs. a proton, for the same conditions in terms of magnetic field and initial particle injection sped we find out that:

$$\frac{r_e}{r_p} = \left(\frac{m_p}{m_e}\right)^2 \tag{4.4}$$

The reaction force in the case of an electron is much larger (about 10^7) than that the one for a proton for the case of equal particle accelerating Lorentz forces.

We could to the above calculations in the frame co-rotating with the particle. From (3.9) we obtain:

$$\tilde{F}' = \gamma^2(u_0) qBu_0 \left(1 + \frac{u_0^2}{c^2}\right) \tag{4.5}$$

From (3.16) we obtain:

$$\tilde{f}' = \frac{p}{u_0} \tag{4.6}$$

The interesting result is that we obtain the same exact result as the one calculated in the frame of the lab, the ratio of forces depends only on the inverse ratio of masses, as shown in (4.4). Even more interestingly, the braking force is the same in both frames.

Summary

We generated a standard blueprint for a general solution to the relativistic electromagnetic theory that gives equivalent of the Lorentz transforms for the case of the transforms between an inertial frame and a uniformly accelerated/uniformly rotating frame. We derived several applications of the transforms: the general form of the relativistic Doppler Effect and of the relativistic aberration formulas for the case of accelerated/uniformly rotating motion. We also gave an explanation of the Mossbauer effect as viewed from a rotating frame. We concluded with an application to the general explanation of the Bremsstrahlung Effect (electromagnetic radiation due to particle acceleration).

References

[1] Moller, C. *"The Theory of Relativity,"* Oxford Press (1960).
[2] Thomas, L. H. "Motion of the spinning electron." *Nature* **117** (1926).
[3] Ben-Menahem, A. "Wigner's rotation revisited." *Am. J. Phys.* **53** (1985).
[4] Ben-Menahem, S. "The Thomas precession and velocityspace curvature." *J. Math. Phys.* **27** (1986).
[5] Kroemer, H. "The Thomas precession factor in spin-orbit interaction." *Am J. Physics.* **72** (2004).
[6] Rhodes, J. A.; Semon, M. D. "Relativistic velocity space, Wigner rotation and Thomas precession". *Am. J. Phys.* **72**, (2005).
[7] Malykin, G. B. *"*Thomas precession: correct and incorrect solutions." *Phys. Usp.* **49** (8): (2006).
[8] Krivoruchenko, M. I. *"*Rotation of the swing plane of Foucault's pendulum and Thomas spin precession: Two faces of one coin" *Phys. Usp.,* **52**. 8 (2009).
[9] Sfarti, A. *"*Hyperbolic Motion Treatment for Bell's Spaceship Experiment," *Fizika A,* **18**, 2 (2009).
[10] Sfarti, A. *"Coordinate Time Hyperbolic Motion for Bell's Spaceship Experiment,"* Fizika A, **19**, 3 (2010).
[11] Sfarti, A. "Relativity solution for "Twin paradox": a comprehensive solution," *IJP*, 86, 10 (2012).
[12] Sfarti, A. *"Generalization of Coordinate Transformations between Accelerated and Inertial Frames – General Formulas of Thomas Precession,"* JAPSI.

[13] Sfarti, A. "Electrodynamics in Accelerated Frames," *Theoretical Physics*, **2**, 4, (2017).
[14] Moller, C. *"The Theory of Relativity,"* Oxford Press (1960).
[15] Thomas, L. H. "Motion of the spinning electron." *Nature*. **117** (1926).
[16] Ben-Menahem, A. "Wigner's rotation revisited." *Am. J. Phys.* **53** (1985).
[17] Ben-Menahem, S. *"The Thomas precession and velocity space curvature."* J. Math. Phys. **27** (1986).
[18] Kroemer, H. "The Thomas precession factor in spin-orbit interaction" *Am J. Physics.* **72** (2004).
[19] Rhodes, J. A.; Semon, M. D. "Relativistic velocity space, Wigner rotation and Thomas precession." *Am. J. Phys.* **72**, (2005).
[20] Malykin, G. B. *"Thomas* precession: correct and incorrect solutions". *Phys. Usp.* **49** *(8):* (2006).
[21] Krivoruchenko, M. I. "Rotation of the swing plane of Foucault's pendulum and Thomas spin precession: Two faces of one coin" *Phys. Usp.*, **52**. 8 (2009).
[22] Sfarti, A. "*Hyperbolic* Motion Treatment for Bell's Spaceship Experiment," *Fizika A,* **18**, 2 (2009).
[23] Sfarti, A. "*Coordinate* Time Hyperbolic Motion for Bell's Spaceship Experiment," *Fizika A,* **19**, 3 (2010).
[24] Sfarti, A. "Relativity solution for "Twin paradox": a comprehensive solution," *IJP*, **86**, 10 (2012).
[25] Sfarti, A. "Generalization of Coordinate Transformations between Accelerated and Inertial Frames – General Formulas of Thomas Precession," *JAPSI*. **8**, 2, (2017).
[26] Sfarti, A. "Electrodynamics in Accelerated Frames as Viewed from an Inertial Frame," *IJPOT*, **3**, 1, (2017).
[27] Sfarti, A. "Electrodynamics in Rotating Frames as Viewed from an Inertial Frame," *IJPOT*, **3**, 1, (2017).
[28] Jackson, J.D. "Classical Electrodynamics," 3rd edition, Wiley, (1998).
[29] Bekefi, G. "Radiation Processes in Plasmas," Wiley, 1st edition (1966).
[30] Ichimaru, S. *"Basic Principles of Plasmas Physics: A Statistical Approach,"* Benjamin/Cummings Pub. Co., (1973).
[31] Moller, C. *"The Theory of Relativity,"* Oxford Press (1960).
[32] Thomas, L. H. "Motion of the spinning electron." *Nature.***117** (1926).
[33] Ben-Menahem, A. *"Wigner's rotation revisited."* Am. J. Phys. **53** (1985).
[34] Ben-Menahem, S. *"The Thomas precession and velocityspace curvature."* J. Math. Phys. **27** (1986).
[35] Kroemer, H. "The Thomas precession factor in spin-orbit interaction". *Am J. Physics.* **72** (2004).
[36] Rhodes, J. A.; Semon, M. D. "Relativistic velocity space, Wigner rotation and Thomas precession." *Am. J. Phys.* **72**, (2005).
[37] Malykin, G. B. *"Thomas precession: correct and incorrect solutions"*. *Phys. Usp.* **49** *(8):* (2006).
[38] Krivoruchenko, M. I. "Rotation of the swing plane of Foucault's pendulum and Thomas spin precession: Two faces of one coin". *Phys. Usp.*, **52.** 8 (2009).

[39] Sfarti, A. "*Hyperbolic* Motion Treatment for Bell's Spaceship Experiment," *Fizika A*, **18**, 2 (2009).

[40] Sfarti, A. "Coordinate Time Hyperbolic Motion for Bell's Spaceship Experiment," *Fizika A*, **19**, 3 (2010).

[41] Sfarti, A. "Relativity solution for "Twin paradox": a comprehensive solution," *IJP*, 86, 10 (2012).

[42] *Sfarti, A.,* "Electrodynamics in Uniformly Rotating Frames," *IJPOT*, **3**, 1, (2017).

[43] Moller C., *"The Theory of Relativity,"* Oxford Press, 1960.

[44] Thomas L. H., "Motion of the spinning electron". *Nature.* **117**, 1926.

[45] Ben-Menahem A., "Wigner's rotation revisited." *Am. J. Phys.* **53,** 1985.

[46] Ben-Menahem S., "The Thomas precession and velocity space curvature." *J. Math. Phys.* **27,** 1986.

[47] Kroemer H., "The Thomas precession factor in spin-orbit interaction" *Am J. Physics.* **72** (2004).

[48] Rhodes J. A., Semon, M. D. "Relativistic velocity space, Wigner rotation and Thomas precession." *Am. J. Phys.* **72**, 2005.

[49] Malykin G. B., "Thomas precession: correct and incorrect solutions."*Phys. Usp.* **49**. *8 ,*2006.

[50] Krivoruchenko M. I., "Rotation of the swing plane of Foucault's pendulum and Thomas spin precession: Two faces of one coin" *Phys. Usp.,* **52**. 8, 2009.

[51] Sfarti A., *"*Hyperbolic Motion Treatment for Bell's Spaceship Experiment," *Fizika A,* **18**, 2, 2009.

[52] Sfarti A., *"*Coordinate Time Hyperbolic Motion for Bell's Spaceship Experiment," *Fizika A*, **19**, 3, 2010.

[53] Sfarti A., "Relativity solution for "Twin paradox": a comprehensive solution," *IJP,* 86, 10 2012.

[54] Sfarti A., *"*Generalization of Coordinate Transformations between Accelerated and Inertial Frames – General Formulas of Thomas Precession," *JAPSI*, **8**, 2, (2017).

[55] Sfarti A., *"*Electrodynamics in Uniformly Rotating Frames," *IJPOT*, **3**, 1, (2017).

[56] Moller, C. "The Theory of Relativity," Oxford Press (1960).

[57] Thomas, L. H. "Motion of the spinning electron." *Nature. 117* (1926).

[58] Ben-Menahem, A. "Wigner's rotation revisited." *Am. J. Phys.* 53 (1985).

[59] Ben-Menahem, S. "The Thomas precession and velocity space curvature." *J. Math. Phys.* 27 (1986).

[60] Kroemer, H. "The Thomas precession factor in spin-orbit interaction" *Am J. Physics*. 72 (2004).

[61] Rhodes, J. A.; Semon, M. D. "Relativistic velocity space, Wigner rotation and Thomas precession." *Am. J. Phys. 72,* (2005).

[62] Malykin, G. B. "Thomas precession: correct and incorrect solutions." *Phys. Usp.49 (8):* (2006).

[63] Krivoruchenko, M. I. "Rotation of the swing plane of Foucault's pendulum and Thomas spin precession: Two faces of one coin" *Phys. Usp., 52. 8* (2009).

[64] Sfarti, A. *"*Hyperbolic Motion Treatment for Bell's Spaceship Experiment," *Fizika A*, 18, 2 (2009).

[65] Sfarti, A. *"Coordinate Time Hyperbolic Motion for Bell's Spaceship Experiment,"* Fizika A, 19, 3 (2010).
[66] Sfarti, A. "Relativity solution for "Twin paradox": a comprehensive solution," IJP, 86, 10 (2012).
[67] Sfarti, A. *"Generalization of Coordinate Transformations between Accelerated and Inertial Frames – General Formulas of Thomas Precession,"* JAPSI, 8, 2, (2017).
[68] Mössbauer, R. L. "Kernresonanzabsorption von γ-Strahlung in Ir^{191}," ["Nuclear resonance absorption of γ-radiation in Ir^{191}] Zeitschrift für Physik A, 15, 1, 124 (1958).
[69] Hay, H. J. et al, "Measurement of the Red Shift in an Accelerated System Using the Mössbauer Effect in Fe^{57}," Phys. Rev. Lett. 4, 165, (1960).
[70] Hay, H. J. in Proc. 2nd Conf. Mössbauer Effect, ed. A Schoen and D M T Compton (New York: Wiley) p 225 (1962).
[71] Champeney, D. C., Moon, P. B. "Absence of Doppler Shift for Gamma Ray Source and Detector on Same Circular Orbit," Proc. Phys. Soc.77, 350 (1961).
[72] Kündig, W. "Measurement of the Transverse Doppler Effect in an Accelerated System," Phys. Rev.129, 2371 (1963).
[73] Champeney, D. C., Isaak, G. R., Khan, A. M. "A time dilatation experiment based on the Mössbauer effect," Proc. Phys. Soc 85, 583, (1965).
[74] Sfarti, A. "Generalization of Coordinate Transformations between Accelerated and Inertial Frames – General Formulas of Thomas Precession," JAPSI, **8**, 2, (2017).
[75] Sfarti, A. "Electrodynamics in Uniformly Rotating Frames as Viewed from an Inertial Frame," IJPOT, **3**, 1, (2017).
[76] Sfarti, A. "Electrodynamics in Uniformly Rotating Frames", Theor. Phys., **2**, 4, (2017).
[77] Kholmetskii, L., Yarman, T., Missevitch, O. V., Phys. Scr. 77, 035302 (2008).
[78] Kholmetskii, L., Yarman, T., Missevitch, O. V., Rogozev, B. I. "A Mössbauer experiment in a rotating system on the second order Doppler shift: confirmation of the corrected result by. Kündig" Phys. Scr. 79, 065007 (2009).
[79] Corda, C. "The Mössbauer rotor experiment and the general theory of relativity," Ann. Phys. 368, 258 (2016).
[80] Jackson, J. D. *"Classical Electrodynamics,"* 3rd edition, Wiley, (1998).
[81] Moller, C. *"The Theory of Relativity,"* Oxford Press (1960).
[82] Sfarti, A. "Relativistic Dynamics and Electrodynamics in Uniformly Accelerated and in Uniformly Rotating Frames - the General Expressions for the Electromagnetic 4-Vector Potential," IJPOT, 3, 2 (2017).
[83] Sfarti, A. *"The General Trajectories of Accelerated Particles in Special Relativity,"* JAPSI, 5, 4, (2016).
[84] Sfarti, A. *"Generalization of Coordinate Transformations between Accelerated and Inertial Frames – General Formulas of Thomas Precession,"* JAPSI, 8, 2, (2017).
[85] Sfarti, A. "Electrodynamics in Rotating Frames as Viewed from an Inertial Frame," IJPOT, 3, 1, (2017).
[86] Sfarti, A. "Electrodynamics in Uniformly Accelerated and in Uniformly Rotating Frames," IJPOT, 3, 2, (2017).

[87] Sfarti, A "The trajectories of charged particles moving at relativistic speeds inside particle separators – a fully symbolic solution," *IJNEST,* 4, 4, (2009).

[88] Sfarti, A., "Lorentz covariant formulation for the laws of physics - particle and elastic body dynamics," *IJNEST,* 6, 4, (2011).

[89] Köhn, C., Ebert, U. "Calculation of beams of positrons, neutrons, and protons associated with terrestrial gamma ray flashes", *Journal Geophys. Res.* (2015), vol. 120, pp. 1620—1635.

[90] Koch, H. W. Motz, J. W. "Bremsstrahlung Cross-Section Formulas and Related Data"; *Rev. Mod. Phys.* 31, 920.

[91] Tessier, F., Kawrakow, I. 'Calculation of the electron-electron bremsstrahlung crosssection in the field of atomic electrons', NIM. *Phys. Res. B* 266, (2007).

[92] Elder, F. R., Gurewitsch, A. M., Langmuir, R. V., Pollock, H. C., "Radiation from Electrons in a Synchrotron" (1947) *Physical Review*, vol. 71, Issue 11, pp. 829-830.

[93] Itzykson, C., Zuber, J.B. *"Quantum field theory"* McGraw Hill, New York. (1980).

[94] Masshoon, B. "Electrodynamics in a rotating frame of reference", *Phys. Lett A* 139, 103 (1989).

[95] Hehl, F.W., et al. "Two lectures on fermions and gravity," *Geometry and theoretical physics*, Bad Honnef Lectures, Springer, Berlin (1991), pp. 56–140.

INDEX

A

actualisation of potential, 3, 13, 14, 15
amplitude, 6, 67, 68, 70, 73, 81, 82, 85, 86, 170, 172, 173, 179, 184, 186, 201, 202, 241, 255, 261, 277, 282
antimatter, 132, 161
approximation of a given current, 170, 171, 174, 176
arbitrary shaped periodic interface, ix, 169
Aristotle, 2, 3, 4, 5, 25
artificial and smart materials, 171
atomic nucleus, 15, 17, 19, 22
atoms, 3, 9, 14, 23

B

beams, vii, ix, 169, 199, 203, 295
blueprint, x, 221, 222, 231, 244, 257, 267, 291
Bohr, 4
bosons, viii, 1, 7, 8, 9, 11, 12, 13, 14, 15, 17, 18, 19, 20, 21, 23, 24, 129
boundary and initial boundary value problems, ix, 169
boundary value problem, ix, 169, 171, 173, 177, 184, 185, 186, 193

C

characteristic frequencies, 177, 180
charge density, 32, 33, 36, 37, 39, 56, 59, 111, 171, 239
classical mechanics, 2, 3, 142, 157, 164, 165, 166
clock synchronization, ix, 111, 128
conjugation, ix, 131, 157, 161, 165, 166, 185
conservation, ix, 15, 22, 131, 165
Coriolis effect, 284
crystal structure, 190, 194, 196
crystals, 188, 196, 197, 199, 203
cyclotron waves, 200, 201

D

defected photonic crystals, 197
derivatives, 33, 34, 35, 59, 81, 104, 152, 209, 217
dielectric constant, 208, 210, 212, 213, 215
differential equations, 37, 72, 74, 77, 79, 80, 82
diffraction, 170, 171, 174, 176, 177, 180, 187, 190, 191, 192, 193, 194, 195, 197, 199, 201, 202, 203
diffraction radiation effects, 171, 205
diffraction radiation generators, 171, 205
dilation, 279, 280
Dirac, vii, 1, 2, 3, 5, 6, 7, 8, 11, 13, 14, 23, 24, 25, 26, 128, 139, 140, 141, 142, 150, 151, 155, 156, 215
Dirac equation, 8, 13, 23, 128, 139, 140, 141, 142, 156
dispersion, ix, 169, 178, 197, 198, 199
displacement, 31, 37, 39, 42, 54, 55, 56, 57, 58, 62, 64, 65, 70, 71, 72, 146
distribution, 18, 87, 94, 95, 176, 200, 215
divergence, 28, 33, 37, 39, 40, 41, 56, 57, 85, 87, 88, 97, 98, 99
duality, vii, ix, 8, 131, 132, 135, 136, 143, 146, 165

E

eigen frequencies, 179, 193
eigen oscillations, 179, 192
Einstein, Albert, viii, 2, 4, 5, 7, 8, 11, 24, 25, 76, 101, 106, 107, 109, 110, 111, 115, 128, 132
electric charge, 6, 9, 10, 12, 13, 24, 28, 29, 37, 59, 87, 89, 90, 98, 161, 232, 246, 269
electric current, 9, 39
electric field, 33, 37, 38, 42, 54, 55, 56, 58, 59, 62, 63, 64, 65, 70, 72, 77, 89, 90, 91, 93, 99, 111, 140, 142, 199
electromagnetic, vii, viii, ix, x, 1, 8, 9, 16, 17, 24, 27, 28, 42, 55, 56, 58, 59, 60, 63, 64, 65, 66, 67, 69, 70, 72, 74, 75, 76, 77, 81, 86, 87, 89, 93, 94, 98, 99, 101, 108, 111, 112, 114, 116, 117, 120, 122, 124, 126, 128, 131, 132, 140, 142, 146, 147, 148,

298 Index

149, 150, 154, 155, 156, 157, 158, 159, 163, 164, 165, 166, 169, 170, 171, 172, 174, 176, 186, 199, 203, 207, 221, 223, 231, 232, 237, 238, 240, 241, 243, 246, 251, 253, 257, 259, 269, 276, 284, 291
electromagnetic fields, vii, 27, 28, 66, 86, 89, 108, 122, 131, 154, 157, 159, 164, 174
electromagnetic waves, 81, 94, 170, 171, 186, 199, 203, 237, 253, 257, 276
electromagnetism, 9
electron, vii, viii, 1, 2, 3, 5, 6, 7, 8, 9, 10, 11, 12, 13, 14, 15, 16, 17, 18, 19, 20, 21, 22, 23, 24, 26, 170, 171, 173, 174, 177, 178, 180, 183, 184, 187, 188, 190, 191, 192, 195, 196, 197, 199, 200, 201, 202, 203, 239, 290, 291, 292, 293, 294, 295
electron beam, 170, 171, 173, 175, 177, 179, 181, 183, 184, 185, 187, 189, 191, 193, 195, 196, 197, 199, 200, 201, 202, 203, 204, 205, 206
electron collectivity, 17, 18, 19, 20
electron pairs, viii, 1, 13, 14, 17, 18
electrons, viii, 1, 3, 6, 8, 9, 10, 11, 12, 13, 14, 15, 16, 17, 18, 19, 20, 21, 22, 23, 24, 29, 54, 90, 97, 98, 100, 172, 176, 178, 192, 194, 203, 238, 239, 290, 295
elementary particle, 2, 6, 8, 16, 24, 132, 136, 165, 166
energy, ix, 3, 5, 6, 8, 10, 11, 21, 23, 24, 66, 86, 87, 88, 89, 90, 91, 92, 93, 94, 95, 96, 97, 98, 99, 100, 131, 134, 146, 148, 149, 150, 157, 163, 164, 166, 170, 172, 176, 177, 180, 186, 187, 192, 193, 194, 209, 214, 215, 219, 238, 239, 284, 290
energy balance equations, 186
energy conservation, 134, 163
energy density, 89, 91, 92, 93, 94, 99
energy transfer, 90, 177
engineering, viii, 27, 86, 94, 95, 96, 97, 98, 99, 100
Epperson, Michael, 4, 5, 25
Euler-Lagrange equations, 157, 159, 160, 164
evolution, vii, viii, 1, 6, 19, 20
excitation, 187, 188, 192, 193, 197, 199, 201, 202, 203
exclusion, 3, 11, 13, 14, 19, 20, 21, 24

F

fermions, 7, 8, 11, 13, 14, 24, 295
field theory, 9, 42, 56, 150, 152, 295
fine structure of diffraction radiation, 199
finite accumulation point, 179
Floquet channel, 173
flow field, 180, 183
force, 9, 29, 30, 35, 38, 42, 162, 163, 213, 237, 238, 286, 287, 288, 289, 290, 291
formation, viii, 1, 6, 9, 10, 11, 12, 18, 21, 23, 192
Fredholm operator equations, 176

G

Galileo, 4, 105
gamma rays, 279
gauge invariant, 140, 142, 143, 209, 219
gauge theory, 213, 220
general relativity, 215, 218
gravitational field, ix, 131, 133, 134, 150, 158, 164, 165, 166, 215
gravitational force, ix, 131, 163, 165, 166
gravity, ix, 33, 131, 150, 156, 166, 209, 295
Group 18 elements, 3

H

Hamiltonian, ix, x, 131, 132, 165, 166, 207, 208, 209, 210
Hartman effect, 9
Heisenberg, 2, 4, 5, 24, 25

I

independent variable, 34, 56, 63, 69
integration, 42, 73, 90, 92, 96, 155, 211, 283

K

krypton, 19, 20, 21

L

Lagrangian density, 209
Late Middle Ages, 4
lepton, ix, 131, 132, 136, 165
light, viii, ix, 8, 27, 64, 101, 102, 106, 107, 108, 109, 110, 111, 112, 115, 122, 124, 125, 128, 129, 132, 143, 146, 171, 197, 225, 236, 239, 241, 253, 255, 257, 261, 276, 277, 281, 283, 284
linear systems, 113
Lorentz lemma, 186

M

magnetic field, 38, 39, 40, 41, 42, 43, 45, 46, 53, 54, 55, 56, 58, 59, 60, 63, 64, 65, 70, 71, 72, 77, 86, 89, 91, 92, 99, 100, 111, 132, 140, 143, 162, 199, 200, 201, 202, 203, 209, 239, 241, 286, 287, 288, 289, 290, 291
mass, 5, 6, 7, 8, 12, 16, 17, 18, 19, 20, 21, 23, 24, 31, 135, 149, 159, 165, 208, 209, 286, 289
matrix, 118, 147, 175, 223, 227, 228, 229, 244, 259, 264, 266, 273
matter, 3, 4, 13, 47, 109, 132, 133, 150, 161, 215, 238, 239

Maxwell equations, viii, 101, 102, 103, 104, 105, 108, 110, 111, 112, 115, 116, 117, 122, 124, 126, 127, 140, 142, 143, 149, 166, 232, 246, 269
Meissner effect, 156
Mendeleev, 2, 23, 24
method of analytical regularization, 171, 174
method of exact absorbing conditions, ix, 170, 171, 184
momentum, ix, 11, 66, 131, 143, 146, 149, 157, 158, 162, 164, 165, 166, 210, 215
monochromatic waves, 185, 192
Mossbauer effect, x, 221, 282, 291, 294

N

neon, 18, 19, 20, 21
noble gases, 2, 21, 22, 23, 24
nucleons, 6, 15, 16, 17, 22

O

operator equation of the first kind, 175
oscillation, vii, 1, 3, 5, 6, 7, 8, 9, 10, 11, 12, 13, 17, 19, 22, 193, 194
Oxford calculators, 4

P

Pauli exclusion, 3, 11, 13, 14, 19, 20, 24
Pauli, Wolfgang, viii, 1, 3, 11, 13, 14, 15, 17, 19, 20, 22, 24, 26
periodic grating, 170, 171, 206
Periodic Table of the Elements, 3, 7, 9, 11, 13, 15, 17, 19, 21, 23, 24, 25
phase transformation, 151, 152, 153, 155, 241, 255, 261, 277, 282
photonic crystal, 184, 188, 190, 196, 197, 198, 199, 203, 204, 205
photons, viii, ix, 1, 6, 8, 9, 11, 12, 13, 15, 16, 17, 22, 24, 131, 150, 166
physics, viii, 2, 3, 4, 6, 24, 27, 29, 81, 132, 134, 150, 165, 166, 196, 295
planar and linear diffraction antennas, 171
plane waves, 170, 172, 173, 178, 194
plasma-like medium, 203
potentia, 3, 4, 6
property of electric charge, 6, 9, 10
protons, viii, 1, 3, 9, 14, 15, 16, 17, 22, 24, 290, 295

Q

QCD, x, 213, 214, 220
QED, 219
quantum chromodynamics, vii, 150, 165
quantum field theory, 7, 210

quantum mechanics, 2, 4, 15, 24, 166
quantum theory, 209
quarks, ix, 131, 136, 165

R

rate of change, 34, 35, 61, 63, 64, 74, 158, 162
real photon, 8, 9
reference frame, viii, 101, 105, 106, 107, 109, 110, 113, 114, 118, 122, 125, 231, 266
refractive index, 170
relativity, ix, 76, 129, 131, 132, 135, 157, 158, 163, 164, 165, 166, 295
Riemann tensor, 147

S

scalar field, ix, 131, 133, 134, 150, 151, 152, 153, 154, 157, 166
space-charge waves, 200
space-time, 112, 128, 133, 135, 139, 141, 142, 150, 165, 215, 219
special theory of relativity, 141
speed of light, viii, 8, 27, 58, 64, 101, 170, 232
spin, 11, 13, 14, 20, 142, 292, 293, 294
spinor fields, 128, 156
standard model, 132
surface waves, 177, 178, 179
symmetry, viii, ix, 11, 34, 56, 62, 64, 101, 102, 104, 108, 109, 110, 111, 112, 115, 117, 118, 122, 125, 128, 131, 132, 134, 136, 139, 141, 143, 146, 147, 150, 151, 153, 159, 161, 165, 166, 193, 219
synchronism conditions, 178, 181, 182

T

The Franck – Condon Principle, 15, 16
total energy, 92, 93, 94, 99
trajectory, 137, 164, 262
transformation, vii, 6, 9, 104, 105, 106, 107, 108,

V

vector, vii, viii, 28, 29, 30, 31, 35, 36, 37, 39, 40, 43, 45, 46, 48, 50, 51, 52, 53, 54, 56, 57, 63, 64, 66, 67, 69, 72, 73, 74, 77, 78, 79, 80, 81, 82, 85, 86, 87, 88, 93, 94, 95, 96, 97, 98, 99, 100, 101, 105, 107, 108, 110, 117, 118, 119, 122, 134, 135, 136, 138, 140, 141, 142, 143, 144, 146, 147, 151, 153, 157, 159, 160, 161, 163, 164, 176, 177, 201, 202, 225, 232, 241, 246, 262, 269
velocity, vii, viii, ix, 28, 29, 33, 34, 55, 59, 68, 69, 70, 72, 74, 75, 77, 83, 101, 102, 105, 106, 107, 108, 109, 110, 111, 112, 114, 115, 118, 119, 121, 122, 124, 125, 126, 129, 131, 138, 157, 162, 163,

164, 166, 170, 171, 172, 177, 178, 181, 196, 197, 198, 201, 202, 238, 239, 244, 256, 258, 263, 264, 266, 279, 285, 292, 293, 294
virtual photon, viii, 1, 6, 8, 9, 10, 11, 13, 14, 15, 24

W

Walter Burley, 4
wave propagation, 58, 172, 173, 201
wave vector, 225, 262
Werner Heisenberg, 4
Whitehead, A. N., 3, 24, 25

X

xenon, 20

Y

Yang-Mills, 213, 218

Z

zepto second, 12

Related Nova Publications

Methods and Instruments for Visual and Optical Diagnostics of Objects and Fast Processes

Author: Gennadiy Sergeevich Evtushenko

Series: Physics Research and Technology

Book Description: This book presents new instruments and methods for studying the dynamics of fast processes. The manuscript consists of two parts: Part I discusses the use of high speed metal vapor brightness amplifiers for object and process imaging, and Part II addresses the plasma parameters of a high-voltage nanosecond discharge initiated by a runaway electron beam.

Hardcover ISBN: 978-1-53613-568-8
Retail Price: $160

Quark Matter: From Subquarks to the Universe

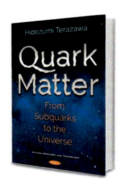

Author: Hidezumi Terazawa

Series: Physics Research and Technology

Book Description: The meaning of "quark matter" is twofold: It refers to 1) compound states of "subquarks" (the most fundamental constituents of matter), which quarks consist of, as "nuclear matter" to those of "nucleons" (the constituents of the nucleus), and 2) compound states of quarks that consist of roughly equal numbers of up, down, and strange quarks, and which may be absolutely stable.

Softcover ISBN: 978-1-53614-151-1
Retail Price: $82

To see complete list of Nova publications, please visit our website at www.novapublishers.com

Related Nova Publications

Polarons: Recent Progress and Perspectives

Editor: Amel Laref

Series: Physics Research and Technology

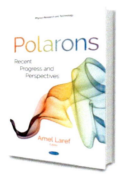

Book Description: This book presents recent research results on the illustrious verge of polaron science, which is broadly applied in condensed matter physics, solid state physics, and chemistry fields.

Hardcover ISBN: 978-1-53613-935-8
Retail Price: $310

Proceedings of the 2017 International Conference on "Physics, Mechanics of New Materials and Their Applications"

Editors: Ivan A. Parinov, Shun-Hsyung Chang, Ph.D., and Vijay K. Gupta

Series: Physics Research and Technology

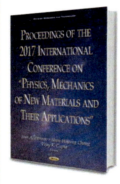

Book Description: The book presents new results of internationally recognized scientific teams in the fields of materials science, physics, mechanics, manufacturing techniques and technologies of advanced materials, operating in diapasons from the nanometer level to the macroscopic level.

Hardcover ISBN: 978-1-53614-083-5
Retail Price: $310

To see complete list of Nova publications, please visit our website at www.novapublishers.com